BIODETERIORATION INVESTIGATION TECHNIQUES

BIODETERIORATION INVESTIGATION TECHNIQUES

Edited by

A. HARRY WALTERS

Visiting Lecturer, Department of Industrial Chemistry, Brunel University, Uxbridge, UK

APPLIED SCIENCE PUBLISHERS LTD
LONDON

APPLIED SCIENCE PUBLISHERS LTD
RIPPLE ROAD, BARKING, ESSEX, ENGLAND

ISBN: 0 85334 696 8

WITH 31 TABLES AND 81 ILLUSTRATIONS

Printed in Great Britain by Galliard (Printers) Ltd Great Yarmouth

List of Contributors

A. E. Apinis

38 Groesffordd, Bryn Crug, Tywyn, Gwynedd, Wales, UK

H. Bizot

Ministère de l'Agriculture, Institut National de la Recherche Agronomique, Laboratoire de Biophysique des Aliments, Chemin de la Géraudière, 44072-Nantes Cedex, France

J. de Jong

IJsselmeerpolders Development Authority, Smedinghuis, Zuiderwagenplein 2, Lelystad, The Netherlands

B. Flannigan

Department of Brewing and Biological Sciences, Heriot-Watt University, Chambers Street, Edinburgh EH1 1HX, Scotland, UK

Björn Henningsson

Department of Forest Products, Royal College of Forestry, Fack, S-104 05 Stockholm 50, Sweden

Waltraut Kerner-Gang

Fachgruppe 'Biologische Materialprüfung', Bundesanstalt für Materialprüfung (BAM), 1 Berlin 45, Unter den Eichen 87, Federal Republic of Germany

R. E. Klausmeier

Naval Weapons Support Center, Crane, Indiana 47522, USA

D. D. Mara
Department of Civil Engineering, University of Dundee, Dundee DD1 4HN, Scotland, UK

J. L. Multon
Ministère de l'Agriculture, Institut National de la Recherche Agronomique, Laboratoire de Biophysique des Aliments, Chemin de la Géraudière, 44072-Nantes Cedex, France

Linda Valentine Oberkotter
Department of Biology, Northeastern University, Boston, Massachusetts 02115, USA

J. L. Osmon
Naval Weapons Support Center, Crane, Indiana 47522, USA

M. Pantke
Fachgruppe 'Biologische Materialprüfung', Bundesanstalt für Materialprüfung (BAM), 1 Berlin 45, Unter den Eichen 87, Federal Republic of Germany

D. G. Parbery
School of Agriculture and Forestry, University of Melbourne, Parkville, Victoria 3052, Australia

Ion Pitis
M.I.C.M. Research Institute for the Electrotechnical Industry, Bd. T. Vladimirescu 45–47, Bucharest 7000—Sectorul 6, Romania

Fred A. Rosenberg
Department of Biology, Northeastern University, Boston, Massachusetts 02115, USA

H. W. Rossmoore
Department of Biology, Wayne State University, Detroit, Michigan 48202, USA

T. Rubidge
Materials Quality Assurance Directorate, Ministry of Defence Procurement Executive, Royal Arsenal East, London SE18 6TD, UK

J. M. SHARPLEY
Sharpley Laboratories Inc., PO Box 846, *Belle Plains Road, Fredericksburg, Virginia* 22401, *USA*

ALICE STANCU
M.E.F.M.C. Research Institute for Building Materials, Calea Grivitei 136, 78122 *Bucharest—Sectorul* 8, *Romania*

F. J. UPSHER
Department of Defence, Materials Research Laboratories, Cordite Avenue, Maribyrnong, Victoria 3032, *Australia*

D. J. A. WILLIAMS
*Department of Chemical Engineering, University College of Swansea, Swansea SA*2 8*PP, Wales, UK*

Editor's Preface

When the time comes that any comparatively new branch of science requires a textbook of specific techniques and methodology, it is fair to say that the subject has really earned full recognition. But hitherto, especially in the case of biodeterioration microbiology, many different types of technical know-how are to be found spread in laboratories all over the world.

This book represents a first attempt to illustrate this fact by gathering together techniques used by workers in eight different countries. The range of subjects covered in biodeterioration studies is extremely wide; therefore the following pages are limited to general approaches in technique, and methods for detecting, assessing and monitoring biodeterioration in bitumens, wood and wood products, pipe-joint rings and sealants, PVC, grain, ferrous metals, optical lenses, floor coverings, aircraft fuels, grassland vegetation, paper and pulp, and metalworking fluids.

Arising from such diversity of materials and types of infection, a common spectrum of technical methods and assessments is gradually emerging. In some instances, of course, there is, as yet, still little evidence of agreed methodological standardisation and result interpretation. Nevertheless, there is a great deal which may be learned from comparing and contrasting in detail both laboratory and field testing techniques, since this works towards the mutual benefit of all concerned with the study of biodeterioration arising from microbial infection.

For microbiologists engaged in other branches of industrial, agricultural, marine and medical work, it is hoped that the methods given will broaden their horizons by providing new ideas, some of which may be worthy of inclusion in both their routine and research work.

It has been a very great pleasure to work with so many self-critical and willing collaborators, and thanks are due to them not only for their contributions, but also for their whole-hearted willingness to co-operate so readily in accepting editorial suggestions—particularly in the case of those overseas workers, several of whom encountered the usual difficulties when writing in scientific English. Thanks are also due to Dr K. Derham, of Applied Science Publishers Ltd, who facilitated the production of this volume.

A. HARRY WALTERS
9 Sunning House
Sunningdale
Berks. SL5 9QN
UK

Contents

1

Methods for Isolation of Biodeterioration Fungi and Their Subsequent Identification

F. J. Upsher

Department of Defence, Materials Research Laboratories, Cordite Avenue, Maribyrnong, Victoria 3032, *Australia*

1. INTRODUCTION

The fungal flora of materials is invariably mixed, often with other organisms such as bacteria and Actinomycetes present, so in order to study the role of any fungus in biodeterioration, or merely to name it, it has first to be isolated. Several methods or combinations of methods can be used, but to obtain the most effective results they have to be selected according to the nature and condition of the flora as assessed by microscopic examination. The isolation methods described here rely on achieving physical separation of fungal spores to give discrete mono-specific colonies without resort to single-spore isolation techniques.

2. METHODS FOR THE EXAMINATION OF FUNGI

Although fungi can be seen on materials using low magnifications, it is necessary to examine microscopic preparations of the flora on slides to find the types present and to note whether they were actually growing or present only as incidental spores. Thus, the methods for examining fungi on materials are the same as those for fungus cultures grown on agar media, though with the latter the main purpose is to identify the organism. Some features of taxonomic importance are noted from visual examination of the colony: colours of the surface and surrounding agar, surface texture, odour and the presence of macroscopic structures such as stromata, sclerotia, ascocarps and pycnidia. The colony is then examined *in situ* on the microscope stage, first with a low magnification (× 50–100), then with a higher power (× 200–250), using transmitted light if the colony is not too opaque. Observation should start at the colony edge, progressing inwards, noting the arrangement of conidia and the apparent manner of their formation, which will be clearest where they are youngest. The presence and morphology of other fruiting bodies will also be noted. Finally, it is necessary to make detailed observations of slide mounts, observing the process of spore formation and the morphology of such structures as conidia, conidiophores, asci and chlamydospores.

For certain Ascomycetes and Coelomycetes it may be necessary to prepare and mount thin sections of the fruiting bodies so that the arrangement and structure of internal elements can be determined, but the majority of biodeterioration microfungi can be identified without sectioning.

Electron microscopy, particularly with the scanning electron microscope, can be of value in studies of biodeterioration fungi since it enables observation of details too small to be resolved with a microscope and it accommodates a great depth of field, structures at different elevations all being in focus together. It is not yet widely used as a tool in the taxonomy of biodeterioration fungi, but details of methods and applications are to be found in recent descriptive reviews (Greenhalgh and Evans, 1971; Hawker, 1971; Heywood, 1971; Williams *et al.*, 1973; Glauert, 1974; Reid, 1974).

2.1. Preparation of Slides for Microscopy

2.1.1. *Adhesive Tape 'Prints'* (Butler and Mann, 1959)

A 5–10 mm length of self-adhesive cellulose tape, held at one end with forceps, is pressed firmly on the fungus and then lifted off so that fragments stick to the adhesive with minimal disarrangement. The tape is then placed, adhesive side up, on a drop of lactophenol on a slide, and another drop put on top. It is then covered with a coverslip, which is pressed down. Excess mounting fluid is soaked up on the edge of a filter paper, taking care not to wet the top of the coverslip. This method is excellent for examining fungi on materials and is suitable for most cultures except those with wet slime spores (*e.g. Fusarium, Cephalosporium*) or those forming a watery exudate. The presence of air bubbles under the coverslip can be advantageous in that the surface detail of spores is more clearly seen when viewed in a bubble.

2.1.2. *Teasing*

Using a dissecting needle, platinum wire or microforceps, a piece of the selected fungus is placed in a drop of mounting fluid on a slide. Fungi not readily wetted can be pre-wetted in 70% ethanol. The fragment is teased with two needles until it is thinly spread, a coverslip is placed on top, then pressed down and excess fluid removed.

2.1.3. *Crushing*

Using a pair of needles or microforceps a selected portion of the fungus, generally with sporing structures, together with a portion of the underlying agar, is placed in a drop of mounting fluid on a microscope slide. It may need pre-wetting with 70% ethanol. A coverslip is carefully placed on top, and if the specimen has enclosed spore-containing bodies they should be viewed before being crushed to observe the morphology of the entire structure. The cover is then gently pressed down, using the back end of a pencil (a harder instrument cracks the glass too easily), until the specimen is spread to a thin film. Excess fluid is soaked up on filter paper. This method can result in disruption and distortion of structures submerged in agar, which may be better observed by placing the agar fragment on a slide and melting the agar away, then replacing it with mounting fluid (Booth, 1971*a*).

2.1.4. *Slide Culture*

These methods are most useful for obtaining mounts of fragile spore structures for micrography. In the simplest method the fungus is inoculated on the edges of an agar block between a slide and coverslip. After a period of incubation in a humid chamber the agar is removed, leaving hyphae with conidiophores adhering to the glass. Both slide and coverslip can then be mounted with lactophenol. Alternatively the fungi can be observed undisturbed, without being wetted. These methods, together with other thin layer culture and special slide culture methods, are described by Booth (1971*a*).

2.2. Mounting Media

The most widely used mycological mounting fluid is Amman's lactophenol (1896), for which the formula is:

Phenol (pure crystals)	10 g
Lactic acid (sp. gr. 1·21)	10 g
Glycerol	20 g
Distilled water	10 g

The phenol and water are warmed until dissolved, then the lactic acid and glycerol are added and mixed.

Evaporation of this solution is so slight that unsealed slides remain satisfactory for long periods and permanent slides may be made by sealing the edges of the coverslip with clear nail varnish or proprietary sealant of similar nature.

Cotton blue (aniline blue, water-soluble) is the stain most often used in lactophenol (0·01–0·05%) and is best for observing hyaline fungi, though it tends to stain the cell contents rather than the walls, making them actually appear smaller. The refractive index of lactophenol is 1·45, which is close to that of fungal cell walls, tending to make surface detail faint. Lactophenol also causes shrinkage and collapse of structures, notably the spores of some Mucorales and Hyphomycetes. For mounting osmotically sensitive types, lactophenol may be used diluted with water to quarter-strength.

Melzer's reagent clears and stains fungal tissues and reveals cell wall detail that can be missed in lactophenol (Dring, 1971) but cannot be used for permanent mounts, for which it is replaced by lactophenol. The formula for Melzer's reagent, as given by Dring (1971), is:

Chloral hydrate	100 g
Potassium iodide	5 g
Iodine	1·5 g
Distilled water	100 ml

2.3. Sectioning

Although it is possible to make some useful sections using a freehand razor, with the specimen sandwiched between Elder pith, other methods are more elaborate, requiring fixing, dehydration and embedding of the fungal tissue. When fungi are contained within a material, preparation will involve the freezing or physical modification of the material to render it suitable for sectioning and the fixation of the fungal component, but it is important that the methods used are not incompatible. Sectioning is then by microtome.

Whereas the preparation of microscopical sections of fungi and materials are occasionally necessary in biodeterioration studies, this topic exceeds the scope of the present chapter; for details of methods, reference should be made to descriptive reviews by Dring (1971) and Greenhalgh and Evans (1971).

3. ISOLATION METHODS

The methods employed to isolate biodeterioration fungi rely on the physical separation of spores and their placement on agar media to form discrete mono-specific colonies. Selection of the appropriate method or methods depends on the condition and apparent composition of the flora. Platinum wire, either straight or as a loop, is used to transfer inocula when there are abundant spores of a single species, or from apparently mono-specific areas. Surface sterilisation enables fungi which are contained within a material to be isolated, without interference by other organisms on the surface. Surface-sterilised specimens may be placed either on a non-nutrient agar medium or homogenised and then plated. Soft materials such as greases and waxes and viscous fluids like oils and plasticisers can be homogenised using a suitable emulsifying agent in the diluent. Swabbing and rinsing are used when there are variable amounts of surface growth, particularly on materials which are not appreciably degraded. Membrane filtration is used to obtain spores from non-viscous fluids such as petroleum fuels. Figure 1 shows the various combinations of methods used in the isolation of biodeterioration fungi.

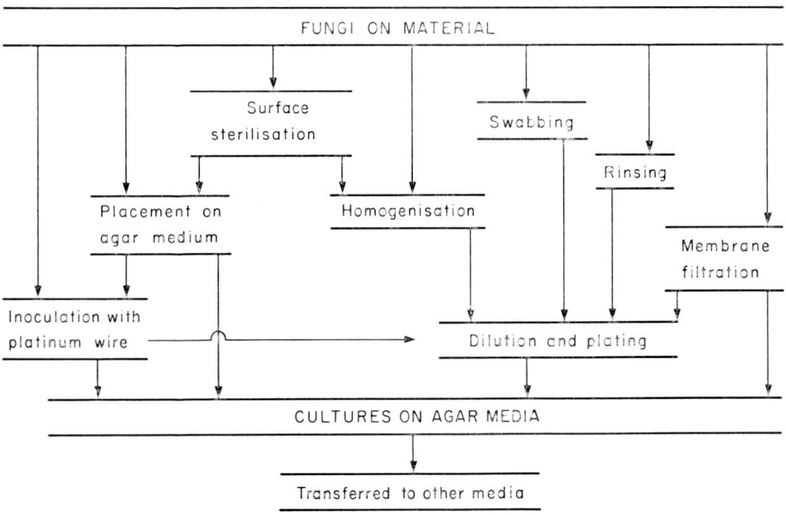

FIG. 1. Flow diagram showing the different methods of isolating fungi from materials.

3.1. Transfer of Inocula with Platinum Wire

This is the most frequently used means of microbial inoculation and is quick, easy and effective as an isolation method. With the aid of a stereo-microscope it can be used to take inocula from small selected areas; even single conidial heads may be sampled. Inocula taken with a wire can be put either directly on to agar media or into sterile diluent for preparing poured plates. A loop is used for all subsequent inoculations on to agar media. Platinum wire is preferred to nichrome since it is corrosion-resistant and more flexible.

Method. The wire, 0·5–0·9 mm diameter and about 50 mm long, held in a metal handle, is used either straight or with a terminal loop, about 3 mm diameter. It should be flamed to red heat, allowed to cool for 15–20 sec, then used dry or wetted with a suitable wetting agent (0·03 % Lissapol N or 0·01 % Tween 80). From a freely sporulating growth a light touch with the loop or the tip will be sufficient, but from a tough, dry or sparse growth some abrasive, agitative effort will be needed. A wet wire or needle is most effective in obtaining spores from perithecia or pycnidia and from dried slime spores. The inoculum is streaked on to the surface of an agar plate, first with a single diametric stroke, then, after flaming, with a number of parallel transverse strokes (Fig. 2). The growth arising from this type of inoculation is densest along the central streak and more sparse along the others. Alternatively, the inoculum can be washed off the wire into 10 ml of sterile diluent, then shaken and plated.

3.2. Placement on Agar Medium

This method is useful when the specimen is old, dry or has been handled or exposed to dry weather so that the spores or other viable units have been removed. Incubation of a fragment of the specimen on a non-nutrient agar enables residual mycelium to grow, deriving its organic nutriment from the specimen material. Any fungi which develop are transferred with platinum wire to other media. If the specimen is surface-sterilised prior to placement on the agar, any subsequent growth will have been derived from previously internal mycelium.

Method. Specimens (10–25 mm) are cut from the material, using sterile instruments, on a clean sterile surface, placed centrally on the surface of a buffered mineral salts agar (BMSA) plate and pressed down to make good contact. Solid, non-porous materials which have been surface-sterilised should be dried between sterile filter papers before placement on the agar, as the presence of free water retards the growth of some fungi.

After incubation for two or more days the specimens may be examined,

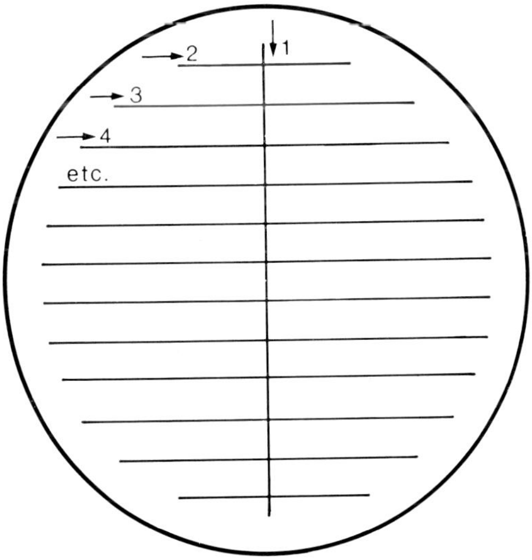

FIG. 2. Diagram showing the order for streaking an agar plate. After the first diametric stroke, the loop is flamed; subsequent transverse streaks are made without flaming.

at first using a stereomicroscope without removing the lid of the dish, so as to avoid contamination. Fungi will have grown on the specimen and on the medium around. Growth-bearing sporing structures may then be examined microscopically and inoculated on to fresh media with a platinum wire.

3.3. Homogenisation

This method is most suitable for soft or fibrous materials such as waxes, greases, gums, papers, boards, ropes, cordage and fabrics, especially when fungal mycelium is visible but no spores are apparent. The numbers of adventitious fungi developing on the isolation plates may be reduced by washing the specimen in sterile BMS solution with a wetting agent, or by surface sterilisation.

Method. A portion of the specimen is aseptically excised and placed in a homogeniser or blender jar with between 10 and 100 vols of sterile BMS with wetting agent. The mixture is blended, preferably at high speed for a short period, until at least half of the specimen is suspended, ensuring that the temperature of the mixture does not exceed about 50°C so that no spores are killed. Taking care to avoid large fragments, the suspension is then plated with or without further dilution.

3.4. Surface Sterilisation

In order to isolate the fungi which are deep in a material, without interference from superficial and adventitious fungi, it is necessary either to remove or kill any viable organisms on the surface. With solid or impervious materials this can be achieved by soaking the specimen in a biocidal solution for 1–5 min. Several solutions are suitable, including formalin (40% formaldehyde in water), 3% hydrogen peroxide, 2% potassium permanganate or 70% ethyl alcohol. Some materials may need pre-wetting with 70% ethanol. After soaking, the specimens are rinsed twice in sterile water. Sections to be placed on agar should be dried between sterile filter papers. Fungi vary in their susceptibilities to sterilants, so to aim at the isolation of the greatest variety of species present, separate specimens should be treated with different sterilants.

Porous or open-structure materials are surface-sterilised with vapour-phase biocides. Exposure to atmospheres saturated with formaldehyde, methanol or propylene oxide vapours for 30 min to 1 hr is sufficient, being careful to avoid excessive penetration of the specimen that would cause the death of fungi contained within. Care should be taken with the methanol and propylene oxide because of their toxic and flammable properties.

After surface sterilisation the specimens may be either placed directly on agar media or homogenised.

3.5. Swabbing

This method is most widely used in the bacteriological sampling of surfaces, particularly in the food industry, and is also useful for fungal sampling. Two types of swabs are commonly used: those made of cotton-wool, from which entrained spores and fragments are washed with BMS solution, and those made of fibrous calcium alginate which dissolves in solutions of sodium hexa-metaphosphate (Calgon). Both are commercially available pre-sterilised as medical supplies, but alternatively may be hand-made by winding a wad of fibre on to the end of an orange stick to form a tip 15 mm × 5 mm, placing it in a test tube which is then closed with a cottonwool plug and sterilised by heating in an oven at 140°C for 2 hr.

Method. The sterile swab is removed from its tube, dipped in BMS solution with a wetting agent, then rubbed firmly on the selected surface. An area of about 100 mm^2 is sufficient. The terminal portion (20–40 mm) of the swab stick is then snapped off into a screw-capped tube with 10 ml of diluent (BMS with wetting agent for cottonwool swabs, 'Calgon' Ringer solution with wetting agent for alginate swabs). The tubes are capped and shaken vigorously, preferably with a mechanical shaker, until the swab is dispersed or dissolved, and the suspension then diluted and plated.

3.6. Rinsing

Rinsing a portion of the specimen yields a suspension of the surface flora, including the adventitious spores. It enables the spores and other propagules to be separated for culturing without using a swab, but may not reveal the presence of immersed, firmly adherent or non-sporulating species.

Method. A portion of the specimen (100 mm^2) is aseptically excised, placed in a screw-capped bottle containing 10 ml of sterile BMS solution with wetting agent and shaken vigorously. The resulting suspension is then diluted and plated.

3.7. Membrane Filtration

Membrane filtration is widely used for collecting particulate matter in low-viscosity liquids. Fungal spores deposited on membrane filters can be grown *in situ* by placing the membrane on a nutrient agar medium, but the spore-load of the suspension is often so great that the colonies become overcrowded and the slower-growing types do not become apparent. Alternatively the spores may be suspended in BMS with wetting agent and plated. Membrane filtration is the basis of methods for estimating the fungal contamination of aviation fuels (Hazzard, 1967; Australian Department of Defence, 1971).

Method. A sterile membrane filter is placed in a pre-sterilised filtration unit and clamped in place, and the unit is then fitted on a filter flask. A filter of pore size 0·8 μm is most suitable for fungal spore separation, but with viscous suspensions or those with a dense particulate content, a larger pore size, up to 5 μm, may be used. The fluid to be sampled is aseptically poured into the funnel of the filtration unit which is then closed or covered with a cottonwool plug or aluminium foil cap. Suction is applied to the flask until the fluid has passed through. If the fluid being sampled is hydrophobic or contains a biocide, the membrane should be washed by passing 2 vols, of about 100 ml, of BMS with wetting agent through it. Using sterile forceps, the membrane is then placed on an agar plate, or alternatively rolled, and placed in a screw-capped tube containing 10 ml of BMS with wetting agent, shaken vigorously, then plated, with dilution if necessary. This plating method cannot be regarded as quantitative since a portion of the spores will remain on the membrane.

3.8. Dilution and Plating

Several of the isolation techniques described above take spores from the specimen and suspend them in an aqueous medium. Aggregates of spores are dispersed by vigorous mechanical agitation assisted by the presence of a surfactant. Spore densities in these suspensions will rarely exceed 10 000/ml and will generally be much lower, but in order to obtain a suitable number of colonies on an isolation plate, *i.e.* between 3 and 30, it is necessary to make a series of decimal dilutions and then to plate from two or more of them. An indication of the dilution needed can be gained by microscopic examination of the suspension on a haemocytometer slide, though a portion of the spores seen will be non-viable.

Thirty colonies on a plate can be excessive, especially if vigorously growing types are present, but growth of these can be restricted by incorporating the dye Rose Bengal in the medium.

Method. The initial suspension is shaken vigorously, then, taking care not to withdraw any large particles, 1·0 ml is removed, using a sterile plugged pipette, and placed in a screw-capped tube containing 9·0 ml of diluent, then shaken. A series of dilutions (1/100, 1/1000, etc.) are prepared by repeating the process, using a fresh pipette at each stage. A convenient reference notation for each dilution uses the logarithmic index, assuming the first suspension to be '−1', successive dilutions −2, −3, etc. This notation also assists in estimating the number of viable spores originally present.

Plates are prepared from this series of dilutions by taking 1·0 ml with a fresh pipette and transferring it to a sterile, pre-labelled petri dish. Several replicates may be taken, each for a different medium. Molten sterile agar at 45–50°C (not too hot to hold in the hand) is then poured on to the suspension in the petri dish and mixed thoroughly.

Since the purpose of the dilution procedure in mycological practice is to bring about a suitable separation of the spores (and not, as in bacteriological practice, to enumerate colonies), the precision of the technique is not too critical and some variations to the basic method can be usefully applied:

(a) 0·1 ml portions may be transferred to the next tube (10 ml) to give an effective 1/100 dilution, or plated, giving a further 1/10 dilution.

(b) 1 ml of suspension can be added to 100 ml of diluent to give a 1/100 dilution.

4. MEDIA

The media used for culturing microfungi must satisfy nutritional and certain physical requirements for germination of spores, and for the growth and maturation of colonies to facilitate identification. They have to provide the inorganic requirements (K, N, P, S, Mg and trace elements Fe, Zn, Cu, Mo, Mn, Bo), the organic nutrients, the major sources being carbohydrates, and to maintain a suitable pH in the range 5–7, but most often between 6 and 6·5. The structural component of mycological culture media is agar, a sulphurylated galactan obtained from various seaweeds, of which a 1·5% (1·0–2·0%) suspension in water forms a firm gel with the useful property of melting at about 95°C but only resetting when cooled below 45°C.

Nutrients are incorporated in the agar in many forms and combinations, but there are two types of media: those with nutrients derived from vegetables, cereals, etc., and the so-called synthetic media made up of a mixture of inorganic salts, with or without a simple organic nutrient which is usually either glucose or sucrose. The latter group more adequately provide the mineral requirements of fungi and are to some extent buffered and find extensive application to physiological and biochemical studies and also in isolation and identification methods.

The media described here support varying amounts of aerial growth and sporulation and favour different types of fungi; hence it is important to use two or more isolation media simultaneously. The nutritional content of media can be increased by the inclusion of various additives (Tables I and II). Solid additives such as cereal grains or plant stems can be either submerged in, or placed on, agar media to induce the formation of conidiophores, coremia and stromata by species that are otherwise tardy. Bacteria are sometimes a nuisance when growing on mycological media, but they can be inhibited by adding an acid to lower the pH or by the inclusion of antibiotics or Rose Bengal.

The method for preparing agar media is standard. The ingredients are mixed in cold distilled water or in the liquid extract and then allowed to stand for 15 min for the agar particles to soak, then heated with constant agitation (*e.g.* magnetic stirrer) to avoid charring, until the agar is thoroughly dispersed. It is then put into vessels for sterilisation and storage (medical flat bottles with metal screw caps, 100–500 ml, are ideal) and sterilised by autoclaving at 15 lb/in^2 (121°C, 105×10^3 N/m^2) for 15 min. Vegetable decoctions are prepared separately before mixing the other ingredients. Thermolabile substances such as antibiotics and yeast nitrogen base (Difco) are prepared as 10 × strength solutions, sterilised by membrane filtration and added by aseptically pipetting 1 vol of additive to 9 vols of the molten medium, previously prepared with only 90% of the prescribed water content. Acids are also added after sterilisation to avoid hydrolysis of the agar and softening of the gel.

In the following formulae the amount of agar is given as 1·5%, but with some grades 1·0 or 1·2% is adequate, though for acid media the amount may need to be increased to 3% to maintain a firm gel. Several of the media or their components are available ready prepared as dry powders from laboratory suppliers and are recommended because of the ease of preparation and consistency of product.

4.1. Agar Media

Buffered mineral salts agar (BMSA):

Ammonium nitrate (NH_4NO_3)	2·7 g
Potassium dihydrogen orthophosphate (KH_2PO_4)	0·9 g
Di-potassium hydrogen orthophosphate (K_2HPO_4)	0·7 g
Magnesium sulphate ($MgSO_4 \cdot 7H_2O$)	0·5 g
Potassium nitrate (KNO_3)	0·5 g
Agar	15 g
Water (distilled)	1 litre

This medium is stable during sterilisation and sheds no precipitate; pH is 6·5. With 3% sucrose included it supports growth similar to that on Czapek-Dox medium but is easier to prepare. BMSA is used for testing the resistance of materials to fungal growth (Standards Association of Australia, 1972).

With the inclusion of 1% finely ground cellulose (chromatographic grade is suitable), cellulose agar (CA) is made, which supports generally a thin growth of cellulolytic fungi but with good sporulation.

Oatmeal agar (OMA) is prepared by including 3% finely ground oats. This encourages good sporulation and pigmentation of a variety of fungi and is particularly good for *Chaetomium* species.

Cornmeal agar (CMA):

Cornmeal	30 g
Agar	15 g
Water (distilled)	1 litre

The cornmeal is boiled for 1 hr, then filtered through muslin. Agar is added to the extract, the volume made up to 1 litre and the medium prepared as above. Cornmeal agar is nutritionally weak and supports restricted growth of a variety of fungi, but is particularly suitable for the observation of darkly pigmented fungi.

Malt agar (MA):

Malt extract (desiccated)	30 g
Peptone (mycological)	5 g
Agar	15 g
Water (distilled)	1 litre

The pH should be corrected to 6·0 or 6·5 using M NaOH solution.

There are several variations to this medium, all of which support luxuriant growth of a wide variety of fungi, especially of the perfect states of *Aspergillus* and *Penicillium*. MA may be acidified to pH 3·5 with lactic acid to supress bacteria. By including 20% or 40% sucrose it becomes suitable for the isolation and cultivation of osmophilic fungi.

Potato carrot agar (PCA) (Langeron, 1952):

Potato (washed, peeled, grated, not new)	20 g
Carrot (washed, peeled, grated)	20 g
Agar	20 g
Water (distilled)	1 litre

The potato and carrot are boiled in water for 1 hr, then sieved or blended. The volume is then made up to 1 litre, the agar added and the medium prepared as above. This is a nutritionally weak medium which supports thin growth of a wide variety of fungi but with good sporulation.

Potato dextrose agar (PDA) (Lacy and Bridgmon, 1962):

Potato (dehydrated, powdered)	22 g
Dextrose (glucose)	20 g
Agar	15 g
Water (distilled)	1 litre

Formerly this medium was made with fresh potatoes, but favourable and more consistent results are obtained and the medium is easier to prepare using dehydrated potato. It is widely used for isolation and culture methods and is specified for the isolation of moulds and yeasts from butter (American Public Health Association, 1953).

Tap water agar (TWA):

Agar	15 g
Tap water	1 litre

Useful for inducing sporulation in some otherwise reluctant fungi by inoculating a plate with a fragment of a nutrient agar bearing the culture.

V8 juice agar (V8A):*

V8 juice (Campbells)	500 ml
Water	500 ml
Agar	15 g

The pH is adjusted to 6·0 using M NaOH solution.

V8 juice is rich in minerals and vitamins, though relatively low in carbohydrates. The medium supports luxuriant growth of a variety of fungi, generally with good sporulation and often with well-developed aerial structures such as coremia.

Yeast malt agar (YMA):

Malt extract (desiccated)	30 g
Yeast extract (desiccated)	5 g
Agar	15 g
Water	1 litre

(*See* under Malt Agar for comments.)

4.2. Additives for Agar Media

The growth of bacteria on mycological media can be directly controlled by incorporating antibiotics or the dye Rose Bengal or by increasing the

* V8 vegetable juice is the registered trade mark of Campbells Soups. The product contains blended juice from tomatoes, carrots, celery, beet, parsley, lettuce, watercress and spinach.

acidity of the medium. Other additives may be used to modify the nutrient qualities of media, and hard vegetable materials used to encourage sporulation. Tables I and II list the most useful additives and indicate their compatibilities with the main media.

4.3. Diluting Fluids

The main function of these solutions is to act as inert carriers without causing either death or proliferation of suspended spores. Distilled water and some tap waters cause osmotic disruption of cells, but satisfactory results are obtained with buffered mineral salts solution (BMSS) which is precipitate-free and compatible with other additives. 'Calgon' Ringer solution is used for dissolving calcium alginate swabs (Higgins, 1950).

TABLE I

ANNOTATED LIST OF ADDITIVES FOR AGAR MEDIA

	Concentration	*Comments*
Antibacterial additives:		
Chloramphenicol (chloromycetin)	30–100 μg/ml	Can be autoclaved
Streptomycin	30–50 μg/ml	Sterilised by membrane filtration
Aureomycin (chlortetracyline)	30–50 μg/ml	Sterilised by membrane filtration
Rose Bengal	30–60 μg/ml	Restricts growth of spreading fungi; autoclavable
Sulphuric, hydrochloric, lactic and tartaric acids	Adjust pH to 3·5–5 as required	Acids depress the variety of fungi that grow, and also soften agar gels
Sporulation additives:		
Lupinus stems	Stems cut to 1 cm lengths; sterilised by autoclaving; all may be either immersed in agar or placed on the surface	
Lime (*Tilia*) stems		
Rice grains (brown)		
Straw		
Nutrients:		
Yeast extract (desicc.)	0·1–2%	
Yeast carbon base (Difco)	0·67%	
Glucose	0·5–2%(–20%)	
Sucrose	0·5–2%(–40%)	
Starch	1–2%	
Cellulose	1–2%	
Peptone	1–3%	
Oatmeal	1–3%	

TABLE II

COMPATIBILITIES OF VARIOUS ADDITIVES WITH MEDIA

	TWA, BMSA	*PCA, CMA*	*MA, YMA*	*PDA*	*V8A*
Antibacterial additives:					
Chloramphenicol (chloromycetin), Streptomycin, Aureomycin (chlortetracycline)	+	+	+	+	+
Rose Bengal	+	+	—	+	—
Sulphuric, hydrochloric, lactic and tartaric acids	—	—	+	+	*
Sporulation additives:					
Lupinus and *Tilia* stems, Rice grains, straw	+	+	—	—	—
Nutrients:					
Yeast extract, Yeast carbon base, Peptone	+	+	—	+	—
Glucose, sucrose	+	+	†	—	—
Starch	+	—	—	—	—
Cellulose	+	—	—	—	—
Oatmeal	+	—	—	—	—

+ Compatible. — Not recommended.
* V8A is naturally acidic, pH 4–4·5, and can be prepared at that pH with 3 % agar.
† For osmophilic fungi.

To prepare these solutions, the salts are dissolved in distilled water, bottled in the required quantities (9, 10, 90 or 99 ml) and sterilised by autoclaving at 15 lb/in^2 (105×10^3 N/m^2) for 15 min.

Buffered mineral salts solution (BMSS):

Ammonium nitrate (NH_4NO_3)	2·7 g
Potassium dihydrogen orthophosphate (KH_2PO_4)	0·9 g
Dipotassium hydrogen orthophosphate (K_2HPO_4)	0·7 g
Magnesium sulphate ($MgSO_4 \cdot 7H_2O$)	0·5 g
Potassium nitrate (KNO_3)	0·5 g
Water (distilled)	1 litre

'Calgon' Ringer solution (CRS):

Sodium hexametaphosphate (Calgon) ($(NaHPO_3)_6$)	10·0 g
Sodium chloride (NaCl)	2·25 g
Potassium chloride (KCl)	0·105 g
Calcium chloride ($CaCl_2$)	0·12 g
Sodium bicarbonate ($NaHCO_3$)	0·05 g
Water (distilled)	1 litre

4.4. Additives for Diluting Fluids

Surfactants are incorporated to improve the wetting properties of diluting fluids and to aid the dispersal of microbial aggregates or to prepare emulsions and suspensions. Two surfactants which have been shown not to be fungitoxic are commonly used, both being non-ionic polyethylene oxide derivatives: Lissapol N is used at 0·03% and Tween 80 at 0·01%. The concentrations may be increased tenfold (0·3% and 0·1%) for preparing emulsions.

5. INCUBATION

5.1. Temperature

Most biodeterioration fungi are mesophilic, growing in the temperature range 10–40°C, with optima around 25°C. They are most often incubated at 25°C, but in testing methods and in physiological investigations they are generally incubated at higher temperatures.

Thermophilic fungi are significant in biodeterioration where elevated temperatures persist. For isolation they are incubated at 45 or 50°C.

Psychrophilic fungi are less often associated with the deterioration of materials but can develop in refrigerated food stores, and in Arctic conditions. Their optimum temperature for growth is below 10°C and incubation should be at 5°C.

5.2. Humidity

Almost all microfungi grow best when the relative humidity exceeds 90%. This level is subtended by the agar medium in petri dishes, but in forced air circulation incubators, refrigerated incubators or at elevated incubation temperatures (35–50°C) the agar can dry out before growth has reached maturity. Under such conditions it is necessary to maintain a high humidity within the incubator by exposing open pans of water (containing copper sulphate (1%) or other non-volatile biocide), but when a high humidity is maintained, condensation can cause difficulties.

5.3. Illumination

Most biodeterioration fungi will grow and sporulate freely in the darkness of a closed incubator, but some, especially those with phytopathogenic tendencies, require stimulation with light, particularly with near-UV radiation in the wavelength range 320–420 nm, to produce spores.

A recommended method to induce sporulation is to expose 4-day-old cultures in plastic petri dishes, horizontally, 400 mm from 'black light' tubes (Osram, Philips Sylvania or other brands), for several days with a 12 hr on/12 hr off cycle (Leach, 1967). The periodic rise in temperature induced by the lights may also in some cases be stimulatory. Cool white fluorescent tubes emit sufficient near-UV radiation to be effective and can be used if the other lamps are not available (Leach, 1971).

6. IDENTIFICATION AND INTERPRETATION

Having obtained pure cultures or, at least, discrete mono-specific colonies producing characteristic sporing structures, there comes the task of precise identification. Familiarity soon permits recognition of the more common genera on sight, but reference to taxonomic literature will be necessary to distinguish between the genera which are superficially similar or to recognise genera encountered less frequently. To identify isolates as species will generally require reference to more specialised works, often monographs dealing with a single genus or group of genera. Listed here are some valuable works which provide means of identification to the level of genus, and monographs of certain common genera. These works will, in turn, refer to other relevant works:

Ainsworth, G. C., *Dictionary of the Fungi*, 5th ed. (1961).

Ainsworth, G. C., Sparrow, F. K. and Sussman, A. S., *The Fungi*, vol. IV A: *A Taxonomic Review with Keys: Ascomycetes and Fungi Imperfecti* (1973). Contains the following contributions:
'Introduction and keys to higher taxa' by G. C. Ainsworth.
'Eurotiales' by D. I. Fennell.
'Sphaeriales' by E. Muller and J. A. von Arx.
'Loculoascomycetes' by E. S. Luttrell.
'Hyphomycetes' by W. B. Kendrick and J. W. Carmichael.
'Coelomycetes' by B. C. Sutton.

Barron, G. L., *The Genera of Hyphomycetes from Soil* (1968).
Booth, C., *The Genus Fusarium* (1971*b*).
Cooney, D. G. and Emerson, R., *Thermophilic Fungi* (1964).
Ellis, M. B., *Dematiaceous Hyphomycetes* (1971).
Raper, K. B. and Fennell, D. I., *The Genus Aspergillus* (1965).
Raper, K. B. and Thom, C., *A Manual of the Penicillia* (1949).
Smith, G. C., *An Introduction to Industrial Mycology*, 6th ed. (1969).
von Arx, J., *The Genera of Fungi Sporulating in Pure Culture* (1970).

There is also the question of interpretation, *i.e.* putting together the observations made on the fungal flora on the material with those obtained by the isolation techniques. Certain inferences can be drawn from these data, regarding the role of the fungal flora and the possible effects on the deterioration of the material:

(a) A species can be recorded as growing on a material when it is observed sporulating there and is subsequently isolated, though no statement can be made on the basis of this evidence concerning its ability to attack or derive nutriment from that material.
(b) When a species, which had not been observed earlier, is isolated in abundance by inoculation by platinum wire, placement on agar or swabbing, the material should be re-examined. The area that was originally sampled may not have been the same as that which was observed, or the species may have been introduced as a contaminant.
(c) Species growing from surface-sterilised specimens on non-nutrient agar would almost certainly have been present in the material.
(d) Species growing from samples of material placed on non-nutrient agar and from homogenised materials, but which had not been observed in the microscopic examination of the material, may have been growing vegetatively, without producing spores, or the species may have lost any recognisable structures. The material should then be re-examined for any further indications.
(e) Swabbing and rinsing invariably produce a mixture of species containing those which were growing and those present as incidental spores. Numerical representation of the species may indicate a distinction, but microscopic examination of the specimen together with other culture methods can confirm the types colonising it.
(f) On plates derived from membrane filtration, when the colonies of one or a few species greatly outnumber the others, it is probable that the major species were growing somewhere in the system that was sampled.

REFERENCES

Ainsworth, G. C. (1961). *Dictionary of the Fungi*, 5th ed. Commonwealth Mycological Institute, Kew, Surrey.

Ainsworth, G. C. (1973). Introduction and keys to higher taxa. In: *The Fungi*, vol. 4A: *A Taxonomic Review with Keys: Ascomycetes and Fungi Imperfecti* (ed. G. C. Ainsworth, F. K. Sparrow and A. S. Sussman), pp. 1–7. Academic Press, London.

American Public Health Association (1953). *Standard Methods for the Examination of Dairy Products*, 10th ed., pp. 171–2. APHA, New York.

Amman, J. (1896). Conservirungsflüssigkeiten und Einschlussmedien für Moose, Chloro- und Cyanophyceen. *Z. Mikroskopie*, **13**, 18–21 (quoted by Smith, 1969, p. 252).

Australian Department of Defence (1971). Def(Aust)208B. Turbine fuel, aviation, kerosine type. SEATO Code Number F35, Avtur. Dept. of Defence, Australia.

Barron, G. L. (1968). *The Genera of Hyphomycetes from Soil.* Williams & Wilkins, Baltimore.

Booth, C. (1971*a*). Introduction to general methods; Fungal culture media. In: *Methods in Microbiology*, vol. 4 (ed. C. Booth), pp. 1–94. Academic Press, London.

Booth, C. (1971*b*). *The Genus Fusarium.* Commonwealth Mycological Institute, Kew, Surrey.

Butler, E. E. and Mann, M. P. (1959). Use of cellophane tape for mounting and photographing phytopathogenic fungi. *Phytopathol.*, **49**, 231–2.

Cooney, D. G. and Emerson, R. (1964). *Thermophilic Fungi.* Freeman & Co., San Francisco.

Dring, D. M. (1971). Techniques for microscopic preparation. In: *Methods in Microbiology*, vol. 4 (ed. C. Booth), pp. 95–112. Academic Press, London.

Ellis, M. B. (1971). *Dematiaceous Hyphomycetes.* Commonwealth Mycological Institute, Kew, Surrey.

Fennell, D. I. (1973). Plectomycetes: Eurotiales. In: *The Fungi*, vol. 4A (ed. G. C. Ainsworth, F. K. Sparrow and A. S. Sussman), pp. 45–68. Academic Press, London.

Glauert, A. M. (1974). Fixing, dehydration and embedding of biological specimens. In: *Practical Methods in Electron Microscopy*, vol. 3 (ed. A. M. Glauert), pp. 1–207. North-Holland Publ. Co., Amsterdam.

Greenhalgh, G. N. and Evans, L. V. (1971). Electron microscopy. In: *Methods in Microbiology*, vol. 4 (ed. C. Booth), pp. 515–65. Academic Press, London.

Hawker, E. L. (1971). Scanning electron microscopy of fungi and its bearing on classification. In: *Scanning Electron Microscopy* (ed. V. H. Heywood), pp. 237–50. Academic Press, London.

Hazzard, G. F. (1967). The detection of microorganisms in petroleum products. *J. Inst. Petrol.*, **53** (524), 267–74.

Heywood, V. H. (1971). Characteristics of the scanning electron microscope and their importance in biological studies. In: *Scanning Electron Microscopy* (ed. V. H. Heywood), pp. 1–16. Academic Press, London.

Higgins, M. (1950). A comparison of the recovery rate of organisms from cotton wool and calcium alginate swabs. *Mon. Bull. Min. Hlth and Pub. Hlth Lab. Serv.*, **9**, 50–1.

Kendrick, W. B. and Carmichael, J. W. (1973). Hyphomycetes. In: *The Fungi*, vol. 4A (ed. G. C. Ainsworth, F. K. Sparrow and A. S. Sussman), pp. 323–509. Academic Press, London.

Lacy, M. L. and Bridgmon, G. H. (1962). Potato dextrose agar prepared from dehydrated mashed potatoes. *Phytopathol.*, **52**, 173.

Langeron, M. (1952). *Précis de Mycologie*, 2nd ed., p. 425. Masson et Cie, Paris.

Leach, C. M. (1967). Interaction of near-ultraviolet light and temperature on sporulation of the fungi *Alternaria*, *Cercosporella*, *Fusarium*, *Helminthosporium* and *Stemphylium*. *Can. J. Bot.*, **45**, 1999–2016.

Leach, C. M. (1971). A practical guide to the effects of visible and ultraviolet light on fungi. In: *Methods in Microbiology*, vol. 4 (ed. C. Booth), pp. 609–64. Academic Press, London.

Luttrell, E. S. (1973). Loculoascomycetes. In: *The Fungi*, vol. 4A (ed. G. C. Ainsworth, F. K. Sparrow and A. S. Sussman), pp. 135–219. Academic Press, London.

Muller, E. and von Arx, J. A. (1973). Pyrenomycetes: Meliolales, Coronophorales, Sphaeriales. *Ibid.*, pp. 87–132.

Raper, K. B. and Fennell, D. I. (1965). *The Genus Aspergillus*. Williams & Wilkins, Baltimore.

Raper, K. B. and Thom, C. (1949). *A Manual of the Penicillia*. Williams & Wilkins, Baltimore.

Reid, N. (1974). Ultramicrotomy. In: *Practical Methods in Electron Microscopy*, vol. 3 (ed. A. M. Glauert), pp. 217–338. North-Holland Publ. Co., Amsterdam.

Smith, G. C. (1969). *An Introduction to Industrial Mycology*, 6th ed. E. J. Arnold, London.

Standards Association of Australia (1972). Australian Standard 1157: *Methods of Testing Materials for Resistance to Fungal Growth*, Part 1. SAA, Sydney.

Sutton, B. C. (1973). Coelomycetes. In: *The Fungi*, vol. 4A (ed. G. C. Ainsworth, F. K. Sparrow and A. S. Sussman), pp. 513–82. Academic Press, London.

von Arx, J. A. (1970). *The Genera of Fungi Sporulating in Pure Culture*. J. Cramer, Lehre.

Williams, S. T., Veldkamp, C. J. and Robinson, C. S. (1973). Preparation of microbes for scanning electron microscopy. In: *Scanning Electron Microscopy*, vol. 3 (ed. O. Johari and I. Corvin), pp. 735–41. ITT Institute, Chicago.

2

Testing of Biodeteriorated Bitumens

ION PITIS* and ALICE STANCU†

* *M.I.C.M. Research Institute for the Electrotechnical Industry, Bucharest, Romania.* † *M.E.F.M.C. Research Institute for Building Materials, Bucharest, Romania*

1. INTRODUCTION

Bitumen used for building insulation, or for metallic pipes and cables buried in the earth, is exposed to the aggressive action of microorganisms such as bacteria, algae, actinomycetes, fungi and so on.

Measurement of the degree of attack by soil microorganisms on the bitumen used for buildings or roads, considered as a conclusive test, is very severe and complex, but it guarantees that the bitumen used for such purposes is highly resistant to attack under working conditions. The laboratory microbial biodeterioration methods used do not give sufficient information alone on the behaviour of bitumen under practical circumstances, and therefore it is necessary to include chemical and mechanical tests, as well as physical ones, in order to estimate the structural changes which may arise in biodeteriorated bitumen.

Research on the resistance of bitumen to soil biological agents has now evolved testing techniques based on:

(a) methods for isolating and identifying the microorganisms concerned in the attack;
(b) chemical and rheological methods to identify the essential changes in bitumens deteriorated under working conditions.

2. TESTING METHODS

The testing of biodeterioration involves:

(a) Laboratory tests for indicating microbial action on the surrounding medium (fungi and earth microorganisms).

(b) Testing under natural conditions of exposure of samples buried in the earth.

The structural and rheological modifications are evaluated by analysing the bitumens before and after exposure to microbial action, as follows:

(a) Macroscopical analysis (matt, glowing or porous appearance).
(b) Analysis for hydrocarbon groups.
(c) Infrared spectrum analysis.
(d) Thermic differential analysis.
(e) Rheological tests:
  (i) Ring and ball.
  (ii) Flexibility on a rod or wood roll.

There are many other methods which may complete the data concerning the structural modifications of the bitumens under conditions of biodeterioration. The most obvious tests have been chosen, especially those which give the most conclusive results in testing before and after biodeterioration on bitumens treated or non-treated with fungistatic substances.

The description of the testing methods is followed by examples which demonstrate their efficiency.

3. TESTING IN LABORATORY CONDITIONS OF EXPOSURE TO SOIL MICROORGANISMS

3.1. Theoretical Basis of the Method

The bitumen samples are laid on a stratum of nutritive medium containing mineral salts and sugar. They are seeded with a spore suspension in a watery solution of mineral salts and sugar, in order to reproduce the conditions under which bitumin exhausts more nutritive medium as it becomes dirtier.

3.2. Apparatus and Materials

A drying oven regulated at 30°C ± 2°, and humidity between 95% and 100% without air circulation.

Testing fungi are *Aspergillus niger* van Tieghem, *Penicillium funiculosum* Thom, *Paecilomyces varioti* Bainier, *Trichoderma viride* Pers. ex. Fr. and *Chaetomium globosum* Kunze. Other species of fungi may also be used.

The testing medium is Czapek-Dox modified nutritive agar-agar medium, with sugar:

$NaNO_3$	2·0 g
KH_2PO_4	0·7 g
K_4HPO_4	0·3 g
KCl	0·5 g
$MgSO_4 \cdot 7H_2O$	0·5 g
Sugar	30·0 g
Distilled water	1 litre
Agar-agar	20·0 g

The watery solution of mineral salts has the same composition as Czapek-Dox modified nutritive medium without the agar-agar.

3.3. Preparation of the Modified Nutritive Czapek-Dox Medium with Sugar

Into a 2 litre Erlenmeyer flask place 200 ml of distilled water, then add the salts (except the phosphates) and sugar. Warm slowly, then add 20 g of agar-agar, and continue heating until the agar-agar is entirely dissolved. Then weigh out the phosphates and dissolve in 200 ml of distilled water, in a 500 ml Erlenmeyer flask. Sterilise both flasks for 30 min at 120°C. After sterilisation mix the two solutions aseptically.

3.4. Samples

At least five samples are needed for each type of test. It is suitable to have samples with 40 × 40 × 40 cm dimensions. The samples may be from bitumen alone, or bitumen applied on a metal or fibreglass tissue support. The bitumen is applied on the support by immersion into warm bitumen at about 130°C.

3.5. Technique

Place into sterilised petri dishes a sufficient quantity of liquid nutritive agar-agar medium with mineral salts and sugar, so that a layer of at least 3–4 mm is obtained. When the nutritive medium has solidified, the bitumen sample is placed on its surface (one sample for each petri dish). All the surface of the medium and also the surface of the sample must be seeded by spraying with a spore suspension prepared as shown in Section 3.2. Then the petri dishes have to be closed and put into the drying oven.

In order to control spore viability, two petri dishes containing the nutritive medium are sprayed with the same spore suspension and these are incubated for the same time as the dishes containing the bitumen samples.

If after 3 or 4 days no mould growth can be observed, the test has to be repeated, using other samples and a fresh spore suspension.

The test period is 28 days or more, if necessary, in which case the spraying has to be repeated with the nutritive solution only, *i.e.* without spores.

If a clear development of moulds is observed by the naked eye before 28 days, the test is finished at that point.

3.6. Evaluation and Interpretation of Results

At the end of the testing the samples are removed from the drying oven and examined both visually and under the microscope (×50).

The assessment of mould development is recorded as follows:

0 = when no growth is visible even under the microscope;
1 = growth scarcely visible by the naked eye, but clearly visible under the microscope (in this case the samples are incubated further up to a total of 56 days);
2 = light growth, which covers at least 25% of the surface of the samples;
3 = medium growth, which covers 25–50% of the surface of the samples;
4 = considerable growth, covering 50% and more of the surface of the samples;
5 = very marked increase, which covers all the surface of the samples.

The evaluation of the resistance of the bitumen to biodeterioration is as follows:

Nature and intensity of the increase	*Appreciation*
0	Strong fungitoxic effect
0 + halo of inhibition	Strong fungitoxic effect due to diffusion
1	Partial fungitoxic effect
2–5	Fungitoxic effect decreasing up to zero

4. TESTING BITUMEN BIODETERIORATION IN SOIL UNDER LABORATORY CONDITIONS

4.1. Main Basis of the Method

The method consists of submitting the sample of bitumen to biodeterioration in soil, under prearranged conditions of moisture and temperature, and determining the microorganism attack based on changes in functional and physico-chemical characteristics.

4.2. Apparatus and Materials

Adjustable thermostat for 30°C ± 1° and 95% ± 5% relative moisture of the air.

Incubator for non-toxic material, with dimensions corresponding to the size and number of samples for the test. The incubators may be of wood, or absorbent or non-glazed ceramic. Incubators made from impermeable materials (glass, plastics, metal, etc.) are unsuitable.

Testing soil, specially prepared.

Cotton tissue of 0·125 kg/m^2 density (or tarpaulin).

Iron plates.

Fibreglass tissue.

Usual laboratory apparatus and materials.

4.3. Preparation of Samples

Since samples of the bitumen which is already in practical use cannot be taken for tests, the bitumen samples are prepared by immersion of the support (iron plates or fibreglass tissue) into warm bitumen at 150°C ± 1°, or by immersion into bitumen suspension at 20°C ± 1°.

These samples may be used for mycological tests and also for other kinds of tests in order to estimate the structural and rheological changes in bitumen, including the reciprocal reaction between the bitumen and the support (corrosion aspects, biodeterioration, etc.).

The supports are:

Iron plates of 0·5–1 mm thickness, used for testing the soil microorganisms action and also the capacity of the bitumen in protecting the support surface against corrosion.

Fibreglass tissue.

4.4. Dimensions and Number of the Samples

To ascertain the changes in behaviour and in the physico-chemical properties of the bitumen, we use a quantity of samples with the shape and adequate dimensions for the specific tests to be done before and after treatment in the conditioned soil:

150 mm × 50 mm for the test of flexibility on a roll.

70 mm × 50 mm for thermic differential analysis.

The number of samples depends on the number of test periods used, but periods of 1 day, 28 days, 3 months and 6 months are obligatory. A 12 month term is optional.

The tests are made on samples of bitumen before and after biodeterioration. For every series of tests five samples are obligatory. Another series of five reserve samples is needed in case of accidents. Hence, for every test 25 samples and 25 reserves are required *in toto*, together with 10 samples for the optional term of 12 months, if needed.

For the ring and ball test, three rings are necessary, to obtain an average of three results. These three rings are filled with bitumen, before and after being biodegraded in soil; the samples are taken from 0·5–1 kg of bitumen, before or after being laid in conditioned soil and taken out for tests after exposure for the set periods.

4.5. Preparation of the Soil for Testing

The soil used for tests must have a composition of equal parts of soil (as used for growing flowers), leaf compost, well-fermented cattle manure and river sand (particles under 1 mm).

The characteristics of the soil must include a pH between 6 and 7·5, a carbon/nitrogen ratio between 8 and 15 and moisture at 30% ± 1%.

The pH is determined by potential measurement on a solution of the soil, using a pH-meter with glass electrodes.

The measurement of nitrogen is determined using current methods such as the Kjeldahl.

The measurement of organic carbon is determined using current methods (such as the simplified method of Anne) acting upon the soil in definite conditions, with a known quantity of oxygen (chromic acid in sulphur medium), then measuring the excess. When calculating the results it is assumed that the consumed oxygen is proportional to the carbon contained in the sample.

4.5.1. *Soil Moisture*

The moisture content of the soil is determined on a dry sample of about 50 g dried at about 100–105°C, until the weight is constant (about 4 hr). If the moisture is higher than 30% the soil is laid in thin layers, then dried up to the desired moisture. For drying the soil, heating is not permitted since it may cause destruction of the microbial flora. If the humidity of the soil is less than 30% it is necessary to add a quantity of water, as follows:

Let x = the number of ml of water necessary to be added to 100 g of soil;
30 = the desired percentage of moisture used;
B = the percentage of soil mosture used;
C = the weight of the soil used for the test (in grammes).

Then
$$x = \frac{30(100 - B)}{70} - B$$
and
$$x' = \frac{x \cdot C}{100}$$

4.5.2. *Verification of Soil Microbial Attack*

The soil, prepared as shown in Section 4.5, is put into the incubator and settled lightly. The humidity of the soil must be 30% ± 1%. Five samples of cotton tissue have to be buried vertically in the soil for three-quarters of their length. The starched cotton tissue samples must be washed with water in order to remove the starch, and then boiled in water, until a drop of I-KI solution, put on the cotton tissue, no longer turns blue. The tissue samples are then slowly wrung out to the weight of 150–180 g/m^2, compared with the initial weight of 100 g of the dry tissue samples, *i.e.* 100–125 g/m^2.

The samples of cotton tissue are taken out of the incubator after 8 days and put into a drying oven at 30°C ± 1°. After this period the tissue samples must have completely lost their stretching strength. The stretch strength is determined by the current methods used for tissue tests. If the samples still have a stretch strength, the soil must be changed for a more active one, and the test repeated with other samples.

When the cotton tissue is completely destroyed—*i.e.* without stretch strength—the soil is considered to be good for bitumen biodeterioration and can be used for bitumen testing.

4.6. Methodology of the Incubation Tests

For incubation tests, petri dishes containing nutritive medium (Section 3.5) were used, and also incubator pots with conditioned soil related to given practical conditions. The incubator pots are filled with the conditioned soil (Section 4.5). The pots must be covered with glass plates. Before testing, the microbiological aggressivity has to be verified and the humidity of the soil adjusted (Sections 4.5.1 and 4.5.2).

If replication of the results is good, 0·5–1 kg of bitumen or 10 samples of bitumen are laid on the required supports for each kind of test, for burial in the soil. The distance between the samples has to be at least 50 mm and from each sample to the wall of the incubator at least 80 mm.

The soil moisture content has to be kept constant during burial, so that one has to verify it and add more water, if necessary, in order to maintain 30% ± 1% humidity.

The samples consist of (a) bitumen (without any treatment or with antifungal treatment) laid on fibreglass tissue support, and (b) bitumen laid on iron plates.

The time for incubation is 2–6 weeks or more.

4.7. Evaluation of Bitumen Biodegradation

The evaluation of biodeterioration of the bitumen when taken out of the incubator pot follows the same rules as the evaluation in the petri dishes (Section 3.6).

5. TESTING UNDER SIMULATED PRACTICAL CONDITIONS FOR SOIL AGGRESSIVITY

5.1. Basic Principles of the Method

This test takes a long time. It allows the determination of the main properties of the building bitumen used for insulation, or of bitumen used for underground pipe and cable protection.

5.2. Samples

As one cannot use bitumen already employed under practical conditions for testing in the laboratory for the physical and chemical changes, iron plates of dimensions 200 mm × 300 mm and 0·1–0·5 mm thickness are immersed in warm bitumen. When taken out, the iron plates are covered with a thin film of bitumen and when this bitumen cools, the sample is ready.

When a high degree of resistance against soil microorganisms is desired, fibreglass tissue supports are used instead of iron plates.

5.3. Working Method

The samples are buried vertically in the soil, very close together, at about 50 cm distance from the place where the same bitumen has been employed. The depth of the buried samples must be at least 60 cm, or at the depth compatible with working conditions. The place where the samples have been buried must be carefully chosen. The number of samples must be according to Section 4.4 and sufficient for all the desired test periods.

5.4. Effects on Biodeteriorated Surfaces

After incubation the samples are taken out of the soil at the end of the set period and examined, in order to estimate the degree of bitumen biodeterioration and the development of the microorganisms. The adherent soil on the samples must be cleaned with water spray and a soft brush. The samples are then sterilised by immersing them for 1 hr in a 1% HgCl solution. They are then washed, first with running water, then with distilled water, and wiped with filter paper, following which they are left to dry in the open air for about 1 hr. They are then examined visually and also under the microscope. Every change in the appearance of the bitumen has then to be recorded (*e.g.* matted, glossy, porous). The surface attack of the microorganisms must also be described, in order to obtain a total percentage evaluation of the biodeteriorated surface.

6. TESTING FOR STRUCTURAL MODIFICATIONS OF THE BIODEGRADED BITUMEN

In order to test for structural modifications of the biodegraded bitumen, one has to use, at the same time, chemical and physical methods.

6.1. Hydrocarbon Group Composition

6.1.1. *Basic Principle of the Method*

Bitumen is a colloidal complex containing hydrocarbon groups, namely oils, resins and asphaltenes. These components can be separated using specific organic solvents.

6.1.2. *Reactives and Materials*

Technical benzene.
Petroleum ether.
Acetone.
Erlenmeyer flasks.
White band filter paper.

6.1.3. *Technique and Interpretation of Results*

(i) *Asphaltene content.* Weigh 5 g of bitumen into an Erlenmeyer flask of 300 ml capacity, add 150 ml of petroleum ether, cover mouth of flask with a glass plate and leave at room temperature for 24 hr. Filter solvent through a white band filter (1), which has been previously weighed. The filter has to be washed with about 50 ml of petroleum ether. The filtrate (1) contains the oil and resin components. The flask with the residue and also the filtrate (1) must be put into a drying oven at 80°C and dried to a constant weight. The residue contains the asphaltenes and impurities.

In order to separate the impurities, the asphaltenes are dissolved in hot benzene. The solution must be filtered through the same filter (1). The filtration can be made under vacuum but the filter has to be washed with hot benzene until the filtrate becomes colourless.

The impurities from the flask and the filtrate (1) are weighed and the asphaltene content is calculated as follows:

$$\text{Percentage of asphaltenes} = \frac{G_1 - G_2}{G} \,.\, 100$$

where G_1 = the impurities plus asphaltene contents, G_2 = the impurities content and G = the bitumen (all weights in grammes).

(ii) *Resins content.* The filtrate (1) is poured into an Erlenmeyer flask (2) (300 ml capacity), previously weighed. The filtrate (2) is distilled on a water bath so that the petroleum ether is retrieved. The residue in the flask (2) is dried in the drying oven at 80°C and then left to cool. Next it is covered with 50 ml of acetone and stirred for 15 min. The resins will sink to the bottom of the flask (2). The upper filtrate is then separated off. The resins have to be washed 5–6 times with acetone. All the acetone used for dissolving the oils and washing the residue (2) is collected in an Erlenmeyer flask (300 ml capacity), previously weighed. The residue from the flask (2) is dried in the drying oven at 130°C until the weight is constant and the resin content then calculated:

$$\text{Percentage of resins} = \frac{G_3}{G} \,.\, 100$$

where G = the bitumen and G_3 = the resins (in grammes).

(iii) *Oils content.* The oils dissolved in acetone from the flask (3) are distilled on a water bath, retrieving the acetone. The residue is dried in a drying oven at 130°C to a constant weight. The oils content is calculated as follows:

$$\text{Percentage of oils} = \frac{G_4}{G} \,.\, 100$$

where G_4 = the oils and G = the bitumen (in grammes).

6.2. Infrared Spectrum Analysis

6.2.1. *Basic Principles of the Analysis*

The infrared ray spectrum analysis utilises waves with lengths greater than the wavelength of the visible spectrum, between 0·8 and 1 Å. Within these limits lies the rotation spectrum and also the vibration spectrum of the molecules which may be excited by electromagnetic waves, specific for the infrared rays involving dipole moment changes.

The excited rotations and vibrations, according to the characteristics of the molecular structure, determine the absorption of some wavelengths from the infrared ray spectrum, corresponding to the absorption bands.

The infrared ray spectrograms are specific for every substance and vary according to the climatic conditions when the substances are exposed to accelerated weathering cycles.

6.2.2. *Apparatus and Materials*

Any type of spectrograph may be used, for instance the Karl Zeiss Jena type UR-20 spectrograph.

White band filter paper.

Bitumen 4% solution in CCl_4, evaporated on a KBr film.

6.2.3. *Technique and Interpretation of Results*

The film of prepared bitumen is radiated with infrared rays and the spectrogram is recorded. The infrared ray spectrum of the bitumen may be ill-defined in bands. A bitumen spectrogram has high amplitudes up to 95%, between the spectrum bands of 2860 cm^{-1} and 2960 cm^{-1}, corresponding to the cyclic aliphatic hydrocarbons, and high amplitudes up to 50% between 1380 cm^{-1} and 1470 cm^{-1}, corresponding to the aliphatic hydrocarbons.

The aromatic hydrocarbons are indicated by the 3030 cm^{-1} band. The oxidised structures are placed between 1600 cm^{-1} and 1700 cm^{-1}. The upper limit of this interval is specific to the cetone and acid group $C = 0$.

Note that the non-biodegraded bitumens, as opposed to the biodegraded bitumens, show in a different position in the infrared ray spectrum bands and changes of the amplitude in the intervals are limited by the absorption bands. The band of 1700 cm^{-1} is removed towards 1800 cm^{-1}, corresponding to the increase of the oxidised compounds.

6.3. Thermic Differential Analysis

6.3.1. *Principle of the Method*

The thermic differential analysis used is an improvement on the traditional method, since it allows study of the phase modifications by means of the recordings for time and temperature.

During constant cooling or warming of solid bodies, it is postulated that if inside a solid body a chemical or physical reaction takes place due to the variation of temperature, a quantity of thermic energy is absorbed or released.

6.3.2. *Apparatus and Materials*

Derivatograph apparatus.
Inert reference material (to temperature variations), *e.g.* Al_2O_3.
Bitumen sample (100 g).

6.3.3. *Technique and Interpretation of Results*

The bitumen samples are cooled or warmed under similar conditions. The temperature of the oven, and any differences of temperature between the bitumen sample and the reference inert material (Al_2O_3), are recorded. When there is no chemical or physical reaction, the thermic flux is transmitted equally through the bitumen sample and through the inert reference material, the difference in temperature being equivalent to zero. Thus a reference graph for time and temperature can be plotted. If this phase is changed, or if inside the sample there is a chemical reaction followed by absorption or elimination of heat, the gradient of temperature relative to the reference material changes also. In this case the temperature variation is recorded as a difference in electromagnetic intensity and there is a deviation from the reference basic line. The degree of deviation is determined by the temperature gradient between the sample and the reference inert material, showing, at the same time, the nature of the thermic process, *i.e.* whether endothermic or exothermic. The results depend on the speed of the heating or cooling of the material.

The temperature variation ranges between 20° and 500°C. The temperature variation and the differences of temperature are recorded graphically.

The interpretation of results is as follows:

Volatilisation interval	°C
Degradation interval	°C
Thermic effect	endo- or exo-
Loss in weight up to 500°C	%
Structural changes:	
Physical, interval	°C
Chemical reaction coupled with physical, interval	°C
Maximum amplitude of the thermic effect	°C

Note that the biodegraded bitumens, in contrast to the corresponding bitumens which have not been degraded, have intervals of volatilisation and of degradation at lower temperatures. They also have greater losses in weight.

7. TESTING RHEOLOGICAL BITUMEN CHANGES

7.1. The Melting Point of the Bitumen, Determined by the Ring and Ball Method

7.1.1. *Basic Principle of the Method*

When a special ball goes through the melting bitumen which fills the inside of a ring, the temperature corresponds to the melting point. The temperature is recorded when the ball touches the reference plate of the ring and ball apparatus, placed near the bottom of the flask which contains the ring and ball metal support. The ring and ball test is a conventional method of testing.

7.1.2. *The Apparatus*

The apparatus (*see* Figs. 1 and 2) consists of:

Berzelius-type glass container.
Metal support, made of three vertical bars on which are fixed two horizontal plates.
Perforated plate provided with two or three holes, 19·2 mm diameter, on which are laid the rings, and another orifice with a 6 mm diameter for a thermometer.
Metal plate support for fallen balls.
Metal lid.
Polished iron balls weighing 3·45–3·55 g.
Supporting hooks.
Mercury thermometer for a 0–200°C interval, with divisions of 0·5°C.
Apparatus for centring the balls on the bitumen surface.
Metallic or china capsule.

The following materials are also necessary:

Polished metal plate.
Sieve or wire tissue.
Glycerine.
Liquid starch or a mixture of 1·3 vols of glycerine with 1 vol of starch.
Thin cigarette paper.

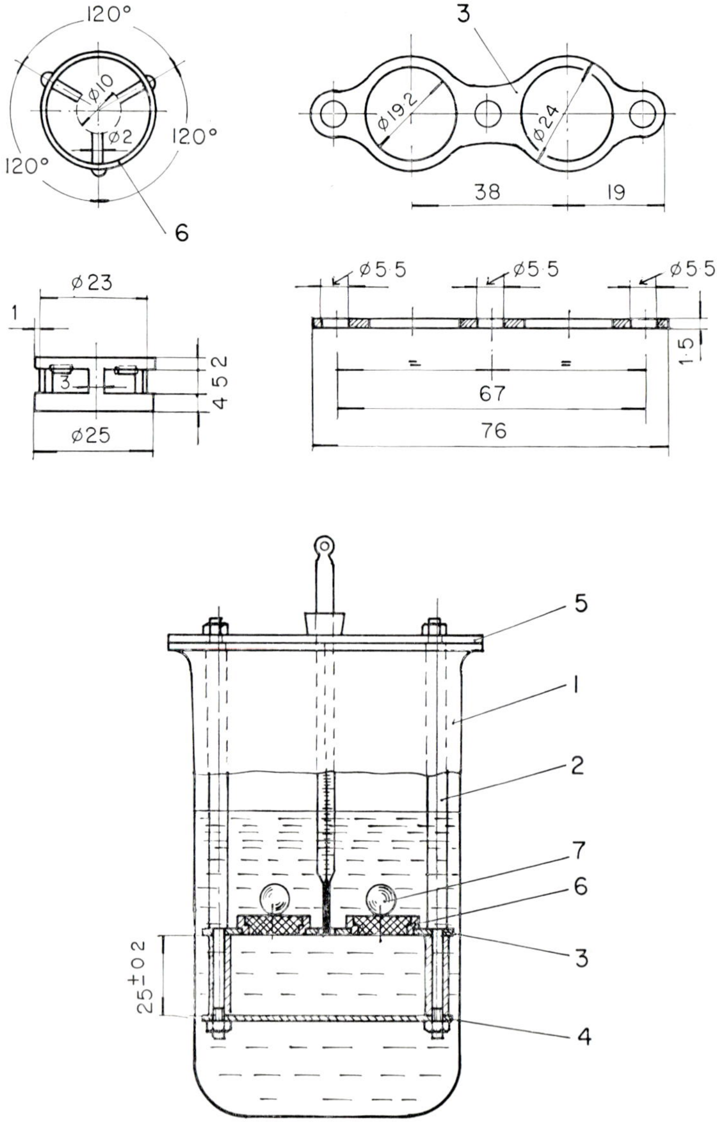

FIG. 1. Ring and ball apparatus with two rings and balls (all dimensions in millimetres). (1) Erlenmeyer flask. (2) Metal support. (3) Brass plate with three or four holes (two or three on which to lay the rings, the third or fourth for the thermometer). (4) Brass plate support for fallen balls. (5) Metal lid. (6) Brass rings. (7) Polished steel balls weighing 3·45–3·55 g.

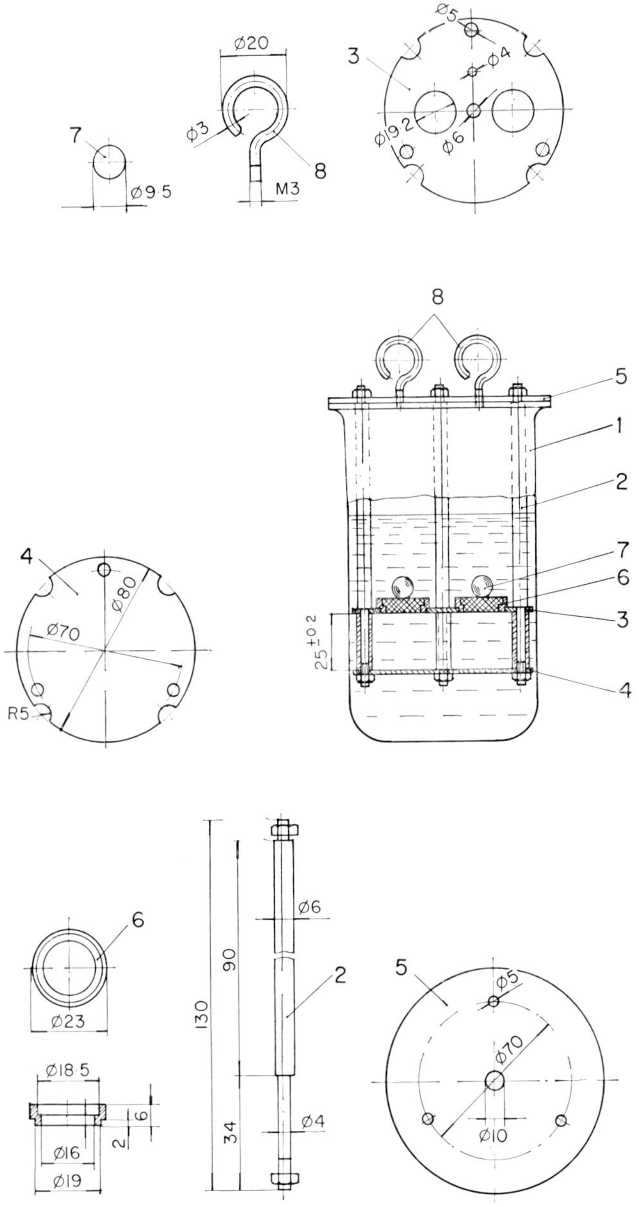

FIG. 2. Ring and ball apparatus with three rings and balls. (1)–(7) as Fig. 1. (8) Supporting hooks.

7.1.3. *Technique and Interpretation of Results*

The bitumen sample, of about 50 g, is melted slowly on an air bath, in the capsule, at a temperature 75–100°C higher than the presumed melting point.

With the bulb end of the thermometer one carefully mixes the material during 15 min of heating, avoiding the introduction of air into the bitumen. Then all is pressed through the sieve. The prepared bitumen is then poured into the balls, which have been previously warmed to the temperature of the bitumen and placed on a metal plate covered with a thin film of liquid starch, to prevent the bitumen adhering. In order to cover the metal plate, a mixture of glycerine and starch may be used or thin cigarette paper soaked in glycerine. When pouring the bitumen in the ring, introduction of air into the bitumen must be avoided.

The rings with bitumen are left for about 30 min at room temperature. With bitumens the melting point of which is less than 26°C, the ring with the bitumen must be left at least 30 min at a lower temperature, *i.e.* at least 8°C less than the melting point.

After cooling, the excess bitumen is removed with a warm knife. If the centring apparatus is missing, one has to make a stamp mark in the centre of the bitumen by pressing slowly with the ball. This is enough to ensure the immobility of the ball and the centring of the ring. The rings and the centring apparatus are placed on the upper horizontal support plate. The thermometer is so placed that its bulb end with the mercury is at the same level as the back surface of the ring and at an equal distance from each of them.

The apparatus with all the accessories is introduced in the glass container, the distance between the back plate of the apparatus and the bottom of the container being 20–30 mm. However, modifications must be made when the bitumen has a melting point above or below 80°C.

(i) *Bitumen with a melting point below* 80°*C.* Fresh distilled water at 5°C $\pm$ 1° is poured into the container so that the level of the water is about 50 mm above the level of the rings with bitumen. The temperature of the bitumen has to be kept at 5°C $\pm$ 1° for 15 min (if necessary the bath has to be kept in a pot with ice). The balls, previously cooled at 5°C $\pm$ 1°, are laid in the centre of the rings, on the stamp or in the centring apparatus, taking care that there is no air between the balls and the bitumen. Then the bath has to be warmed, so that the inside water temperature increases by 5°C $\pm$ 0·5° every minute, starting from 20°C. This speed of temperature increase is mandatory if good results are expected. The bath has to be kept away from air currents. When the ball touches the back plate of the apparatus, the temperature then corresponds to the melting point of the bitumen. The melting point for each ring must be recorded.

(ii) *Bitumen with a melting point above 80°C.* The test is the same as for bitumen with a melting point below 80°C. There are only a few differences. Instead of distilled water one has to use glycerine, which is poured into the container at 32°C ± 1° at the beginning of the test. The balls must also be at the same temperature (32°C ± 1°) before being put on the apparatus support.

The result of the test is expressed as the average of the melting points determined in the same time. The difference between the two results should preferably be within 1°C. The maximum differences between the tests are included in Table I. The results must be recorded within limits of 0·5°C.

TABLE I

Melting point	*Maximum differences allowed between tests*	
	In the same laboratory	*In different laboratories*
≤50°C	1°C	2°C
>50°C	2°C	4°C

7.2. The Flexibility Test

7.2.1. *Basic Principle of the Method*

The behaviour of the bitumen is assessed by the flexibility test by bending a bitumen film of 1–1·5 mm, which is tested at 0°C and 2°C.

7.2.2. *Apparatus and Materials*

Water pot for ice.
Wood roll with half-circular section.
Fibreglass tissue.
Aluminium sheet.

7.2.3. *Technique and Interpretation of Results*

A fibreglass tissue support is covered with a 1–1·5 mm bitumen film by immersion in warm bitumen at 150°C ± 10°. (The dimensions of the samples are given in Section 4.4.) Alternatively one may determine the flexibility of samples using an aluminium support.

For the fibreglass tissue the samples are tested perpendicularly.

(i) *Flexibility at* 0°*C*. The samples are laid in water with ice for 30 min, then taken out and within 5 sec bent slowly on the binding roll. Non-degraded bitumen film should not be fissured or detached from the support.

From the extraction of the samples out of the water with ice to the bending on the roll, the interval of time has to be about 15 sec. The radius of the apparatus is fixed according to official recommendations.

(ii) *Flexibility at* 20°*C* ± 2°. The samples are put into water at 20°C ± 2° for 10–15 min, then every sample is bent slowly on the roll. The bitumen surface must be carefully inspected for every fissure or detachment. The result is recorded as an average of five tests.

Flexibility behaviour is recorded as ‘good’ or ‘bad’. The biodegraded bitumens, when tested for flexibility, fissure and detach themselves from the support when bent on the roll.

3

Evaluation Techniques for the Resistance of Pipe-Joint Rings and Sealants to Microorganisms

WALTRAUT KERNER-GANG

Fachgruppe 'Biologische Materialprüfung', Bundesanstalt für Materialprüfung (BAM), 1 Berlin 45, Unter den Eichen 87, Federal Republic of Germany

While pipe-joint rings are made of rubber (natural or synthetic rubber or mixtures of both), sealants may be produced from a variety of materials. As we cannot discuss all sealant materials we shall restrict the description to microbiological tests for bituminous fillers and plastic sealants.

1. PIPE-JOINT RINGS

When water pipes, drainage pipelines and bye-washes are laid, the sealing of the pipe joints is particularly important. While formerly hemp and bituminous bandages were used and the remaining gaps were filled with a plastic filler, today rubber pipe-joint rings are used, initially those made of solid rubber and later those made of cellular rubber which is produced by adding an inflating agent during the vulcanising process.

Several authors have investigated the microbiological resistance of rubber and rubber pipe-joint rings and have come to different conclusions. More than 60 years ago Söhngen and Fol (1914) pointed out that natural rubber may be attacked by Actinomycetes. ZoBell and co-workers (1942 and later) and Blake and co-workers (1949 and later) reported that vulcanised rubber mixtures may be deteriorated by different micro-organisms. Rook (1955) showed that vulcanised rubber was susceptible to attack by Streptomycetes. This was confirmed by Nette *et al.* (1960).

Large-scale investigations carried out in the Dutch Keuringsinstituut voor Waterleidingsartikelen (KIWA, 1961) showed that natural rubber was attacked even when certain fungicides were compounded into the rubber. Deterioration was found to be caused by Streptomycetes. Synthetic rubber, on the other hand, was not attacked by these organisms (Leeflang, 1963). Heederik (1966), too, reported on damage to pipe-joint rings in sewage mains in different countries and Hills (1967) described similar results as they have been reported by KIWA for New Zealand.

These negative findings contrast with more positive experience in other countries. Dunkley (1964) estimated the service life of natural rubber pipe-joint rings in Great Britain to be 50 years. Dickenson (1965) tends to the opinion that natural rubber is resistant to microorganisms; he contended that the unusual composition of Dutch drinking water and the special design of the pipe-joint rings were responsible for observed deteriorations in the Netherlands.

An inquiry conducted by 36 construction and drainage boards in the Federal Republic of Germany by the Committee of Standards for Water Supply revealed that no microbiological attack of pipe-joint rings in sewer pipes was observed (Küch and Kerner-Gang, 1973).

Cundell and Mulcock (1973*a*) inspected 583 natural rubber rings in sewage lines in New Zealand and found that the proportion of rings subject to microbial attack was 63%. With regard to conflicting results reported in the literature, they investigated (Cundell and Mulcock, 1973*b*) the resistance of vulcanised rubber to microbial deterioration and confirmed the susceptibility of natural rubber to deterioration by soil- and water-borne microorganisms. Synthetic rubber was not degraded with the exception of styrene–butadiene rubber. The criteria used for determining the relative resistance of vulcanised rubber were the appearance of the rubber surface, its loss in tensile strength and the respiratory activities of the organisms growing on the rubber. In another investigation (Cundell and Mulcock, 1975) scanning electron micrographs were used to compare the patterns of damage produced by the effects of ozone, heat and microorganisms.

As the results described in the available literature are partly conflicting, the microbiological resistance of solid and cellular rubber rings was studied at the BAM; the results were evaluated by means of the scanning electron microscope (Kerner-Gang, 1973). It was found that solid rubber rings made of natural or synthetic rubber or a mixture of both were, in general, more resistant to microbiological deterioration than cellular rubber rings. Solid rubber rings of synthetic rubber were found to be most resistant, whereas cellular rubber rings of a mixture of natural and synthetic rubber proved to be most susceptible.

The test methods used at the BAM for comparative studies of the microbiological resistance of four solid and three cellular rubber specimens will be outlined below.

1.1. Tests with Pure Cultures on Agar Without a Carbon Source

This investigation is made to determine whether the chosen test organisms are able to utilise the test material as the sole carbon source. Specimens that are colonised at the end of the specified test period may be considered as 'susceptible' and must not be subjected to further tests.

Method of test. 3–5 mm thick discs of rubber rings to be tested are placed on a 3–4 mm plate of a commonly used mineral salts agar without a carbon source, *i.e.* one specimen in the centre of each petri dish; it is then inoculated with 0·5 ml of an aqueous spore suspension by spreading the suspension evenly over the specimens and the agar with a pipette. The dishes are subsequently incubated for 8 weeks at 24°C ± 1°.

Table I shows those microorganisms used at the BAM and the rate of deterioration of three cellular rubber rings made of natural rubber, synthetic rubber, and a mixture of natural and styrene–butadiene rubber.

TABLE I

DETERIORATION OF CELLULAR RUBBER RINGS EXPOSED IN PETRI DISHES ON AN AGAR WITHOUT A CARBON SOURCE AFTER A TEST PERIOD OF 8 WEEKS

Test organisms	*Cellular rubber rings of*		
	natural rubber	*synthetic rubber*	*mixture of natural and styrene–butadiene rubber*
Aspergillus flavus	++	(+)	++
Aspergillus terreus	+	—	++
Cladosporium herbarum	(+)	—	++
Humicola grisea	+	—	++
Paecilomyces varioti	(+)	—	++
Penicillium citrinum	+	—	++
Pen. funiculosum	(+)	(+)	++
Pen. ochro-chloron	+	(+)	++
Phialophora fastigiata	+	—	+
Trichoderma viride	+	(+)	++
Streptomyces griseus	(+)	—	++

— = no growth.
(+) = growth at the outer edge.
+ = moderate growth.
++ = significant growth.

Further test organisms (fungi or Streptomycetes) which are considered to yield informative results may additionally be included in certain tests. At least eight of the organisms listed in Table I should, however, be used. Two to three parallel tests should be carried out with each fungus.

This method can be considered as a screening test for determining the resistance or susceptibility of rubber rings and for selecting those materials which should undergo a more rigorous test. This would, for example, apply to the synthetic rubber mentioned in Table I.

1.2. Long-term Tests

It is difficult to recommend further test methods as the conditions of use will ultimately decide the test method to be used. Alternative test methods are the Leeflang Test Bath and the soil-burial method in which the specimens are buried in moist soil rich in microorganisms for 1–2 years at 28–30°C or the material is stored for some time in the sludge of a sewage treatment plant.

Leeflang (1963) devised a test method to gauge the relative resistance of rubber pipe-joint rings for water and sewer mains to microbiological deterioration. This method has also been used by other authors in the same or a modified way. Rubber strips of approximately 50 cm^2 are put into dishes together with a deteriorated piece of rubber to serve as a source of infection. A slow flow of unchlorinated tap water at 20°C is maintained through the dish, which is protected from light to prevent algal growth. Unless the test material is attacked earlier, the test will be extended over 12 months.

This method is certainly very suitable, the only disadvantage being that attacked material is required which is not readily available in every laboratory.

After a long-term test the material should in any case be examined by scanning electron microscopy to ensure that the surface was actually deteriorated and not just superficially overgrown by the fungi. Figures 1 and 2 show attacked specimens under the SEM.

1.3. Evaluation of Preservatives

It is known that vulcanised rubber may inhibit microbial growth. Some authors attribute this effect to the use of vulcanising catalysts (Haldenwanger, 1970). On the other hand, preservatives may also be incorporated to prevent microbial attack.

Method of test. The effectiveness of microbiocidal preservatives is tested according to the method described in Section 1.1, with the exception

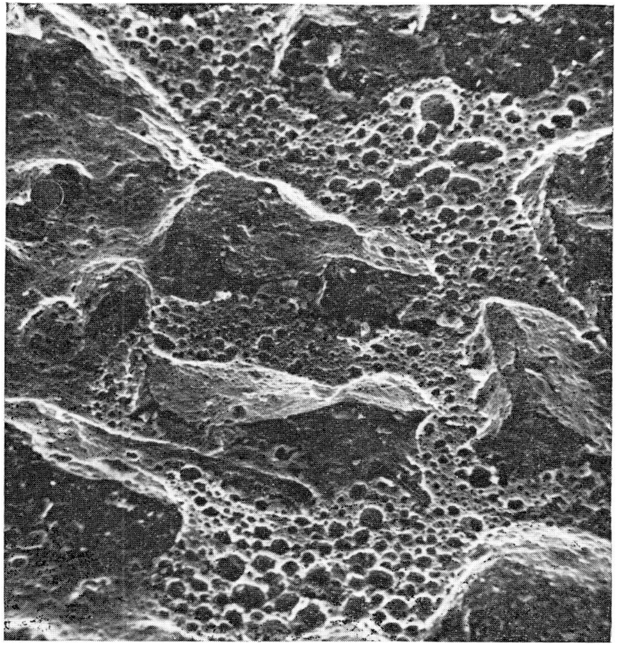

FIG. 1. Superficies of a solid rubber ring (synthetic rubber) after 2 years' exposure in the sludge.

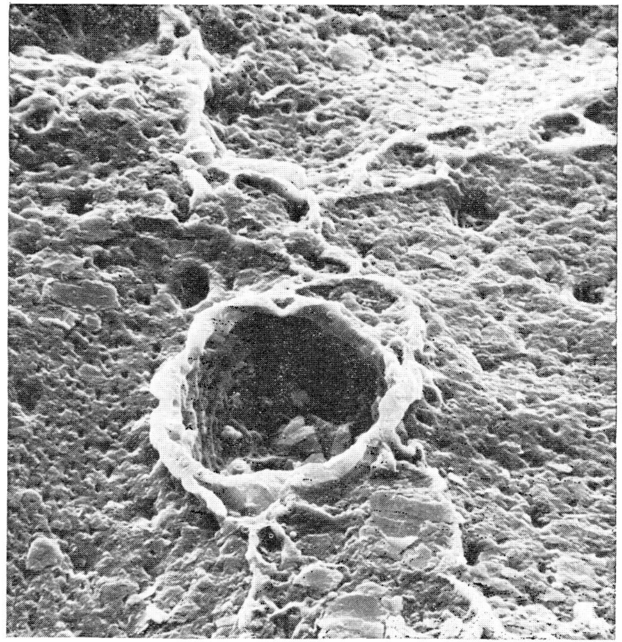

FIG. 2. Superficies of a cellular rubber ring (mixture of synthetic and natural rubber) after 15 months' exposure to fungi.

that a carbon source is added to the agar to enable the fungi to develop more vigorously (*e.g.* mineral salts agar supplemented with 30 g of glucose/litre, or malt agar). The effectiveness of the formulation may be determined by the zone of inhibition produced around the specimen.

When bacteriological media are used, this method may also be applied to determine the bactericidal effectiveness of certain substances when the optimum temperature for bacteria is maintained (*e.g.* 37°C for *Staphylococcus aureus*). A test period of 3–5 days is mostly sufficient.

Zyska *et al.* (1972) used the soil-burial method to study the effectiveness of several microbiocides incorporated in rubber for cable coverings. In parallel tests the effects of preserved rubber powder on soil micro-organisms were measured in the Warburg apparatus.

2. SEALANTS

2.1. Bituminous Sealants

A method was developed at the BAM to evaluate the resistance of bituminous sealants to mildew. This method presupposes the use of suitable test fungi (Kerner-Gang, 1967). Of 21 fungal species and strains investigated for their behaviour towards seven bituminous sealants, six fungi were selected for the tests described below.

Method of test. The bitumen to be tested is liquefied under heat and poured into petri dishes in such a way as to cover an area of a diameter of 4–5 cm, 3 mm thick. Subsequently the remaining area between the bitumen and the edge of the dish is filled with a mineral salts agar according to

TABLE II

COMPOSITION OF THE CULTURE MEDIA ACCORDING TO ASTM

(amount per 1000 ml aqua dest.)

Components	*Mineral salts agar*	*Mineral salts solution*
KH_2PO_4	0·7 g	0·7 g
K_2HPO_4	0·7 g	0·7 g
$MgSO_4 \cdot 7H_2O$	0·7 g	0·7 g
NH_4NO_3	1·0 g	1·0 g
NaCl	0·005 g	0·005 g
$FeSO_4 \cdot 7H_2O$	0·002 g	0·002 g
$ZnSO_4 \cdot 7H_2O$	0·002 g	0·002 g
$MnSO_4 \cdot 7H_2O$	0·001 g	0·001 g
Agar–agar	20 g	—

ASTM (Table II). The reaction of the fungi both on the border area between bitumen and agar and on the bitumen itself is observed. The bitumen and the agar surfaces are then inoculated by means of a pipette with 0·5 ml of a spore suspension which is produced as follows.

The spore mixture comprises *Aspergillus flavus*, *A. ustus*, *A. versicolor*, *Dactylium fusarioides*, *Penicillium citrinum* and *Trichoderma viride*. When the fungal cultures are well sporulated, in general 2–4 weeks after inoculation, an aqueous spore suspension of each species is produced by transferring the spores from inclined agar tubes into sterile water. Each suspension is then filtered through sterile glass wool and centrifuged aseptically. The supernatant liquid is discarded and the residue is resuspended

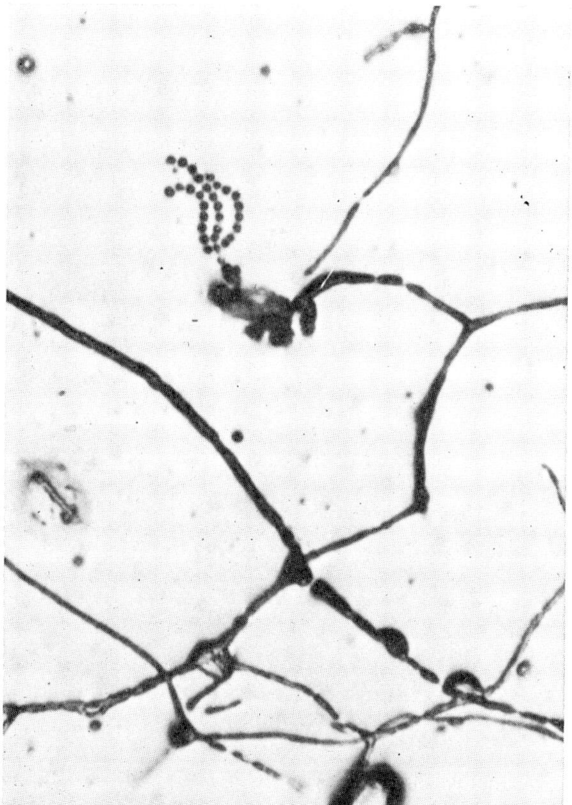

FIG. 3. Mould growth on bituminous sealants after a 4-week incubation period.

in a sterile mineral salts solution (Table II). Finally, equal amounts of all six suspensions are mixed into the spore suspension.

After inoculation the dishes are incubated for 4 weeks at 24°C. After 1 and 2 weeks the dishes are inspected. After 4 weeks, fungal growth on the bitumen is assessed if necessary with the help of a magnifying glass (Fig. 3).

2.2. Plastic Sealants

The test method described below was used at the BAM to evaluate sealants for socket joints for their resistance to mildew attack. Elastic material on a polyurethane basis and rigid material on a phthalate basis were tested.

Method of test. The tests are carried out as described in Section 1.1 for testing rubber pipe joints. But instead of the rubber, specimens of the sealant of 6–10 cm^2 are placed on the agar. Figure 4 shows fungal growth of *Humicola grisea* on the rigid material after 3 weeks.

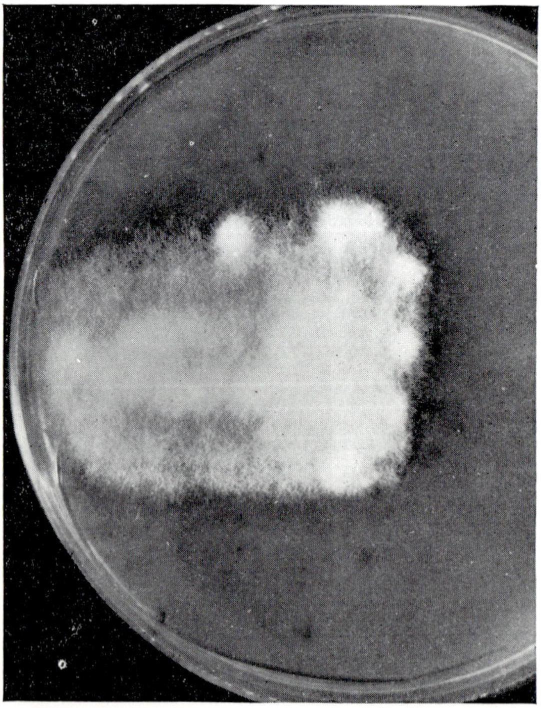

FIG. 4. Fungal growth on plastic sealants after a 3-week incubation period on a carbon-free agar. Test fungus: *Humicola grisea*.

REFERENCES

Blake, J. T. and Kitchin, D. W. (1949). Effect of soil microorganisms on rubber insulation. *Ind. Eng. Chem.*, **41**, 1633–41.

Cundell, A. M. and Mulcock, A. P. (1973*a*). Microbiological deterioration of natural rubber pipe-joint rings. *Material u. Organismen*, **8**, 165–77.

Cundell, A. M. and Mulcock, A. P. (1973*b*). Measurement of the microbiological deterioration of vulcanised rubber. *Material u. Organismen*, **8**, 1–15.

Cundell, A. M. and Mulcock, A. P. (1975). The biodegradation of vulcanised rubber. *Devel. Ind. Microbiol.*, **16**, 88–96.

Dickenson, P. B. (1965). Verwendung von Dichtungsringen aus Naturkautschuk in unterirdischen Rohrleitungen. *Kautschuk-Fortschr. u. Entw.*, **18**, 87–92.

Dunkley, W. E. (1964). Die Verwendungsdauer des Kautschuks in Rohren und Leitungen. *Kautschuk-Fortschr. u. Entw.*, **17**, 98–102.

Haldenwanger, H. H. M. (1970). *Biologische Zerstörung der makromolekularen Werkstoffe*. Bd. 15 der Reihe Chemie, Physik und Technologie der Kunststoffe (ed. K. A. Wolf). Springer Vlg, Berlin.

Heederik, J. P. (1966). Prüfen von Betonrohren in den Niederlanden. *Beton-Ztg.*, **32,** 635.

Hills, D. A. (1967). The degradation of natural rubber pipe-joint rings. *Rubber J.*, **149,** 12, 13, 15, 17, 77.

Kerner-Gang, W. (1967). Zur Prüfung von Bitumen auf Widerstandsfähigkeit gegen Pilzbefall. *Material u. Organismen*, **3**, 161–77.

Kerner-Gang, W. (1973). Beständigkeit von Abdichtungsmaterialien auf elastomerer Grundlage gegenüber Mikroorganismen. *Material u. Organismen*, **8**, 17–37.

KIWA (1961). *Aantasting van Rubberringen voor Waterleidingsbuizen.* s'Gravenhage, 1961.

Küch, W. and Kerner-Gang, W. (1973). Untersuchung elastomerer Dichtungselemente für Rohrverbindungen in Entwässerungskanälen und -leitungen auf mechanische, chemische und biologische Eigenschaften. *Rohre—Rohrleitungsbau—Rohrleitungstransport*, **12**, 92–109.

Leeflang, K. W. H. (1963). Microbiological degradation of rubber. *J. Amer. Water Works Assoc.*, **55**, 1523–35.

Nette, J. H., Pomorzewa, N. W. and Koslowa, J. J. (1960). Die Zersetzung des Kautschuks durch Mikroorganismen. *Sowjetwiss. naturwiss. Beiträge*, No. 7, 770–5.

Rook, J. J. (1955). Microbiological deterioration of vulcanised rubber. *Appl. Microbiol.*, **3**, 302–9.

Söhngen, N. L. and Fol, J. G. (1914). Die Zersetzung des Kautschuks durch Mikroorganismen. *Zbl. Bakteriol.*, II Abt., **40**, 87–98.

ZoBell, C. E. and Grant, C. W. (1942). The bacterial oxidation of rubber. *Science*, **96**, 379–80.

Zyska, B. J., Rytych, B. J., Zankowicz, L. P. and Fudalej, D. S. (1972). Microbiological deterioration of rubber cables in deep mines and the evaluation of some fungicides in rubber. In: *Biodeterioration of Materials*, vol. 2 (ed. A. H. Walters and E. H. Hueck-van der Plas), pp. 256–67. Applied Science Publishers, London.

4

Test Methods for Evaluation of Susceptibility of Plasticised PVC and Its Components to Microbial Attack

M. PANTKE

Fachgruppe 'Biologische Materialprüfung', Bundesanstalt für Materialprüfung (BAM), 1 Berlin 45, Unter den Eichen 87, Federal Republic of Germany

1. INTRODUCTION

In the past decades plastics have enjoyed an ever greater popularity due to the diversity of their fields of application. Based on the countless possibilities of adapting plasticised PVC to a multiplicity of end-uses by the development of new technologies and above all by compounding special ingredients, PVC has assumed a prominent role. Properties such as elasticity, flexibility at low temperatures, impact resistance, resistance to migration, insulating properties and thermal stability may be controlled by compounding the basic polymer with certain plasticisers, stabilisers, antioxidants, lubricants, filling agents and pigments.

The chemical constitution of the different components is diverse and some of them may support microbial growth. By way of their enzymes and metabolites, microorganisms may moreover deteriorate plasticised PVC materials. Deterioration may extend from surface growth which may easily be wiped off, permanent discolorations, to mechanical and electrical failures or even the destruction of the entire product.

Microbial attack of plasticised PVC products or its components has been described in a large number of publications, *e.g.* Lee (1962), Wessel (1964, 1968), Haldenwanger (1970), Pantke (1971) and Becker and Gross (1974).

2. TEST METHODS

The problem of the resistance of plasticised PVC to microbiological deterioration is primarily of economic importance. The service performance of such a material under certain conditions of use can best be predicted by choosing a test method that simulates the conditions under which the material will be used most effectively. Tests are carried out with finished or semi-finished products or preferably with films. The lower the film thickness, the quicker is the development of microbial attack. Moreover, those non-polymer components may be tested that are supposed to support or inhibit microbial growth.

The test methods used for assessing the biosusceptibility of plastics may be classified into tests in liquid media, solid media, soil burial and durability tests.

According to the damage observed, it was mainly tests with fungi and bacteria which were developed. For liquid culture tests pure cultures are mostly used, whereas a natural microflora (*e.g.* soil suspension) is used in the soil burial and durability tests.

The testing time of the different methods varies. The soil burial method may take weeks or even months. With natural weathering tests at least one or more years are required. Culture tests in liquid and solid media take much less time, from a few days to a few weeks. Short-term tests in liquid media can give a valuable indication of the likely performance of a material after only 24 hr, although the value of the data obtained should be considered with caution.

As it is impossible to review the great number of test methods described in the literature, those methods will be selected that may be carried out in a microbiological laboratory with the usual technical equipment. Special or rarely applied methods will only be mentioned. The collection of test methods presented is of course not complete, but it is hoped that the description of a number of commonly applied methods as well as the relevant literature will serve as practical instructions and as an element of orientation.

3. TEST ORGANISMS

The causal agents of deterioration and the organisms used to evaluate the resistance of plastics are strictly microorganisms. This term comprises different groups. Mould fungi have almost always been isolated from attacked plastics. They are used in the majority of test methods. In addition, bacteria, yeasts and Actinomycetes are also known to degrade plastics and are used in some tests.

It should be stressed again that the handling of microorganisms presupposes the necessary training and skill as the work sometimes involves the handling of pathogens. Work should therefore only be carried out by microbiologically trained staff.

3.1. Test Fungi

When microbiological damage to plastics is examined, quite a number of organisms, mostly fungi, are isolated. Some of them are used in a variety of test methods. In addition, several test organisms have been adopted from other fields of materials testing, *e.g.* textile testing. Early standard methods specified only the genus and the species to be utilised. As the physiological characteristics and hence the aggressiveness of different strains may vary considerably, however, the strains were later also specified (Table I). It is advisable to obtain the specified test organisms from culture collections. When the test fungi are frequently required and kept in culture, checks should be made at certain intervals to ensure that no characteristics have changed.

Apart from the fungi mentioned in Table I, an attempt was made to find other organisms which have specific plastic-degrading properties. Mills *et al.* (1971) mention *Talaromyces emersonii*, for example, a thermophilic Ascomycete. This organism may often be isolated from soil and urban refuse. It proved to be very adaptable to a variety of plasticisers. In further investigations Mills and Eggins (1974) studied the growth of 19 thermophilic fungi on a number of PVC plasticisers as well as their ability to produce esterases.

In 1970 Osmon *et al.* described studies to follow the deterioration of plasticisers and plasticised PVC by yeasts. Further tests are reported by Osmon and Klausmeier (1971) on *Candida lipolytica* CBS 6114. This non-pathogenic test organism degrades the same plasticisers and plasticised PVC formulations as the mixed fungal culture specified in ASTM D 1924 or *Pseudomonas aeruginosa*, which will be discussed later.

A large proportion of 28 Actinomycetes isolated from soil were able to degrade six representative organic ester plasticisers (Klausmeier and Osmon, 1976).

As early as 1961 Klausmeier and Jones pointed out that it was necessary to obtain further test organisms by means of the so-called enrichment culture. Cavett and Woodrow (1968) described an enrichment culture technique to isolate plastic-degrading organisms. River mud near the discharge of a plastic factory was enriched with various plasticisers for a period of 2–3 weeks. 1 g portions were placed in 100 ml amounts of sterile FA No. 5 mineral salts medium (Table II) containing 1% of the same

TABLE I

TEST FUNGI USED

	ASTM D 1924	*Berk* et al. (1957)	*Burgess and Darby* (1965)	*ISO TC* 61	*NF* $_X$ 41–513, 514, 515	*Wälchli and Zinkernagel* (1966)
Alternaria solani		×				
A. tenuis			×			×
Aspergillus amstelodami					×	×
A. flavus	ATCC 9643	×	×		×	×
A. niger	ATCC 9642	×	×	×		×
A. oryzae		×	×			
A. sydowii						×
A. terreus		×				
A. ustus		×				
A. versicolor	ATCC 11730	×	×			×
Chaetomium globosum			×	×	×	×
Cladosporium herbarum			×			×
Curvularia geniculata		×				
Fusarium sp.		×				
Glomerella cingulata		×				
Memnoniella echinata					×	×
Metarrhizium glutinosum			×			
Mucor sp.		×				
Myrothecium verrucaria		×			×	×
Paecilomyces varioti		×	×	×	×	×
Penicillium brevi-compactum					×	×
P. chrysogenum		×	×			
P. citrinum		×				
P. cyclopium					×	×
P. frequentans		×				
P. funiculosum	ATCC 9644	×	×	×	×	×
P. glaucum						×
Pullularia pullulans	ATCC 9348	×	×			
Stachybotrys atra		×	×		×	×
Stemphylium consortiale		×				
Sterigmatocystis nigra					×	×
Trichoderma sp.	ATCC 9645	×				
Trichoderma T-1					×	×
T. viride			×	×		×

TABLE II
CULTURE MEDIA USED

Components	*FA No. 5 medium* (*Berk* et al. 1957) *and culture medium according to ASTM* 1924	*Czapek-Dox agar, modified culture medium according to ISO R* 846*	*Basal medium for bacteria, yeasts and Actinomycetes* (*Pankhurst* et al., 1972)	*Culture medium according to NF X* 41-513
	Parts by weight (*g*)			
Sodium nitrate ($NaNO_3$)		2·0	2·0	
Ammonium nitrate (NH_4NO_3)	1·0			3·0
Potassium dihydrogen phosphate (KH_2PO_4)	0·7	0·7	1·0	1·0
Potassium monohydrogen phosphate (K_2HPO_4)	0·7	0·3	0·4	
Potassium chloride (KCl)		0·5	0·5	0·25
Magnesium sulphate ($MgSO_4 \cdot 7H_2O$)	0·7	0·5		0·5
Sodium chloride (NaCl)	0·005			
Ferrous sulphate ($FeSO_4 \cdot 7H_2O$)	0·002	0·01	0·01	
Zinc sulphate ($ZnSO_4 \cdot 7H_2O$)	0·002			
Manganese sulphate ($MnSO_4 \cdot 4H_2O$)	0·001			
Saccharose		(30·0)*		
Agar-agar	20·0	20·0		15·0
pH adjusted to	6·0–6·5		7·2–7·5	6·2–6·4

* Method B with, Method A without sugar

plasticiser used in the mud enrichment. The media were then incubated at 20°C on a shaker for 2–3 days and 1 ml transferred to a fresh culture solution. This process was repeated for 2–3 weeks. As soon as fungal growth was visible, the organisms were isolated and transferred to solid media.

3.1.1. *Production of the Spore Suspension*

Depending on the test method used, single or mixed fungal cultures are utilised for assessing the potential susceptibility of plastics. The preparation of the spore suspension varies accordingly. The method described here has been applied in a somewhat modified form by Klausmeier (1970*a*).

Well-sporulated subcultures grown at 28°C for 14–21 days are utilised to prepare the spore suspension. 10 ml of sterile aqua dest. containing 0·05 g/litre of a non-toxic wetting agent, *e.g.* Tween 80, is poured into the respective culture tubes. A sterile inoculating wire is used to scrape the spores gently from the fungal mat. All spore charges obtained from one fungal species are poured into a sterile 125 ml glass-stoppered Erlenmeyer flask containing 45 ml of sterile aqua dest. and 10–15 glass beads, 5 mm in diameter. The flask is then shaken vigorously for a few minutes to liberate the spores from the fruiting bodies and to break up the spore clumps. In order to remove mycelial fragments, the suspension is filtered through a sterilised glass funnel containing a thin layer of sterile glass wool. The filtered spore suspension is centrifuged aseptically and the supernatant liquid discarded. The residue is resuspended in 40–50 ml of sterile water and again centrifuged. In this manner the spores obtained are washed three times. The final washed residue is diluted with a sterile mineral salts solution (Table II) to contain 1 million ± 200 000 spores/ml. This is determined by means of a counting chamber (a small clump of spores is counted as one spore). When a mixed spore suspension is required, equal volumes of the spore suspensions of the individual organisms are mixed. Spore suspensions should be prepared fresh for each test the same day. For further tests they may, if necessary, also be kept in a refrigerator at 3–10°C for up to 4 days. To check the viability of the spores, some slanted agar tubes are also inoculated in each test. The agar of the tube must be overgrown within a few days.

3.2. Bacteria

Apart from the organisms mentioned which are used to evaluate the susceptibility of plastics, bacteria should not be omitted. Booth *et al.* (1968), for example, isolated 62 bacterial cultures from attacked plastic material. 23 of these bacteria, mainly belonging to the genera *Brevibacterium* and *Pseudomonas*, caused weight losses of up to 20% under laboratory conditions. *Pseudomonas aeruginosa* NCTC 8060 is mentioned by several authors as a test organism. Burgess and Darby (1965) wrote in the introduction to their paper: '*Pseudomonas aeruginosa* was chosen as a test organism because of its ability to attack certain PVC materials and in full knowledge of Bejuki's forceful warning on its potential pathogenicity.' Hitz *et al.* (1967) used the same strain for their 'tube test', whereas Cavett and Woodrow (1968) used a fluorescent *Pseudomonas* strain CBCC 1237 for respirometric measurements. Klausmeier and Jamison (1973) carried out weight-loss measurements of plasticised PVC deteriorated by *Pseudomonas aeruginosa* ATCC 13388.

4. LIQUID CULTURE TESTS

In the test methods described below, the observed growth of the organisms is used to determine the susceptibility or resistance of a PVC material. Methods to determine changes in the test material will be discussed in Section 8.

4.1. Evaluation of the Resistance of Plasticisers and Other Liquid Substances to Fungi (according to NF X 41-513)

3·5 g of the liquid to be tested is poured into an Erlenmeyer flask containing 70 g of a mineral salts solution (Table II). It is sterilised for 5 min at 120°C and inoculated with 1 ml of a mixed spore suspension (Table I). If the substance to be tested is not thermostable it is added after the sterilisation of the nutrient solution. The inoculated flasks are incubated for 6 days at 30°C and 95% relative humidity on a shaker together with three controls containing castor oil instead of the substance to be tested. For evaluation, *see* the following section.

4.2. Determination of the Fungistatic Effects of Substances (according to NF X 41-513)

Mixtures of growing proportions of the substance to be tested and castor oil are produced and added to the nutrient solution. The intensity of fungal growth in both this and the above test is rated as follows:

0 = No growth.
1 = Growth restricted to some mycelial clusters.
2 = Moderate growth, mycelial clusters scattered in the liquid.
3 = Considerable growth, dense homogeneous mycelium.
4 = Heavy, homogeneous growth, equal or superior to that observed in the control with pure castor oil.

4.3. Determination of the Mycelial Dry Weight

Stahl and Pessen (1953) used the weight of mycelial pellets in shake flasks as a criterion of the degree of susceptibility of plasticisers to microbiological deterioration. This presupposes that the mycelium grown in shake culture may be separated easily from the test specimens. Pankhurst *et al.* (1972) filtered the liquid of each flask through a weighed membrane filter and washed it several times with distilled water. After filtration the membrane filters were dried and reweighed.

Sometimes it is impossible to weigh fungal mycelium, *e.g.* when the material to be tested is a powder or cannot be separated from the mycelium. In such cases the material to be tested is incorporated in a solid medium and examined by a method described in Section 5.

4.4. Enumeration of Bacteria

The bacterial cell count in the culture liquid may be used to assess the degradability of plasticisers. It may be measured by counting them in a counting chamber, by the agar-dilution test or by the MPN (Most Probable Number) method. All three methods involve a lot of work, which may be avoided by measuring the turbidity of the culture liquid produced by growing numbers of bacteria.

4.5. Determination of the Degree of Turbidity

This method, which is called the 'tube test method' to determine the bacterial deterioration of plasticised PVC, was described by Hitz and Zinkernagel (1967). Test specimens are sterilised with ethylene oxide and placed into test tubes each containing 20 ml of a sterile mineral salts solution (FA No. 5 medium without agar—Table II). Dumbbell samples of 75 mm length and 0·5 mm thickness proved to be suitable for subsequent mechanical testing. Half the test tubes serve as sterile controls. Four drops of a 0·1 % phenyl-Hg-borate solution are added to each of these test tubes. The remaining tubes are inoculated with *Pseudomonas aeruginosa* NCTC 8060. The cell suspension is prepared as follows. Bacteria are harvested from a 24 hr agar slant culture (Brain Heart Infusion Agar) by means of a platinum loop and are introduced into 10 ml 1/15 M of a sodium phosphate buffer (pH 6·7). The solution is centrifuged three times for 5 min each at 2900 g and resuspended each time in a fresh buffer solution. The bacterial density is measured in a counting chamber and adjusted to 10^6 bacteria/ml. The tubes are inoculated with three drops of this cell suspension. The density of bacterial growth may be determined by means of a nephelometer. The contents of each culture tube are emptied under aseptic conditions into sterilised nephelometer tubes. With progressing bacterial density the suspension must be diluted 1:10 with sterile mineral salts solution. Bacterial suspensions of the same density produced in parallel tests, which have not, however, been washed and centrifuged, yielded similar results.

4.6. Determination of Oxygen Consumption

A screening test for new PVC formulations which takes only a few days will certainly be preferred to conventional methods, such as the soil burial method, which may yield results only after several weeks or even months. The data obtained from such a screening test must, however, be critically evaluated. On the other hand such rapid tests are a very valuable means of roughly assessing the likely performance of materials, especially with regard to the time saved.

Pankhurst *et al.* (1972) described a method to determine the oxygen uptake as follows:

> 'Oxygen uptake by *Pseudomonas aeruginosa*, *P. resinovorans* NCIB 9280 and *Serratia marcescens* NCIB 9523 in the presence of polymeric substrates was measured at 30°C by the direct method of Warburg (Umbreit *et al.*, 1957). Agar cultures were harvested with a platinum loop, suspended in 0·15 M KCl and centrifuged for 10 min at 4500 rpm. The supernatant liquid was discarded and the sediment resuspended in 0·15 M KCl and centrifuged again. The final sediment was resuspended in equal proportions of 0·15 M KCl and M/20 phosphate buffer (pH 7·6). The main compartment of each Warburg flask contained cell suspension (2·5 ml) and polymeric substrate (0·004 g). The centre well contained 20% aqueous sodium hydroxide (0·2 ml) and a piece of fluted filter paper. Thermobarometer flasks contained distilled water (2·5 ml) only. Tests were duplicated and lasted about 48 hr.'

The organisms used for these tests were not 'adapted' and yielded astonishing results contrasting with earlier tests (Burgess and Darby, 1965; Cavett and Woodrow, 1968). The authors conclude that 'in respirometric tests only the ability of resting cells to oxidise the substrate is measured, not their ability to grow on it'. For this reason the authors believe that growth tests carried out over a longer period which offer the test organisms the possibility of adapting themselves are more affirmative and reproducible. Despite all the limitations, the determination of the oxygen uptake is doubtless a valuable means of screening substances for their behaviour to microbial attack. The influence of endogenous respiration must not, however, be overlooked in the evaluation of manometric tests. Cavett and

Woodrow (1968) describe a radiotracer technique to measure the endogenous respiration of test organisms. The authors also suggest, however, using DOS (di-2-ethylhexylsebacate) as a reference standard.

4.7. Determination of Endogenous Metabolism

The fact that some *Pseudomonas* species, which do not possess lipid or carbohydrate reserves, utilise endogenous nitrogen-containing compounds (protein and RNA) as their energy source under starvation conditions was used by Sharpe and Woodrow (1972) to measure the endogenous metabolism by determining the ammonium concentration in the suspension. When metabolisable material is present in the medium, the production of ammonium ions should decrease with increasing ability of the organisms to metabolise this material. The test organism used was *Pseudomonas fluorescens* CBCC 1237, which proved to be very active against a number of commercial plasticisers. A subculture was made of bacteria cultured in Heart Infusion Broth by inoculating them into 100 ml of a mineral salts solution (FA No. 5 medium without agar—Table II) and shaking them for 18 hr at 30°C. The harvested cells were washed three times in 0·1 M phosphate buffer, pH 7·0, and used at a concentration of $5{\cdot}0 \times 10^9$ cells/ml.

Pieces of untreated PVC film are wiped with paper tissues, dried for 24 hr over $CaCl_2$ at 45°C and weighed. One piece of each is placed into 10 ml of the bacterial suspension and incubated at 30°C under shaking. The test has to be duplicated. The controls either contained no film or an addition of 0·02% $HgCl_2$. After 48 hr, 1·5 ml samples are acidified with 1·5 ml of 10% trichloroacetic acid and centrifuged. The ammonia content is determined with Nessler's reagent using a spectrophotometer at 480 nm. The 'ammonia values' are calculated from the following formula:

$$\text{Ammonia value }(\%) = \frac{C_t - C_0}{(C_t{}^0 - C_0{}^0)} \times 100$$

where C_0, C_t = ammonia concentrations initially and at time t, and $C_0{}^0$, $C_t{}^0$ = ammonia concentrations initially and at time t in the endogenous control tube.

5. TESTS ON SOLID MEDIA

5.1. Tests with Fungi on Agar Media, with Visual Examination (according to ISO R 846)

Both procedures described here are modifications of Methods A and B of the ISO Recommendation R 846. They were described by Hitz *et al.*

(1967). The medium (Czapek-Dox agar) used in procedure A contains only mineral salts, whereas that used in procedure B is supplemented with 30 g/litre saccharose. Procedure A is intended to determine whether a plastic may serve as a nutritive substance for fungi; procedure B yields information on possible fungitoxic properties of the material.

Procedure A, an adaptation of ISO Method A (modified). 18 specimens are made of each plastic to be tested. If possible, square specimens of an edge length between 30 and 50 mm and a maximum thickness of 4 mm should be produced. 2 × 3 specimens of each type of plastic are placed in two petri dishes on 3–4 mm agar plates and inoculated with a mixed spore suspension of five fungi (Table I). Six further specimens are kept as zero controls. The remaining six specimens are also placed in groups of three on two agar plates and sprayed with an $HgCl_2$ solution (1:100). The sterile controls are incubated together with the inoculated specimens at 30°C ± 2° and 95–100% relative humidity. After 4 weeks the inoculated specimens are reinoculated with a mixed spore suspension and incubated for another 4 weeks. The reinoculation is the main difference compared with the original ISO method.

Fungal growth is then rated as follows:

0 = No growth is apparent under the microscope.

1 = Fungal growth is hardly visible to the naked eye but is quite apparent under the microscope.

2 = The growth on the surface is visible to the naked eye but does not cover more than 25% of the test surface.

3 = The growth on the surface is visible to the naked eye and covers more than 25% of the test surface.

The evaluation of the resistance of plastics to the growth of fungus is determined according to the degree of fungal growth. Table III gives the evaluation of the resistance of plastics.

Procedure B, an adaptation of ISO Method B (modification II). The sugar-containing agar is sprayed with a mineral salts spore suspension of the five test fungi and incubated for 3 days at 30°C ± 2° and 95–100% relative humidity. The specimens are then placed on the fungal mat developed. (The number of specimens and their arrangement is identical with that described for procedure A.) After an incubation period of 4 weeks the specimens are reinoculated as described for procedure A and

TABLE III

Intensity of fungal growth	Evaluation of testing material
0	Plastics are not nutritive media for microorganisms (they are inert or fungitoxic)
1	Plastics contain nutritive substances or are contaminated to such a small degree that they permit only a slight growth
2–3	Plastics are not resistant to a fungal attack and contain nutritive substances suitable for the development of microorganisms

incubated for another 4 weeks. The intensity of growth is rated as follows:

0 = No fungal growth is apparent even under the microscope. Inhibition zone absent or present; if present, the dimension in millimetres should be stated.
1 = Fungal growth is hardly visible to the naked eye but is quite apparent under the microscope.
2 = Slight growth covering less than 25% of the test surface.
3 = Medium growth covering 25–50% of the test surface.
4 = Considerable growth covering 50% or more of the test surface.
5 = Heavy growth covering the entire test surface.

The evaluation of test results is made according to Table IV.

5.2. Tests with Bacteria on Agar Media (according to ASTM D 1924)

A report drawn up by Hitz *et al.* (1967) describes a co-operative test carried out by Task Group A of the ISO TC 61 WG 6. In addition to the fungal tests mentioned in Section 5.1, two ASTM methods to evaluate the

TABLE IV

Intensity of fungal growth	Evaluation of testing material
0	Strong fungitoxic effect
0 or zone of inhibition	Strong fungitoxic effect due to diffusing substance
1	Not quite a complete fungitoxic effect
2–5	Decreasing effectiveness to complete absence of fungitoxic effect

resistance of plasticised PVC to bacterial attack were used. In procedure A the specimens are placed on the surface of the agar; in procedure B the contact is more extensive as the test specimens are completely immersed in the agar.

ASTM procedure A (adapted). Sterile mineral salts agar (FA No. 5 medium—Table II) is melted and then cooled to 45°C and inoculated with a cell suspension of *Pseudomonas aeruginosa* NCTC 8060 to obtain a concentration of about 50 000 cells/ml. The agar is poured into sterile petri dishes in such a manner as to provide an agar layer of about 1 cm. When the agar has solidified, groups of three specimens are placed in each of the two dishes. 2 × 3 further specimens are placed in two petri dishes on uninoculated media which are additionally disinfected by spraying them with an $HgCl_2$ solution (1:100). The remaining six specimens are kept as zero controls. The inoculated specimens are incubated together with the sterile controls for 4 weeks at 30°C ± 2° and 95–100% relative humidity.

Bacterial growth on the surface of the test specimens will hardly be detectable to the naked eye. Visible bacterial growth may, however, be observed on the agar surface around susceptible specimens; growth is registered as (+) and no growth as (−).

ASTM procedure B (adapted). The molten medium is inoculated with bacteria as described in procedure A. The inoculated medium is divided into equal portions. One portion of the seeded agar is poured into two sterile dishes and allowed to solidify. Three test specimens are each placed on the agar surface and covered with the second portion of the inoculated agar. Six sterile and six zero controls are treated as in procedure A. Incubation and evaluation is made according to procedure A. In addition, the weight loss of the specimens is determined and mechanical changes are measured by a method described in Section 8.

5.3. Determination of the Colony Diameter

This method was used by Berk *et al.* (1957) to evaluate the growth of 24 fungal species on 127 chemical components. The organic substances to be tested are added to a sterile FA No. 5 mineral salts agar (Table II) in such quantities as to produce a carbon content of 0·8%. The unsterilised plasticisers are blended with the agar (50–60°C) in a sterile blender and 30 ml portions are poured into sterile 9 cm petri dishes. If organic acids are blended with the agar, the pH has to be adjusted to 6·4 by adding sodium hydroxide. Each agar plate containing one substance is inoculated with one test fungus. The spore suspension is applied with a 2 mm platinum loop. The inoculated plates are incubated in an inverted position for 2 weeks at 29°C ± 1° and subsequently the colony diameter is determined.

Pantke (1973) used a somewhat modified method to determine the growth rate of a number of fungi isolated from attacked plasticised PVC films. A melted mineral salts agar is cooled to 48°C in a water bath after sterilisation and 3% of the plasticiser to be evaluated is added. The liquid is stirred for 30 sec at 20 000 rpm to produce an emulsion which is stable for a short time and then poured into sterile one-way 9 cm petri dishes in 20 ml portions. To obtain more accurate results, five replicate agar plates are inoculated with each plasticiser. Reproducibility may be improved by using a seeded agar instead of a spore suspension. By means of sterile glass tubes, agar pieces of a diameter of 0·5 cm are punched out from the front of the freshly grown subcultures. They are placed into the centre of a solidified plasticiser agar by means of an inoculating wire. A malt agar (3% malt, 1·5% agar) is usually used as a medium for the subcultures. The inoculated pieces also serve as a 'starting assistance'. Another possibility is to grow the subcultures on a mineral salts agar with the plasticiser to be tested to provide the test organisms with the possibility of adapting themselves. Five replicate plates are used as reference standards and viability controls. They contain sugar instead of the plasticiser as a carbon source. The growth of the fungal colony may be measured sufficiently accurately through the agar by using a millimetre scale placed beneath the dish. Unless the cultures are circular, the different extensions are measured and the mean calculated. Measurements are made daily at the same time. To prevent vigorously sporulating cultures from spreading, and thus affecting the results, the plates are not moved. Daily measurements are continued until the fungal colony has reached the edge of the dish. The graph plotting the mean values of the daily measurements shows the growth rate of the fungi on the tested plasticiser.

It may often be observed that the mostly dim plasticiser–agar suspensions around the microbial colony are cleared. Klausmeier (1966) used clearing around a colony growing on plasticiser–emulsion agar as an index of the utilisation of the plasticiser by the organisms. This method cannot, however, be generally applied as some microorganisms are able to utilise plasticisers without clearing the plasticiser–emulsion agar.

5.4. Determination of the Oxygen Consumption

Burgess and Darby (1964, 1965) used the oxygen consumption of fungi and bacteria to assess the activity of microorganisms on plasticised PVC films. The films used had the following composition: suspension PVC 65 parts, plasticiser 35 parts, Ba/Cd stabiliser 2 parts and epoxide stabiliser 1 part. 28 circular holes, 0·5 cm in diameter, are punched according to a regular pattern into discs, 7 cm in diameter, to increase the ratio of edge to surface. 25 ml of FA No. 5 mineral salts agar is poured into Erlenmeyer flasks and after the agar has solidified, the films are placed on the agar. The flasks are inoculated by spraying them with a mixture of different spore suspensions or with a suspension of *Pseudomonas aeruginosa* NCTC 8060. The sterile control flasks containing agar but no plastics are sprayed with a 0·1 % $HgCl_2$ solution. The test and the control flasks are fitted with thimbles containing 1 ml of 40 % KOH and a filter paper wick. One test and one control flask are attached to the side arms of a differential pressure manometer in a room conditioned to 28°C. The CO_2 given off by the microorganisms is absorbed by the KOH. After 24 hr the system is closed. The oxygen consumed by the microorganisms results in a reduced pressure. The values are read at convenient intervals from the manometer over 72 hr.

Pantke (1973) also used the oxygen uptake of mould fungi to assess the susceptibility of PVC plasticisers. As already described in Section 5.3, the plasticisers are suspended in mineral salts agar. Preliminary tests to explore a suitable inoculating technique yielded unsatisfactory results both when an inoculated agar piece was used and when a spore suspension was applied. But when 0·5 ml of a spore suspension containing $2{\cdot}3 \times 10^6$ spores/ml was blended with 10 ml of the plasticiser–agar emulsion, the mould fungi used, mainly *Trichoderma pseudokoningii*, developed quickly and uniformly.

In the above method according to Burgess and Darby the oxygen absorbed produces pressure changes which can be read from the manometer U-tube. Unless the pressure is balanced at certain intervals (*i.e.* discontinuously), the oxygen pressure is continuously reduced during the test and exerts additional influences on the reaction of the organisms. By means of a measuring and regulating device the oxygen absorbed in the method described by Pantke is continuously replaced through electrolysis, thus maintaining the original pressure conditions. The technical details of this apparatus will not be discussed in this paper.

By means of the respirometric technique the susceptibility of a number of commercial PVC plasticisers has been studied. The results agree with those obtained by measuring the colony diameter of microorganisms as described in Section 5.3. Figure 1 shows different growth rates of a test fungus on several types of plasticisers.

The value of the respirometric technique compared with other methods

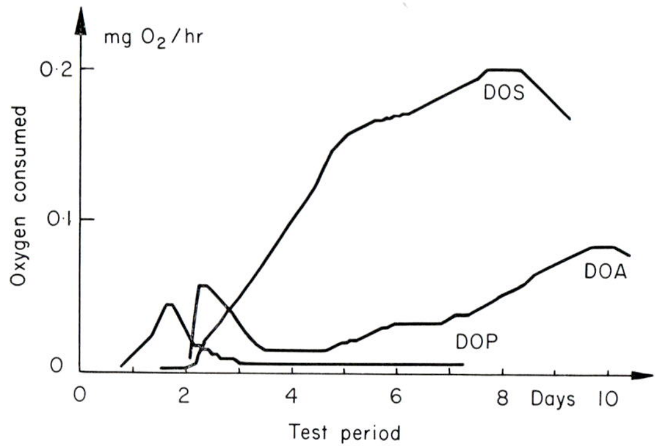

FIG. 1. Oxygen consumption by *Trichoderma pseudokoningii* growing on mineral salts agar containing 3 % di-2-ethylhexylphthalate (DOP), di-2-ethylhexyladipate (DOA) or di-2-ethylhexylsebacate (DOS).

has already been discussed in Section 4.6. It should be added, however, that this technique, as well as being very time-saving, also requires costly equipment.

6. SOIL BURIAL TESTS

Soil burial tests to evaluate the resistance of plastics are described in a number of standards and recommendations. This method is mainly applied for plastics that come into contact with soil, such as protective films for building construction, cable insulating materials and coated fabrics for tents. Exposure times and test conditions will be decided by the potential hazards to which the material is exposed under natural conditions. The test material is buried in soil either under natural field or laboratory conditions.

In an interim report, Decoste (1968) presented a detailed description of test procedures and evaluations of a field study started in 1958. In another interim report on this test programme, Connolly (1972) described the effects observed on a number of different materials: 'The PVC containing certain plasticisers have shown varying degrees of fungal attack. The flexibility retention has ranged from good to poor. Generally there have been no significant changes in tensile strength without an accompanying but much larger change in insulation resistance.' Kaplan *et al.* (1970)

tested the effectiveness of different biocides incorporated in plasticised PVC films. The test specimens were partly weathered before being buried in soil in the field.

In contrast to soil burial in the field, certain parameters, mainly climatic ones, are specified in laboratory tests. Different standard methods (*e.g.* ASTM 684, AATCC 30, DIN 53 933, NF X 41-503 and -514) were developed which were originally used for testing textiles and which were later also adopted for testing plastics. The simultaneous testing of unprotected textile strips in order to control the aggressiveness of the test soils used is common with all methods. But in other respects the methods differ considerably, *e.g.* type of soil (composition, moisture content, pH, C/N ratio), natural flora of microorganisms or subsequent inoculation of a previously sterilised soil; number, shape, pretreatment and installation of the test specimens and the sterile controls; climatic conditions, exposure period and method of evaluation. Wälchli and Zinkernagel (1966) compared methods and results of soil burial tests (NF X 41-514, DIN 53 933, EMPA proposal VII/11 and VITNO-BIO A 1) to assess the resistance of plastics to microorganisms.

The inherent problems of the soil burial methods are demonstrated, for example, by the fact that the DIN 53 933 has been revoked without being replaced. Where good comparability and reproducibility of results are important, soil burial tests cannot be recommended. But as soil burial tests are very rigorous tests, they are valuable in developing and testing new plasticised PVC formulations. The procedure used to evaluate wood blocks against soft rot attack might offer a new possibility for testing plastics. In this procedure the wood blocks are not buried in soil but in vermiculite (Al–Fe–Mg silicate) which has previously been inoculated with different pure cultures of fungi (Kerner-Gang and Gersonde, 1972).

In the methods described in Sections 4 and 5 the evaluation of the degradability of plastics by microbial attack is based on the response of the test organisms. Such reactions, however, are difficult to observe in the soil burial method. For these reasons, the methods described in Section 8, *e.g.* weight loss and loss in tensile strength, are often used.

7. DURABILITY TESTS

By means of durability tests the behaviour of plastic materials under real working conditions can be evaluated. Like soil burial tests, they may be carried out in the field or in the laboratory.

Field tests are mainly carried out in subtropical and tropical areas because of the climatic conditions which are very favourable for microbial growth. The materials to be tested, such as articles of equipment, cables and fabrics for tents, are exposed in the open. The tests extend over several years in order to keep the effects of climatic variations as low as possible.

Based on the temperature and humidity values published by different meteorological stations around the world, Rychtera and Niederführová (1965) plotted a world map distinguishing between four climatic zones with microbiological corrosion hazards. Laboratory mould tests with different test cycles are suggested for testing plastics that will be employed in the four areas. The test cycles should include a pre-critical period (28°C ± 2°, 76% ± 2% relative humidity) and a critical period (28°C ± 2°, 96% ± 2% relative humidity) extending over 28 days altogether. In zone two the critical period is fixed at 3 days, in zone three at 7 days and in zone four at 28 days. By comparing the results with parallel atmospheric mould tests the authors concluded that their method simulated natural processes in the different areas much better than did the methods formerly used.

7.1. Testing in the Tropical Chamber

Standardised climatic conditions are maintained in the tropical chamber to eliminate natural climatic variations. NF X 41-515 will be described as an example. This procedure is used to determine the effects of microorganisms on flat-shaped articles, packaging materials and equipment made of plastic. A closed room is used as the tropical chamber. The indoor furniture, such as shelves, must be made of a material that neither inhibits nor stimulates microbial growth. The daily temperature is adjusted to 24°C for 6 hr and to 30°C for 18 hr and the relative humidity controlled at 90–100%. When fungistatically preserved materials are tested, a constant draught-free air exchange is particularly important. Small chambers are inoculated by spraying them with a spore suspension every 3 months, bigger chambers exceeding 5 m^3 every 6 months (Table I). Trays with soil are additionally introduced into the chambers. The aggressiveness is checked by introducing textile strips into the chamber. When these strips are not completely rotted within 1 month, the chamber must be reinoculated. The materials to be tested are installed as finished articles or as test specimens or placed on soil after having been inoculated with 1 ml of spore suspension per 100 cm^2 of test surface. The final evaluation is made according to the following rating system:

0 = No growth visible under the microscope.
1 = Growth visible under the microscope.
2 = Growth on parts or the entire surface visible to the naked eye.

The test period is 1–6 months.

7.2. Humidity Cabinet Tests

Plastic building elements or films are sprayed with a mixed spore suspension and incubated in a water-vapour-saturated atmosphere without an additional nutritive source at 28°C ± 1° for 4 or 12 weeks. This method was primarily used in electrical engineering (DIN 40 046, Part 10, Test J). Kerner-Gang (1976) presents a description and a diagram of the test arrangement.

8. METHODS OF EVALUATING PHYSICAL PROPERTIES

In Sections 4 to 7 a number of test methods have been described. In all these methods the resistance or degradability of the tested substances is determined by the reaction of the microorganisms. In this section some methods will be discussed in which the damage caused by microbial attack is measured on the material itself.

8.1. Weight-Loss Determination

This is probably the most common method used in the majority of all papers cited here. If possible, films of the plasticised PVC are used from which the specimens are cut (at least five specimens per test). If tensile tests are also envisaged, the specimens should be cut in the respective specified dimensions. They must be notched for identification in a suitable manner. Unless it is agreed to test the material as it is supplied, the specimens are washed in running tap water, rinsed with aqua dest., and dried at room temperature over a suitable drying agent in a desiccator to a constant weight (1–3 days). The weight of the specimens is recorded to the nearest 0·1 mg. Before being installed they are sterilised. Possible side effects caused by the different sterilisation processes are reported by Klausmeier (1970*b*). The prepared specimens are installed and inoculated according to the test method used. Uninoculated or sterilised controls have to be included in the test. They are used to determine physical effects (water retention, swelling, migration of plasticiser, etc.) which are simultaneously observed. At the end of the specified test period the specimens are removed, sterilised, freed from growth, washed as above and reweighed to the nearest 0·1 mg. The percentage weight loss is calculated from the following formula:

$$\text{Weight loss } (\%) = \frac{(M_0 - M_x) - (M_0^{\text{st}} - M_x^{\text{st}})}{M_0} \times 100$$

where M_0, M_x = mass of test specimen before and after the biological test, and M_0^{st}, M_x^{st} = mass of sterile control specimen before and after the biological test.

Most authors consider the weight-loss method to be very sensitive. The results obtained agree well with those obtained from other test methods.

8.2. Mechanical Evaluation

When a plasticised PVC formulation contains susceptible plasticisers, the microorganisms may assimilate and utilise them as a carbon source. This causes a loss in plasticiser which in turn affects the mechanical properties of the material. Mechanical evaluation is made with inoculated specimens and uninoculated controls of certain dimensions and shapes in accordance with nationally or internationally standardised tensile tests for plastics. Design and qualities of the tensile testing machine are specified in the different standards.

Hitz *et al.* (1967) reported on a co-operative test carried out by ISO TC 61. The tensile strength at break was determined and the elongation at break was calculated from the load extension diagram. Tensile strength (TS) and elongation at break (EB) are expressed as percentage change according to the following formula:

$$\%\text{ change} = \frac{\text{TS or EB (inoc. spec.)} - \text{TS or EB (sterile control)}}{\text{TS or EB (sterile control)}} \times 100$$

Doležel (1967) also reported on the evaluation of the results of the co-operative test: ‘Changes in tensile strength are relatively small and no statistically significant differences were found between the tensile strength of samples exposed to fungi and samples tested under sterile conditions. There is no direct relation between the degree of growth and tensile strength changes. Depending on the test method used there is a statistically significant difference between the elongation at break of inoculated specimens and the sterile control. The greatest elongation change is caused by the biological test according to the ISO A modified method.’

Hitz *et al.* introduced a new term ‘biological value’ (BV) to record mechanical changes induced by microbial attack. The stress (S in kg/cm^2) at 33% extension (E) is determined graphically from the load-extension curves:

$$BV = \frac{S \text{ at } 33\%\ E \text{ (inoc. spec.)} - S \text{ at } 33\%\ E \text{ (sterile control)}}{S \text{ at } 33\%\ E \text{ (sterile control)}} \times 100$$

At low extensions, a difference in stress was noted between the inoculated specimen and the sterile control.

8.3. Determination of Electrical Properties

Test methods applied in electrical engineering may also be used to determine whether microbiological attack has caused changes in electrical properties of plastics. Depending on the problems to be studied, one or several properties should be measured: surface and volume resistivity and dielectrical loss factor tan δ (Doležel, 1967). Measurements can be performed either on washed and conditioned specimens, or the attacked specimens can be measured before removing them from the test arrangement. A final operational test is recommended, especially with electrical and electronic equipment which has been tested in a tropical cabinet.

Wasserbauer and Blahnik (1975) presented a review and analysed the reasons for microbial corrosion of electrical equipment. They make suggestions regarding the design of electrical instruments used in areas under extreme climatic conditions.

8.4. Chemical Analyses

A loss in plasticiser substance caused by microbial attack of PVC materials may also be determined by direct chemical analysis. Inoculated films and untreated controls are extracted with diethylether for 6 hr. These extractives are analysed by means of quantitative thin-layer chromatography (Pantke, 1970). The percentage losses of the different components agree well with weight losses determined prior to the chemical analysis.

8.5. Investigation of the Surface Condition

Another means of identifying microbial attack is to evaluate the surface condition of the test material. Damage may even be partly recognised by the naked eye when the specimens are washed. The surface gets rough where it is overgrown. This is even more easily recognisable under a magnifying glass or an incidental light microscope. By means of a scanning electron microscope Pantke (1970) studied plasticised PVC films after soil burial. Figure 2 shows progressive deterioration of the specimen surface related to the time of exposure.

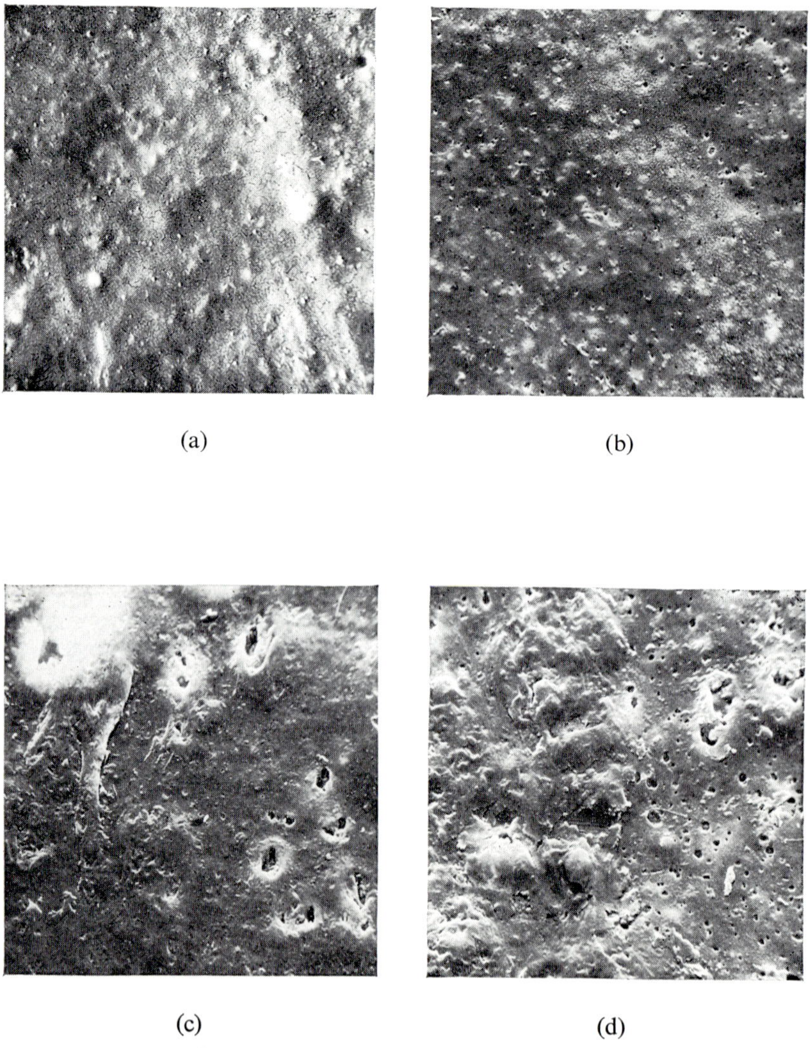

(a) (b) (c) (d)

FIG. 2. Progressive deterioration of the surface of plasticised PVC films after soil burial related to the time of exposure. (a) Condition of material at time of supply; (b) after 2 weeks; (c) after 4 weeks; (d) after 8 weeks.

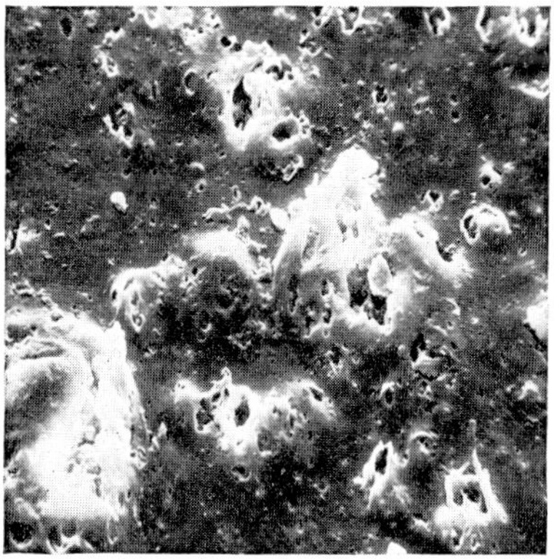

(e)

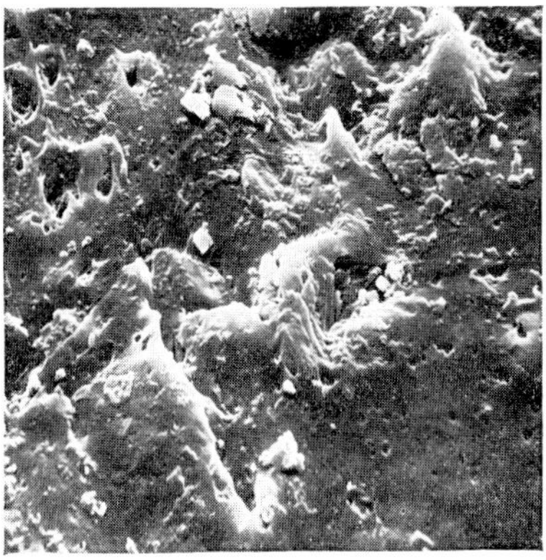

(f)

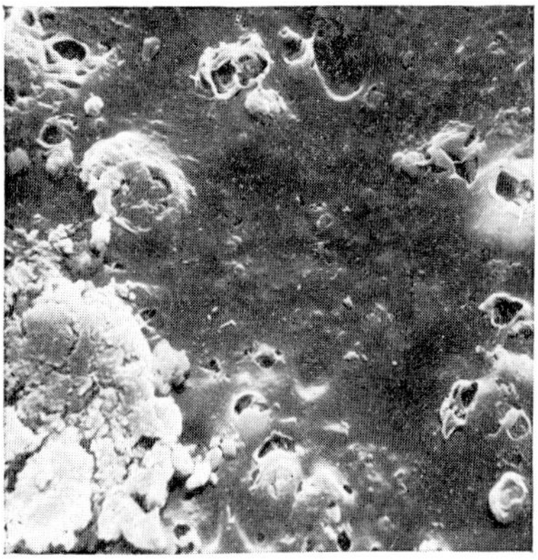

(g)

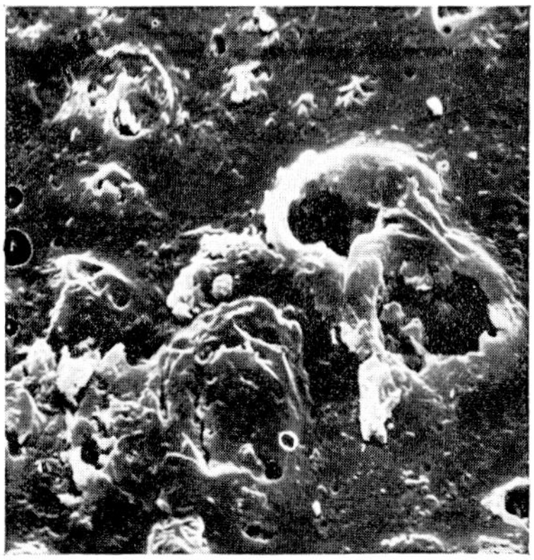

(h)

FIG. 2—*contd.* (e) after 12 weeks; (f) after 18 weeks; (g) after 24 weeks; (h) after 48 weeks.

REFERENCES

American Society for Testing and Materials (1968*a*). Recommended practice for determining resistance of plastics to fungi, ASTM Designation D 1924–63. *Book of ASTM Standards*, Part 27, pp. 625–57.

American Society for Testing and Materials (1968*b*). Resistance of plastics to bacteria, ASTM Designation D 2676-67T. *Book of ASTM Standards*, Part 26, p. 840.

Association Française de Normalisation (1961). *Protection des matières plastiques: Méthode d'essai de résistance des constituants aux microorganismes.* NF X 41-513.

Association Française de Normalisation (1962). *Protection des matières plastiques: Méthode d'essai de résistance des matériels et appareillages aux microorganismes.* NF X 41-515.

Becker, H. and Gross, H. (1974). Über die Widerstandsfähigkeit makromolekularer Werkstoffe gegen mikrobiellen Angriff (Ein Übersichtsbericht). *Material u. Organismen*, **9** (2), 81–131.

Berk, S., Ebert, H. and Teitell, L. (1957). Utilization of plasticizers and related organic compounds by fungi. *Ind. Eng. Chem.*, **49** (7), 1115–24.

Booth, G. H., Cooper, A. W. and Robb, J. A. (1968). Bacterial degradation of plasticized PVC. *J. Appl. Bacteriol.*, **31** (3), 305–10.

Burgess, R. and Darby, A. E. (1964). Two tests for the assessment of microbiological activity on plastics. *Brit. Plastics*, **37** (1), 32–7.

Burgess, R. and Darby, A. E. (1965). Microbiological testing of plastics. *Brit. Plastics*, **38** (3), 165–9.

Cavett, J. J. and Woodrow, M. N. (1968). A rapid method for determining degradation of plasticized PVC by microorganisms. In: *Biodeterioration of Materials*, vol. 1 (ed. A. H. Walters and J. J. Elphick), pp. 162–9. Applied Science Publishers, London.

Connolly, R. A. (1972). Soil burial of materials and structures. In: *Biodeterioration of Materials*, vol. 2 (ed. A. H. Walters and E. H. Hueck-van der Plas), pp. 168–78. Applied Science Publishers, London.

Decoste, J. B. (1968). Soil burial resistance of vinyl chloride plastics. *Ind. Eng. Chem./Prod. Res. Devel.*, **7** (4), 238–47.

Doležel, B. (1967). The resistance of plastics to microorganisms. *Brit. Plastics*, **40** (10), 105–12.

Haldenwanger, H. H. M. (1970). *Biologische Zerstörung der makromolekularen Werkstoffe*. Springer-Vlg, Berlin. 281 pp.

Hitz, H. R. and Zinkernagel, R. (1967). Tube test method for evaluation of biodegradation of plasticized PVC by *Pseudomonas aeruginosa* NCTC 8060. *Int. Biodetn. Bull.*, **3** (1), 21–3.

Hitz, H. R., Merz, A. and Zinkernagel, R. (1967). Determination of the resistance of plasticized PVC to attack by fungi and bacteria by the weight loss method and evaluation of mechanical properties. *Material u. Organismen*, **2** (4), 271–96.

International Organisation for Standardisation (1968). *Recommended Practice for the Evaluation of the Resistance of Plastics to Fungi by Visual Examination.* ISO Recommendation R 846. 10 pp.

Kaplan, A. M., Greenberger, M. and Wendt, T. M. (1970). Evaluation of biocides for treatment of polyvinylchloride film. *Polymer Eng. Sci.*, **10** (4), 241–6.

Kerner-Gang, W. (1976). Evaluation techniques for resistance of floor coverings to mildew. This volume, pp. 95–103.

Kerner-Gang, W. and Gersonde, M. (1972). Untersuchungen zur Prüfung mit Moderfäulepilzen im Vermiculit-Eingrabe-Verfahren. *Material u. Organismen*, **7** (4), 241–58.

Klausmeier, R. E. (1966). The effect of extraneous nutrients on the biodeterioration of plastics. *Soc. Chem. Ind.*, *Monograph* No. 23, 232–43.

Klausmeier, R. E. (1970*a*). Test procedure for the third interlaboratory test on deterioration of plastics. Rep. Int. Biodegradation Res. Group, 70/51. 12 pp.

Klausmeier, R. E. (1970*b*). Effect of sterilization procedures on plastic films. Rep. Int. Biodegradation Res. Group, 70/50. 8 pp.

Klausmeier, R. E. and Jamison, E. I. (1973). A rapid method for estimation of microbial degradation of plastics. *Devel. Ind. Microbiol.*, **14**, 258–63.

Klausmeier, R. E. and Jones, W. A. (1961). Microbial degradation of plasticizers. *Devel. Ind. Microbiol.*, **2**, 47–53.

Klausmeier, R. E. and Osmon, J. L. (1976). Biodegradation of plastics by actinomycetes. In: *Biodeterioration of Materials*, vol. 3 (ed. J. M. Sharpley and A. M. Kaplan), pp. 815–18. Applied Science Publishers, London.

Lee, R. W. H. (1962). Bibliography on resistance of plastics to microorganisms, insects, rodents and other pests. Avail. OTS, AD 601 309. 33 pp.

Mills, J. and Eggins, H. O. W. (1974). The biodeterioration of certain plasticizers by thermophilic fungi. *Int. Biodetn. Bull.*, **10** (2), 39–44.

Mills, J., Barnes, T. G. and Eggins, H. O. W. (1971). *Talaromyces emersonii*—a possible biodeteriogen. *Int. Biodetn. Bull.*, **7** (3), 105–8.

Osmon, J. L. and Klausmeier, R. E. (1971). *Candida lipolytica:* a nonpathogenic test organism for deterioration of plastics. *Devel. Ind. Microbiol.*, **12**, 197–203.

Osmon, J. L., Klausmeier, R. E. and Jamison, E. I. (1970). The ability of selected-yeast-cultures to degrade plasticized-polyvinyl-systems. *Devel. Ind. Microbiol.*, **11**, 447–52.

Pankhurst, E. S., Davies, M. J. and Blake, H. M. (1972). The ability of polymers or materials containing polymers to provide a source of carbon for selected microorganisms. In: *Biodeterioration of Materials*, vol. 2 (ed. A. H. Walters and E. H. Hueck-van der Plas), pp. 76–90. Applied Science Publishers, London.

Pantke, M. (1970). Gravimetrisch, elektronenoptisch und chemisch erfassbare Veränderungen an PVC-Folien nach Erd-Eingrabe-Versuchen. *Material u. Organismen*, **5** (3), 197–215.

Pantke, M. (1971). Literaturübersicht zum Abbau von Weich-PVC durch Mikroorganismen. *Material u. Organismen*, **6** (4), 295–315.

Pantke, M. (1973). Der Sauerstoffverbrauch von Schimmelpilzen als Mass für die Abbaubarkeit von Weich-PVC und dessen Bestandteilen. *Material u. Organismen*, **8** (4), 249–70.

Rychtera, M. and Niederführová, E. (1965). Beziehungen zwischen natürlicher mikrobieller Korrosion und Laborschimmelpilzprüfung. *Werkstoffe u. Korrosion*, **16** (2), 116–24.

Sharpe, A. N. and Woodrow, M. N. (1972). A rapid test for biodegradability by *Pseudomonas* organisms. In: *Biodeterioration of Materials*, vol. 2 (ed. A. H. Walters and E. H. Hueck-van der Plas), pp. 233–7. Applied Science Publishers, London.

Stahl, W. H. and Pessen, H. (1953). The microbiological degradation of plasticizers. I. Growth on esters and alcohols. *Appl. Microbiol.*, **1,** 30–5.

Umbreit, W. W., Burris, R. H. and Stauffer, J. F. (1957). *Manometric Techniques*, 3rd ed. Burgess, Minneapolis.

Wälchli, O. and Zinkernagel, R. (1966). Comparison of test methods for determining the resistance of plastics to microbiological attack. *Material u. Organismen*, **1** (3), 161–71.

Wasserbauer, R. and Blahnik, R. (1975). Microbial corrosion of electrical equipment. Analysis of causes and consequences: a review. *Int. Biodetn. Bull.*, **11** (1), 9–15.

Wessel, C. J. (1964). Biodeterioration of plastics: plastics attacked by fungus, bacteria, termites, insects, rodents, marine organisms. *Soc. Plastics Eng. Trans.*, **4** (3), 193–207.

Wessel, C. J. (1968). The information background in the field of biological deterioration of nonmetallic materials. *Mater. Res. Std.*, No. 9, 10–17.

5

Techniques for Assessing Biodeterioration of Plastics and Plasticisers

J. L. OSMON and R. E. KLAUSMEIER

Naval Weapons Support Center, Crane, Indiana 47522, *USA*

1. INTRODUCTION

The term 'plastic', when referring to a material, is used almost always to refer to a formulation, not to a single compound. These formulations are usually identified by the polymer or resin that comprises the bulk of the formulation (*i.e.* polyethylenes or polythenes, polyvinyl chlorides—sometimes called simply vinyls—and the like). However, the formulation may well contain considerable amounts of other ingredients.

For example, a flexible 'vinyl' formulation (PVC) may contain approximately 65%, by weight, of the polymer. The polymer itself does not represent a uniform chemical entity, but rather a range of compounds, all similar, but varying in molecular weight, degrees of branching, degree of substitution, etc.

To this polymer may be added a mixture of plasticisers in the amount of 25–30%. The types of plasticisers vary with the required final characteristics: cold flexibility, elasticity, dielectric constant, fungus resistance, etc. The plasticisers also are not pure chemical compounds since commercial practices utilise the cheapest possible grade of product available to meet end-use requirements. Some plasticisers are natural oils (castor oil, tung oil, etc.) or are derivatives of these natural oils. Others are synthetic esters, hydrocarbon fractions, and similar viscous liquids or low-melting solids which are miscible with the polymer and provide suitable changes in flexibility.

Other ingredients in the formulation may include stabilisers which are usually metal salts which act as acid scavengers. (During processing, polyvinyl chloride is usually processed near the melting temperature of the mix. At elevated temperatures PVC liberates hydrochloric acid. This free acid has serious consequences on the polymer, the formulation and the equipment.) Additional ingredients may include pigments, materials to produce opacity, agents to provide a glossy finish and other sundry additives.

Similar additives may be found in other plastic formulations, such as absorbers of ultraviolet light to protect polyolefins from actinic degradation. There are fillers ranging from cotton or linen cloth to ceramic or carbon fibres. There are mould release agents, fire retardants and possibly even dirt and debris not intentionally included.

In resin production not all ingredients polymerise even under ideal conditions. If less than stoichiometric amounts of reactants are added, the polymer may well contain unreacted, and possibly biologically susceptible, residues.

Furthermore, a manufacturer may change ingredients in a formulation routinely as a result of availability, favourable pricing, etc., since many of the ingredients may be interchangeable as far as most formulation characteristics are concerned.

Obviously, then, testing 'plastics' for susceptibility to microbial degradation offers a challenge. Fortunately most, but not all, commercially used polymers are themselves inherently resistant to attack from microbes to any significant extent. The susceptible components are usually additives. Microbial deterioration of these additives can alter characteristics such as tensile strength, elasticity (Berk, 1950), dielectric constant (Teitell *et al.*, 1955) and others.

In general, it is better to test entire formulations as prepared for end-use. Human error and manufacturing technology will occasionally transform a formulation prepared entirely from 'resistant' materials into a product which fosters the growth of a veritable jungle of microbes.

However, under certain conditions there may be valid reasons for testing the biosusceptibility of the individual formulation ingredients. But, as discussed above, once these individual ingredients are evaluated, it is paramount that they be tested in the formulation in which they are to be used. In the experiences of the authors, materials rated 'inert' have sometimes proved not to be so, and several plasticisers which may have been relegated to disuse on the basis of simple tests have proved to be biologically non-susceptible when tested in complete formulations.

Finally, it will be noted that in almost all the tests described, the final evaluation is the effect of the microbe on the material rather than the converse. Microbial physiologists can describe many unusual effects of small amounts of biologically active compounds on microorganisms.

Study of these effects is not the purpose of studying biodegradation of plastics. Rather, it is intended to study the effects of microbes on plastics. These tests are intended to meet these goals. For this reason, the discussion of techniques for examining the biodegradability of plasticisers will now follow.

2. METHOD FOR ASSESSING PLASTICISER DEGRADATION

Perhaps the most widely applicable method for assessing plasticiser biodeterioration is the homogenised plasticiser agar plate technique. Briefly, this method consists of mixing the plasticiser to be tested with a molten agar basal medium in a Waring blender such that the plasticiser forms an emulsion which is translucent or opaque. Organisms capable of degrading the plasticiser will form a clear zone or halo around the area of growth when propagated on petri plates of this emulsified medium. This has been shown to be due to the production of extracellular esterases (Reese *et al.*, 1955; Berk *et al.*, 1957; Williams *et al.*, 1969; Klausmeier *et al.*, 1971). Some comments on considerations which impact upon the outcome of this test precede the step-by-step instructions on the technique.

2.1. Selection of Inoculum

The selection of an appropriate organism for use as an inoculum in performing this test is of course somewhat dependent upon the objectives of the experiment. In some cases it may be desirable to use a pure culture when the ability of a specific organism to degrade a plasticiser is in question. If, however, the objective of the study is to determine the resistance of the plasticiser to biological attack from all sources, some sort of mixed inoculum is generally to be preferred. If the end-use of the plasticiser is known, an inoculum derived from the environment in which the plasticiser may be expected to be exposed is an appropriate choice. If the end-use is not known, inocula prepared from soil, humus, domestic sewage, surface waters or a mixture of known organisms have all proved effective. Studies designed to isolate organisms capable of degrading specific plasticisers should also utilise mixed inocula derived from sources such as soil, sewage or surface waters. Appropriate dilutions of such inocula will yield plates bearing well-isolated colonies showing the typical halo indicating degradation of the plasticiser. Pure cultures can easily be isolated for further study or identification from these colonies. A suitable inoculum can be prepared from soil by the following method.

1. Add approximately 1 g of freshly collected soil to 10 ml of distilled water. If determinations are to be made on the basis of the number of organisms per gramme of soil, two soil samples should be precisely weighed. One is used to prepare the inoculum and the other should be dried to a constant weight in a desiccator. The weight of the dried soil sample may then be employed to determine the moisture content of the soil and adjust the calculated number of organisms per gramme of soil to a dry weight basis.

If the soil sample as collected is excessively wet and difficult to handle, the soil may be partially dried in air at room temperature prior to inoculum preparation. Care must be taken not to overdry the soil or some organisms will be killed. Drying to the state at which the soil, when clenched in the hand, will adhere to form a clump which breaks apart when released and allowed to fall is normally sufficient.

2. Shake vigorously. Allow the suspension to stand 10 min or until most particulates have settled out.

3. Decant the supernatant and filter through a coarse filter such as glass wool, fritted glass or a coarse filter paper such as No. 41 Whatman.

4. Appropriate dilutions of the filtrate may be used as a mixed inoculum. The dilution required will depend upon the particular soil sample used and must be determined empirically. The filtrate will normally be turbid due to suspended small soil particles. The turbidity increases with increasing clay content of the soil.

2.2. Selection of a Basal Medium

Since the plasticiser is normally expected to serve as the source of carbon for the growth of the test organisms, the function of the basal medium is to provide an agar base to preserve the emulsion characteristics of the media, and also to provide all other nutrients required for the growth of the test organisms. Normally, one of the commonly used mineral salts solutions is adequate for this purpose (Table I). For some special applications it is desirable to alter the composition of the basal medium in order to produce a selective medium which will allow only the growth of specific types of microorganisms. For example, it might be desirable to omit a nitrogen source from the basal medium in order to isolate organisms which fix atmospheric nitrogen and degrade a particular plasticiser. Antibiotics might be incorporated into the basal medium in order to suppress the growth of certain groups of organisms such as bacteria when the isolation of plasticiser-degrading fungi is the objective. In any case, the primary consideration is that the basal medium shall provide all elements necessary

TABLE I

MINERAL SALTS MEDIA

(all formulae are per 1 litre distilled water)

OECD nutrient salts:	
KH_2PO_4	0·7 g
K_2HPO_4	0·7 g
$MgSO_4$	0·7 g
NH_4NO_3	1·0 g
NaCl	0·005 g
$FeSO_4 \cdot 7H_2O$	0·002 g
$ZnSO_4 \cdot 7H_2O$	0·002 g
$MnSO_4 \cdot 7H_2O$	0·001 g
Crane mineral salts (especially for fungi):	
NH_4NO_3	3·0 g
KCl	0·25 g
K_2HPO_4	2·56 g
$MgSO_4 \cdot 7H_2O$	2·08 g
NaCl	0·005 g
$ZnSO_4 \cdot 7H_2O$	0·002 g
$FeSO_4 \cdot 7H_2O$	0·002 g
$MnSO_4 \cdot 7H_2O$	0·001 g
or	
$MnSO_4 \cdot H_2O$	0·006 g

for the growth of the organisms in question with the exception of carbon, which should be provided by the plasticiser. The primary exception to this rule is discussed in the next section.

2.3. The Addition of Other Carbon Sources

It has been found that certain microorganisms are able to degrade plasticisers only when another nutrient source is present in the medium (Klausmeier, 1966; Klausmeier and Jones, 1961). This phenomenon, known as co-metabolism, is not uncommonly encountered in the biodegradation of the recalcitrant molecules. When working with a mixed inoculum derived from a natural source such as soil, the addition of a carbon source other than the plasticiser, such as glucose or yeast extract, will generally result in an increased number of colonies appearing on the homogenised plasticiser agar plates. It is, however, quite easy to distinguish those organisms which are degrading the plasticiser from those which are merely utilising the extraneous carbon source for growth by the clear zone or halo surrounding the colony. If it is desired to determine whether or not a specific colony can utilise the plasticiser as a sole carbon source, it is a simple matter to transfer a portion of any colony exhibiting a halo to a fresh homogenised plasticiser agar plate which does not contain the extraneous nutrient. Growth and clearing on this plate indicate utilisation of the plasticiser as a sole carbon source. Failure of the organism to grow or clear this second medium indicates that the organism is able to degrade the plasticiser only through co-metabolism. The same procedure may be followed when pure cultures are used as an inoculum. Petri plates containing the plasticiser both in the presence and absence of extraneous nutrient may be inoculated and incubated concurrently to determine the involvement of co-metabolism in the degradation of the plasticiser.

As might be expected, instances have been reported in which a particular organism would form a halo on plasticiser agar medium when the plasticiser was the sole source of carbon, but would not form a halo on plasticiser agar medium when an extraneous carbon source was present (Osmon *et al.*, 1970). In these instances, the organism obviously prefers to grow on the more readily utilisable carbon source without regard to the presence of the plasticiser. In most instances, extended incubation of the agar plate containing extraneous nutrient will result in the formation of the halo while the extraneous nutrient is exhausted from the medium. However, examples may be encountered where the build-up of metabolic products inhibits further growth before the plasticiser can be degraded. In experiments designed to determine the resistance of a particular plasticiser to microbial attack under virtually all conditions, it is advisable to perform the experiment in duplicate using media which contain the plasticiser as a sole carbon source and also media containing an additional nutrient. As to the selection of the nutrient to be added, yeast extract at a concentration of 1% is the choice of the authors. Virtually any readily utilisable source of carbon, such as glucose, will provide the nutrient necessary for co-metabolism; however, many organisms require specific factors in order to grow which may not be derived from either the plasticiser or a simple carbon source such as glucose. Yeast extract contains a large number of these growth factors in small amounts and therefore allows this additional class of organisms to be examined as potential agents of biodeterioration of the plasticiser.

2.4. Incubation

The length of time for which homogenised plasticiser agar plates should be incubated is somewhat dependent upon the objectives of the experiment and conditions of incubation. As is the case with all agar media, the medium in the petri plates tends to dehydrate during incubation. As dehydration progresses, the agar medium becomes thinner and the emulsion gradually breaks down, resulting in clearing of the medium which might be mistaken for clearing due to the growth of the test organism. It is normally quite easy to distinguish between these two situations. However, if extended incubation is attempted, provisions must be made for maintaining a high humidity in the incubation chamber to prevent the drying-out of the agar medium. This may be accomplished by increasing the humidity within the incubator or by enclosing the petri plates in a sealed container in which high humidity can be sustained. It is also possible to extend the usable life of an agar medium by pouring the medium in the petri dish to a greater depth. The increased depth, however, does have an adverse affect on halo formation by the organism since the extracellular enzyme involved in the clearing process must be present in larger concentrations and diffuse through a greater distance in order to effect halo formation. The length of time required for an organism to produce a halo on a plasticiser agar medium is, of course, dependent upon the rate of growth of the organism. In most cases, the authors consider 14 days to be a sufficient incubation period for bacterial cultures and 28 days to be a sufficient length of time when examining fungi. When a mixed inoculum such as a soil suspension is used, it is necessary to examine the plates daily. Less frequent examinations may result in the observation that virtually the entire plate has been cleared, and the identity of the organisms responsible for the clearing cannot be determined due to the overlapping of the haloes produced by the various colonies. The conditions under which the plates should be incubated will of course be determined by the objectives of the experiment. For example, if the objective of the experiment is to determine the resistance of a plasticiser to anaerobes, the plates must be incubated under anaerobic conditions. 55°C might be chosen as the incubation temperature if the isolation or examination of thermophilic organisms was the objective of the experiment. Incubation under various gaseous atmospheres or the use of other selective techniques is also possible. For most applications, however, the authors consider aerobic conditions and a temperature of 20–35°C to be adequate.

2.5. Controls

The use of controls in experiments of this type is generally not necessary. If, however, the determination of whether or not the plasticiser can serve as the sole source of carbon is of critical importance, it is necessary to prepare a control agar medium. The control should be prepared in exactly the same manner as outlined for the homogenised plasticiser agar media but without the addition of the plasticiser. If, upon inoculation, growth is obtained upon the control plate, it must be assumed that either the organism is autotrophic, or the organism is able to utilise agar or its impurities as a source of nutrient. In this case, appropriate further experimentation would have to be performed to determine the possible involvement of co-metabolism in the biodegradation of the plasticiser.

2.6. Procedure

Step 1: *Select an Appropriate Inoculum*

See Section 2.1 for comments in this regard.

Step 2: *Prepare Media*

See Sections 2.2 and 2.3 for comments concerning media selection. Formulae for suitable general-purpose mineral salts media are listed in Table I.

Dissolve the appropriate mineral salts in distilled water. Add 20 g of agar per litre of solution. This mixture should be heated in a double boiler or by steam until the agar melts. The molten agar medium is then poured into a suitable blender such as a Waring blender, and plasticiser is added to give a resulting 2% concentration of the plasticiser.

The blender is turned on and the contents mixed at as high a speed as is practical for a minimum of 2 min. For some plasticisers, longer periods may be required in order to form stable emulsions.

The emulsified medium should be autoclaved at 15 lb/in^2 (105×10^3 N/m^2) for 10 min. Using aseptic technique, pour 15–30 ml of the sterile medium into a sterile petri dish. In a standard 9 cm dish this results in a plate having a depth of 0·5–1·0 cm. See Section 2.4 for comments concerning the proper thickness of the plate.

Step 3: *Prepare Inoculum*

If the inoculum is to be prepared from a mixture of pure cultures held as slant cultures, a suspension must be prepared from the slant. Add a small amount of sterile normal saline (0·85% NaCl) to the slant and gently scrape the surface of the agar with a sterile needle or loop to dislodge the cells or spores. The resultant suspension should be poured into a sterile container and vigorously shaken to break up clumps, centrifuged and the supernatant discarded. Resuspend the cells in sterile saline, centrifuge, decant and again suspend in sterile saline. The cells are now washed and free of nutrient from the agar slant. All cultures to be used should be treated as above, combined and, if necessary, diluted.

Step 4: Inoculate the Plates

If the inoculum has been prepared as in step 3 above, pipette 0·5 ml of the suspension on to the agar and spread the inoculum evenly over the surface with a flame-sterilised bent glass rod.

An alternative procedure, if pure cultures are utilised, is to pick a small amount of growth directly from the agar slant on a sterile loop and streak the agar plate.

Step 5: Incubate the Plates

The plates should be placed upside down in an appropriate incubator and examined daily for evidence of growth. See Section 2.4 for further comments on incubation.

Step 6: Recording and Evaluating Results

Most bacteria will show positive results in 1 week while fungi may require 2 weeks or more. The comments in Section 2.4 are applicable.

A positive result will be evidenced by a clear zone or halo around the microbial growth whether present as a streak or as colonies. The clear zone results from the action of extracellular enzymes which destroy the emulsified plasticiser, rendering the medium transparent. If clearing extends throughout the depth of the medium, the halo will be sharp and unmistakable. If clearing is not complete, the zone will be an indistinct translucent area surrounding the growth. This is best viewed by holding the plate up to a source of light and viewing from the underside of the plate.

Some organisms, notably Actinomycetes and some yeasts, tend to produce a depression in the agar immediately surrounding the growth. This can be mistaken for a translucent zone, a possibility of which the experimenter should be aware.

Normally it is useful to record the presence or absence of growth on each plate (especially if nutrient other than the plasticiser is present) as well as any clearing produced. The system favoured by the authors is to record growth as either + or − and clearing as 'C'. A plate which shows both growth and clearing would therefore be recorded as +C.

Since the extent of clearing may vary under given conditions and a subjective judgement on the part of the experimenter is often required, it is helpful to perform the test in triplicate.

3. THE INFLUENCE OF PLASTICISER MIGRATION ON THE BIODEGRADATION OF PLASTIC FORMULATIONS

It has been shown that a plasticiser which is easily biodegraded in the pure state may not be subject to biodegradation when incorporated into a plastic formulation (DeCoste, 1968; Osmon *et al.*, 1972). The reason for this is that in most cases, for the plasticiser to be attacked, it must be mobile in the formulation. That is, it must migrate to the surface of the plastic sample for degradation to occur. If a relatively resistant resin base such as polyvinyl chloride is used, neither the organisms nor the enzymes are able to penetrate a significant distance into the plastic formulation. Therefore, the plasticiser must migrate to the surface in order to be degraded. Butylene glycol polyadipate (BGA), for example, is easily degraded by a wide variety of microorganisms. However, polyvinyl chloride formulations which incorporate BGA as a plasticiser are quite resistant to microbial attack. This has been shown to be due to the stability of the plasticiser in the formulation. BGA does not migrate to the surface and therefore cannot be attacked by the microorganisms. Other plasticisers such as some of the phthalate esters migrate readily through the plastic formulation but are resistant to microbial attack. Therefore, while some loss of plasticiser may be expected due to leaching or blooming, under most conditions the diffusion equilibria are such that the bulk of the plasticiser remains stable in the formulation. Only when a plasticiser which is readily degraded by microorganisms and diffusible within the plastic formulation is used, do we find significant biodeterioration of plasticised formulations. Of course, if the resin itself is biodegradable, eventually all the plasticiser will be subjected to microbial attack regardless of its mobility within the formulation.

4. METHOD FOR ASSESSING THE BIODEGRADABILITY OF PLASTIC FORMULATIONS

4.1. General Comments

In assessing the biodegradation of a plastic sample, two methodologies are involved. The first involves the means by which the sample is exposed to the test organisms. The second concerns the methods used to evaluate the effect of the organisms on the sample (*i.e.* the biodegradability of the sample).

A variety of techniques have been devised for bringing the plastic sample into contact with appropriate microorganisms. Soil burial, suspension in water and hanging from wires in a moist atmosphere have all been employed. This discussion will concern itself, however, only with methods involving exposure of the sample on an agar medium in a petri plate. Although there are variations, all agar plate methods for testing plastic samples follow the same basic routine. An agar medium is prepared and petri dishes are prepared using this medium. Secondly, the plastic sample is placed on the surface of the agar medium and inoculated with an appropriate test inoculum. The sample is then incubated under appropriate conditions for a predetermined length of time. At the end of incubation the sample is evaluated to determine the extent of biodegradation. The discussions in Section 2 regarding the selection of inoculum, the selection of a basal medium, the use of extraneous nutrients and the conditions of incubation are all applicable to this type of test.

One of the most obvious techniques for evaluating the biodegradability of a plastic sample is a visual rating of the amount of growth present on the sample following inoculation and incubation. Unfortunately this method is also one of the most unreliable. A visual rating is a subjective judgement on the part of the experimenter. Several authors have reported on the fallibility of visual ratings of growth for assessing deterioration of plastic formulations. Also, if extraneous nutrient is incorporated into the agar medium to examine co-metabolism effects, a visual rating becomes meaningless. Even if extraneous nutrient is not present, and the amount of growth is carefully and objectively measured, the measurement reflects an effect of the plastic upon the microorganisms, and not the desired objective of measuring the effect of the organisms on the plastic. The same can be said of techniques which measure other growth parameters such as optical density, oxygen uptake or production of metabolic substances (Kaplan, 1964; Hitz *et al.*, 1967; Sharpe and Woodrow, 1972; Klausmeier and Jamison, 1973). Furthermore, the amount of growth is dependent upon the type of organism involved in the deterioration. A non-filamentous organism may cause much more severe deterioration of the plastic than a filamentous organism such as a fungus, although the fungi may cover a considerably larger proportion of the surface area of the sample. In the opinion of the authors, therefore, while it may be interesting to note the amount of growth on a sample as heavy, moderate, light, etc., these observations should not be given undue significance when determining the amount of degradation which has taken place. Measurement of the physical characteristics of the plastic sample itself is considered to be the most reliable index of biodeterioration. Characteristics such as the flexibility, elasticity or tensile strength of the sample can, in some cases, reliably indicate the degree of biodeterioration of plastic formulations. However, the single measurement which has been found to correlate most closely with all other indices of plastic degradation by microorganisms is weight loss (Hazeu, 1967; Hitz *et al.*, 1967). The application of the weight loss technique, as practiced in this laboratory, to the agar plate method of assessing biodegradability of plastics is a modification of the method proposed by Burgess and Darby (1964).

4.2. Procedure

Step 1: *Preparation of the Sample*

See Section 5 for comments concerning sample shape and size.

It is normally good practice to clean the plastic sample prior to subjecting it to microbial resistance testing. Cleaning with an aqueous detergent solution followed by a thorough rinsing with distilled water is normally sufficient for the removal of dirt and grease or oil contaminants. Since most plastics are hydrophobic, it is usually quite simple to blot the sample dry following the distilled water rinse. Rubber gloves may be worn by the experimenter to prevent recontamination of the sample by fingerprints, or the clean sample may be handled with a forceps.

The sample should be desiccated for at least 24 hr and weighed to the nearest 0·1 mg on an analytical balance. If large numbers of small samples are being tested, it is convenient to hold each one in a numbered paper envelope. Data can be recorded by sample number.

Step 2: *Sterilising the Samples*

If pure cultures are employed as the test inoculum, the sample should be sterilised prior to testing. If the formulation to be examined is not harmed by heat, autoclaving at 120°C for 5 min will sterilise the sample. Most plastic formulations, however, cannot tolerate this amount of heat; therefore, sterilisation by means of ethylene oxide must be employed. If a mixed inoculum such as a soil suspension is to be used, sterilisation of the samples is not required. The procedure for sterilising plastic samples by ethylene oxide exposure will be dictated by the type of steriliser employed. Follow manufacturers' directions. The following general considerations have proved useful to the authors:

(a) Do not exceed a temperature of 54°C.
(b) Small paper envelopes are convenient for enclosing samples, especially small pre-weighed samples of extruded films.

(c) Samples should not be left in the steriliser for longer than manufacturers' recommendation. Damage to the plastic sample may result even if the heat and electric power have been turned off.
(d) Samples should be aerated in room air for at least 8 hr after sterilisation to allow residual ethylene oxide to dissipate prior to testing.

Step 3: *Select and Prepare Media*

Select a basal medium following the procedure and comments detailed in Section 2.2.

For this test the medium should be prepared by dissolving the inorganic salts (and any carbon source used) in distilled water. To the solution add 20 g of agar per litre and autoclave 15 min at 15 lb/in^2 (105×10^3 N/m^2). After autoclaving, pour 20–30 ml of the molten medium into each 9 cm petri dish to be used. The depth of the medium is not critical in this test but should be sufficient to prevent drying of the agar during the incubation period.

Step 4: *Select and Prepare Inoculum*

The procedure here is the same as that detailed in Section 2.1. The resultant inoculum must be a suspension. The concentration of the inoculum is not critical so long as a minimum of 10^4 cells/ml are maintained (Osmon and Klausmeier, 1971).

Controls are necessary. If sterilised samples are employed, duplicate controls for each formulation or incubation condition should be aseptically placed on sterile agar plates identical to the test plates and incubated concurrently. If non-sterile samples are used, or difficulty is encountered in maintaining sterility throughout an extended incubation period, the control sample may be dipped in an aqueous solution of 0·1 % $HgCl_2$ to prevent microbial growth. This seldom, if ever, interferes with other experimental considerations. Situations may, however, be encountered where the use of $HgCl_2$ would be ill-advised.

Step 5: *Inoculate the Samples*

Three basic techniques are available for inoculating plastic samples used in agar plate biodegradation studies.

(i) *Spreading*. The inoculum may be pipetted on to the surface of the agar medium prior to the placement of the sample and spread over the surface of the plate by means of a glass rod. The plastic sample is then placed on the surface of the inoculated agar medium. If a pure culture inoculum is to be used, the pipetting and spreading of the inoculum and placement of the sample must be done observing aseptic technique.

(ii) *Spraying*. The sample may be sprayed with a suspension of organisms after having been placed on the surface of the agar medium. Again, if pure culture inoculum is employed, the atomiser must be sterilised and aseptic techniques observed during the inoculation procedure.

(iii) *Dipping*. The plastic sample may be dipped into a suspension of the organisms used for inoculation and then placed on the surface of the agar medium. Normally, droplets of the suspension will adhere to the plastic sample in sufficient quantity to provide for adequate inoculation. Aseptic technique must, of course, be observed when working with pure culture inocula.

Step 6: Incubation

See comments in Section 2.4 concerning incubation conditions.

Plates should be incubated right side up and examined periodically to note the presence of microbial growth. It is often useful to pull samples at various times (*i.e.* 1, 2, 3 and 4 weeks) for measurement of weight loss.

Step 7: Measurement of Weight Loss

When removed from incubation, the plastic samples should be submerged in a 0·1% solution of $HgCl_2$ for at least 10 min to halt any further microbial action. (*Caution:* This solution is toxic. Avoid contact with skin and eyes or prolonged inhalation of vapours.) The samples are then thoroughly rinsed with water.

It is usually necessary to remove microbial growth from the samples at this point. This is best accomplished by thorough rubbing with the fingers or, if the sample is irregular, gentle scrubbing with a soft brush under running tap water.

The sample is then rinsed with distilled water, blotted dry and placed in a desiccator for at least 24 hr. The sample is then weighed and the weight recorded.

Weight loss is usually expressed as a percentage calculated as follows:

$$\text{Loss (\%)} = \frac{\text{Initial weight} - \text{Final weight}}{\text{Initial weight}} \times 100$$

The percentage loss is then corrected for the amount of loss of the control samples.

Step 8: Evaluating Results

In evaluating weight loss data from plasticised samples, a corrected loss of less than 2% is not usually considered significant. A loss of greater than 10% is very significant, while intermediate values require discretion in evaluation.

A knowledge of the formulation is helpful. For example, in a simple resin/plasticiser formulation containing 30% plasticiser, a weight loss of 1·5% represents a loss of 5% of the plasticiser. The same loss in a formulation containing 10% plasticiser represents a 15% loss of plasticiser. A judgement of the effect of the plasticiser loss on the performance of the formulation may be necessary. It should also be noted that effective plasticiser loss may be greater than weight loss in some cases (Klausmeier, 1972).

When testing any plastic formulation it is important to determine also the fraction of ultimate weight loss which the measured loss represents. A plot of weight loss versus time using data gathered at intervals throughout the incubation period should approach horizontal for an accurate measurement of biodegradability. Care must be taken to ensure that the stabilisation of weight loss is not due to extraneous causes such as desiccation or exhaustion of some mineral from the agar, but, indeed, reflects the inability of the organisms further to degrade the plastic. The same type of graph may also be used with caution to predict the effective life of a plastic in a given usage environment.

5. THE INFLUENCE OF SAMPLE SHAPE AND SIZE

It has been shown that the shape and size of the plastic sample can greatly influence the results of biodeterioration testing (Osmon *et al.*, 1972). This is due to two causes. First, a plastic sample placed on the surface of a moist agar medium effectively prevents the diffusion of oxygen to that portion of the medium covered by the sample. Growth of the test organisms, especially non-filamentous organisms, is therefore confined largely to the perimeter of the sample. The second cause is the aforementioned migration of plasticiser through the plastic formulation. Obviously, the greater the perimeter, the greater is the area of the sample exposed to attack. Minimising the distance from any point in the plastic sample to the perimeter of the sample will maximise the rate at which biodegradation proceeds. The accompanying graphs (Figs. 1 and 2) illustrate the effect of sample shape on the rate of biodeterioration of plasticised polyvinyl

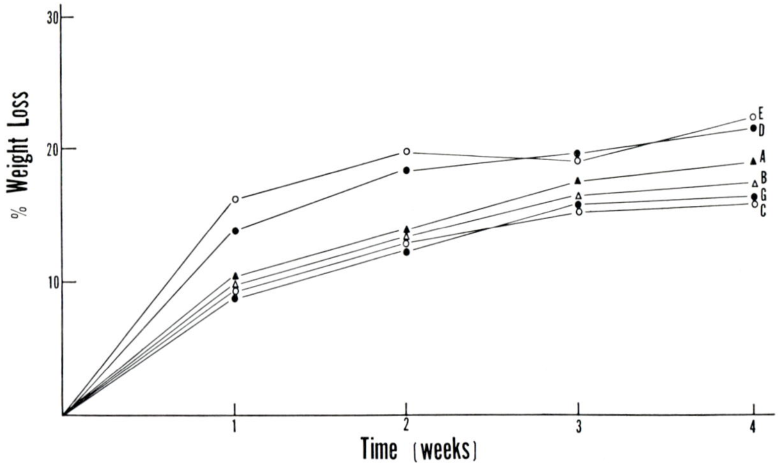

FIG. 1. Weight loss of plastic samples of various shapes incubated with *Pseudomonas*.

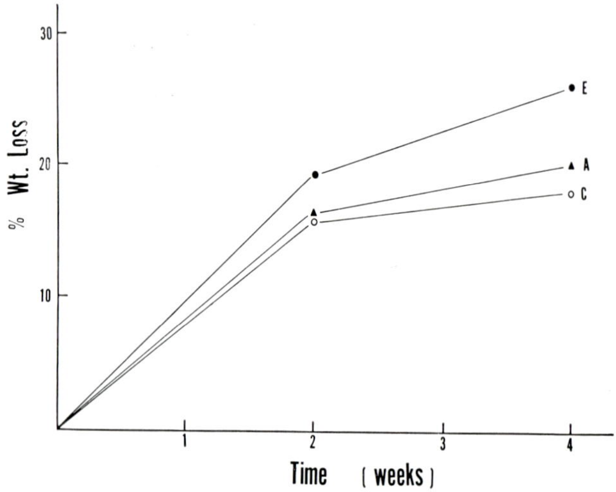

FIG. 2. Weight loss of plastic samples of various shapes incubated with fungi.

chloride formulations. Each of the samples represented in the graph has an identical surface area but a different shape and, therefore, a different perimeter (*see* Table II). A positive correlation is obvious between perimeter and biodegradation as shown by weight loss. When the ratio of perimeter to surface area is maximised, the rate of biodegradation is also maximised. The authors prefer to use samples having a very narrow width and of a small thickness. When working with extruded films, the authors' laboratory uses a manually operated press with a die which excises samples 1 mm wide by 100 mm long from the plastic film. By using this technique, a large number of samples can be prepared in a short period of time. It is recognised that when testing certain types of preformed items, no choice is available regarding sample size or shape. However, in most instances it is possible to cut or pare samples for testing from the original sample of material. The best results are achieved when neither the width nor the thickness of the sample to be tested exceeds 1 mm. The length of the sample can be any length which is convenient to handle.

TABLE II

DIMENSIONS AND SHAPE OF SAMPLES

Code	*Shape*	*Perimeter (cm)*	*Area (cm^2)*
A	Rectangle 3 × 10 cm	26	30
B	Rectangle 6 × 5 cm	22	30
C	Square 5·5 cm	22	30
D	Rectangle 7 × 5 cm with 1 × 5 cm cutout	36	30
E	Rectangle 10 × 7 cm with 5 × 8 cm cutout	60	30
G	Circle with radius of 3·09 cm	19·4	30

REFERENCES

Berk, S. (1950). Effect of fungus growth on the tensile strength of polyvinyl chloride films plasticized with three plasticizers. *ASTM Bull.*, **168,** 53–5.

Berk, S., Ebert, H. and Teitell, L. (1957). Utilization of plasticizers and related organic compounds by fungi. *Ind. Eng. Chem.*, **49,** 1115–24.

Burgess, R. and Darby, A. E. (1964). Two tests for the assessment of microbial activity on plastics. *Brit. Plastics*, **37,** 32–7.

DeCoste, J. B. (1968). *Ind. Eng. Chem., Prod. Res. Devel.*, **7,** 238.

Hazeu, W. (1967). Results of the first interlaboratory test on plastics. *Int. Bicdetn. Bull.*, **3,** 15–19.

Hitz, H. R., Merz, A. and Zinkernagel, R. (1967). Determination of the resistance of plasticized PVC by fungi and bacteria by the weight loss method and evaluation of mechanical properties. A report on the cooperative test of the Task Group A (Biological Attack) of the ISO Technical Committee 61, Working Group 6. *Material u. Organismen*, **2,** 271–96.

Kaplan, A. M. (1964). Comments on the respirometric technique for assessment of microbiological susceptibility of materials. *Devel. Ind. Microbiol.*, **6,** 191–201.

Klausmeier, R. E. (1966). The effect of extraneous nutrients on the biodeterioration of plastics. In: *Microbiological Deterioration in the Tropics*, SCI Monograph 23, pp. 232–43.

Klausmeier, R. E. (1972). Results of the second interlaboratory experiment in biodeterioration of plastics. *Int. Biodetn. Bull.*, **8,** 3–7.

Klausmeier, R. E. and Jamison, E. I. (1973). A rapid method for estimation of microbial degradation of plastics. *Devel. Ind. Microbiol.*, **14,** 258–63.

Klausmeier, R. E. and Jones, W. A. (1961). Microbial degradation of plasticizers. *Devel. Ind. Microbiol.*, **2,** 47.

Klausmeier, R. E., Williams, P. L., Osmon, J. L., Jamison, E. I. and Kanzig, J. L. (1971). The effect of added nutrient on plasticizer deterioration. *Devel. Ind. Microbiol.*, **12,** 204–11.

Osmon, J. L. and Klausmeier, R. E. (1971). *Candida lipolytica:* a non-pathogenic test organism for deterioration of plastics. *Devel. Ind. Microbiol.*, **12,** 197–203.

Osmon, J. L., Klausmeier, R. E. and Jamison, E. I. (1970). The ability of selected yeast cultures to degrade plasticized polyvinyl systems. *Devel. Ind. Microbiol.*, **11,** 447–52.

Osmon, J. L., Klausmeier, R. E. and Jamison, E. I. (1972). Rate-limiting factors in biodeterioration of plastics. In: *Biodeterioration of Materials*, vol. 2 (ed. A. H. Walters and E. H. Hueck-van der Plas), pp. 66–75. Applied Science Publishers, London.

Reese, E. T., Cravetz, H. and Mandels, G. R. (1955). Activity of fungi on oils. *Farlowia*, **4,** 409–21.

Sharpe, A. N. and Woodrow, M. N. (1972) A rapid test for biodegradability by *Pseudomonas* organisms. In: *Biodeterioration of Materials*, vol. 2 (ed. A. H. Walters and E. H. Hueck-van der Plas), pp. 233–7. Applied Science Publishers, London.

Teitell, L., Berk, S. and Kravitz, A. (1955). The effects of fungi on the direct current surface conductance of electrical insulating materials. *Appl. Microbiol.*, **3,** 75–81.

Williams, P. L., Kanzig, J. L. and Klausmeier, R. E. (1969). Evaluation and production of esterases on plasticizer substrates by fungal species. *Devel. Ind. Microbiol.*, **10,** 177–81.

6

Evaluation Techniques for Resistance of Floor Coverings to Mildew

WALTRAUT KERNER-GANG

Fachgruppe 'Biologische Materialprüfung', Bundesanstalt für Materialprüfung (BAM), 1 Berlin 45, Unter den Eichen 87, Federal Republic of Germany

1. INTRODUCTION

Microbiological test methods for evaluating the rotproofing of floor coverings vary greatly depending on the different types and applications. They must be differentiated between those applied to (a) *synthetic* floor coverings (with and without backing) and (b) *textile* floor coverings, the pile material of which may consist of natural or synthetic fibres with plastic coated or foamed backing.

The structural differences between plastic and textile must be taken into account when their mildew resistance is determined. There are a great variety of test methods for these two materials (*e.g.* Haldenwanger, 1970; Becker and Gross, 1974; Kühne, 1975; Pantke, 1976). Those test methods that are successfully being applied at the BAM will be described here; but the evaluation of the hygienic treatment of hospital carpetings and places where pathogenic microorganisms occur abundantly will not be considered. Several test methods for evaluating the permanence of bacteriostatic preservatives for floor coverings have, for instance, already been given by Mebes (1973).

2. SYNTHETIC FLOOR COVERINGS (PVC) WITH AND WITHOUT BACKING

Damage caused by mildew growth on synthetic (polyvinyl chloride) floor coverings has been reported by Kerner-Gang (1969). When the backing of these floor coverings (often used for sound and heat insulation and made from jute or textile fabric) is not sufficiently preserved, fungi may easily develop under certain moisture conditions. The pattern of such biodeterioration may, however, vary. Under certain conditions a number of mould fungi produce pigments, for instance, especially when their metabolic processes differ from the normal state. Without changing the synthetic molecules chemically, these pigments often penetrate the synthetic material, colouring its surface yellow or reddish. This pigmentation cannot be wiped off from the surface. Certain bacteria, too, may produce similar discolorations.

In our experience there are two test methods which are most suitable for these types of floor coverings: (1) the evaluation of resistance of natural fibre backing to cellulose-decomposing fungi, and (2) the behaviour of the synthetic upper material to discolouring fungi.

2.1. Evaluation of the Resistance of Carpet Backings of Natural Fibres to Cellulose-decomposing Fungi

For carpet producers it is often important to carry out mildew resistance tests with felt or other fabrics, for which the growth test in accordance with DIN 53 931 (DNA, 1969) is recommended.

Method of test. Petri dishes are prepared with 20 ml of a mineral salts–cellulose agar in each (Table I(A)). By means of a Drigalski spatula, 0·5 ml of an aqueous spore suspension is evenly distributed over the agar surface.

TABLE I

COMPOSITION OF THE CULTURE MEDIA MENTIONED
(quantities in g/1000 ml aqua dest.)

Component	*A*	*B*	*C*	*D*	*E*
Ammonium nitrate	2	—	—	—	—
Cellulose powder	10	—	—	—	—
Iron sulphate	—	—	0·01	0·01	0·01
Magnesium sulphate	1	—	0·5	0·5	0·5
Malt extract	—	50	—	—	—
Potassium chloride	—	—	0·5	0·5	0·5
Potassium phosphate, prim.	—	—	0·7	0·7	0·7
Potassium phosphate, sec.	1	—	0·3	0·3	0·3
Potassium tartrate	1	—	—	—	—
Saccharose	—	—	—	—	30
Sodium nitrate	—	—	2	2	2
Solution of heavy metal salts	1 drop	—	—	—	—
Agar-agar	20	15	20	—	20

The following fungi are used individually:

Chaetomium globosum ATCC 6205
Penicillium funiculosum QM 391
Trichoderma sp. ATCC 9645 (T1)

These fungi are allowed to develop for 24 hr at 29°C ± 1°. After that period the test specimens are placed on the agar surface. The specimens should have a diameter of 50 mm or be square (50 mm × 50 mm). At least three replicate specimens should be used with each of the fungal species. With finished textiles one unfinished control specimen should also be included. The dishes are now incubated for 2 weeks at the above temperature.

Evaluation. At the end of the test period the intensity of fungal growth is rated as follows (if necessary with a magnifying glass) (Fig. 1):

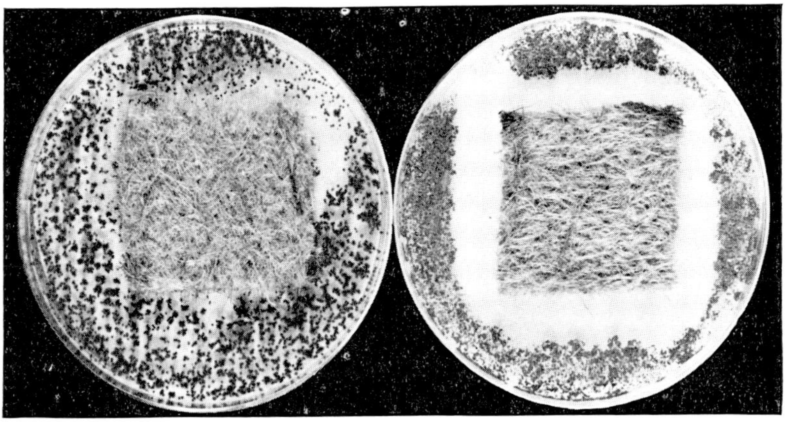

FIG. 1. Evaluation of jute felt with *Chaetomium globosum.* (*left*) Untreated, rating 1. (*right*) Preserved with a fungicide, rating 0.

Surface growth:

00 = No fungal growth apparent.
0 = Inhibition zone around the test specimen.
1 = Growth is observed at the edges of the test specimen.
2 = Growth covers the edges of the test specimen but less than 25% of the test surface.
3 = Growth of some colonies covering 25–75% of the test surface.
4 = Considerable growth covering 75% or more of the test surface.
5 = Heavy growth covering the entire test surface (100%).

Intensity of fungal growth:

+ Slight growth, largely mycelium only.
+ + Moderate growth, partly sporulated.
+ + + Heavy growth, many spores.

When leach-resistant finishes are applied to the specimens, their resistance to leaching may be evaluated by the leaching procedure according to DIN 53 930 prior to the fungal test.

2.2. Evaluation of the Resistance of Synthetic Floor Coverings to Discolouring Fungi

The behaviour of polyvinyl chloride floor coverings (with and without backing) is tested as follows at the BAM.

Method of test. Petri dishes are prepared with 20 ml of a malt agar (Table I(B)). Their surface is inoculated with 0·5 ml of a spore suspension evenly distributed by means of a Drigalski spatula. The fungi may develop for 48 hr at 29°C ± 1°. Round test specimens with a diameter of 5–8 cm are then placed on the mycelial mat. The dishes are then incubated at the above temperature. Inspection is made after 4, 8 and 12 weeks. Should the agar be observed to dry up, 1–2 ml of sterile water should be added to every dish at each inspection.

The following fungal species were found to cause discoloration of light-coloured polyvinyl chloride floor coverings:

Aspergillus versicolor	Red
Chaetomium alba-arenulum	Dark grey to black
Dactylium fusarioides	Pink to red
Epicoccum nigrum	Pink to orange
Penicillium funiculosum	Red, partly yellow to orange
Penicillium janthinellum	Yellow, partly greenish
Penicillium purpurogenum	Red
Penicillium vermiculatum	Chrome yellow, partly also red
Penicillium wortmannii	Yellow, some areas pink

It is recommended that five or six of these species should be used, and that they should be strains which already show clear pigmentation in the test tubes. At least three test specimens should be used for every species and, if possible, another specimen of a material which is easily discoloured (Fig. 2).

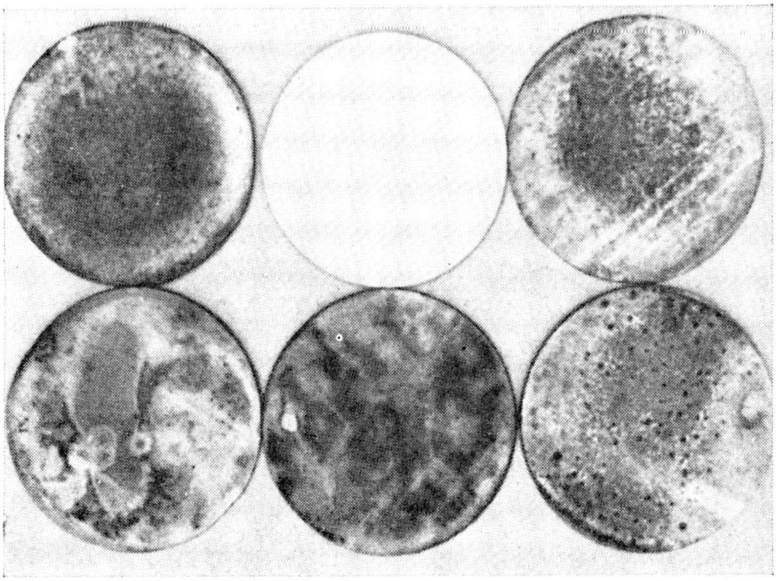

FIG. 2. Discoloration of light-coloured polyvinyl chloride floor coverings caused by several mould fungi within 12 weeks under laboratory conditions. Control: upper row, centre (about half the natural size).

When the test specimens are brittle, in addition to being discoloured, this indicates that plasticiser components are undergoing microbiological attack.

3. TEXTILE FLOOR COVERINGS

Based on the ever-growing popularity of textile floor coverings for so-called wet rooms and hobby rooms, resistance to microbiological attack is necessary because of the high water-retaining capacity of such coverings. Even purely synthetic floor coverings may contain components such as plasticisers, finishes, etc., on which mildew may develop. At the BAM a number of textile floor coverings were tested for their resistance to microbial attack by different test methods (Kerner-Gang and Meckel, 1972). These coverings had either been qualified as outdoor coverings or were recommended for wet rooms. The pile of the floor coverings tested consisted either of synthetic or natural fibres (*e.g.* polyamide, polyester, viscose fibres or wool), and the backing was coated, foamed or dotted with adhesive. Valuable results were obtained with specimens suspended freely, and also with the agar method according to ISO R 846 (or ISO/DIS 846 (1975) revised).

3.1. Test with Freely Suspended Test Specimens

This test method was originally developed in electrical engineering (DIN 40 046, Part 10, Test J) and has been adopted for these tests. Specimens sprayed with a fungal spore suspension are suspended in an atmosphere saturated with water vapour. The fungi are given no additional food source.

Method of test. Specimens measuring about 5 cm × 5 cm are fastened to the lid of a test vessel with perlon strings (Fig. 3). They are sprayed with an aqueous spore suspension containing the spores of the following species and strains of moulds:

Aspergillus niger ATCC 6275
Aspergillus terreus PQMD 82j
Aureobasidium pullulans ATCC 9348
Paecilomyces varioti IAM 5001
Penicillium funiculosum IAM 7013
Penicillium ochro-chloron ATCC 9112
Scopulariopsis brevicaulis 5146
Trichoderma viride IAM 5061

FIG. 3. Mildew test with suspended test specimens.

The lids with the specimens are placed on the test vessels, the bottoms of which are covered with water (*e.g.* 100 ml for 1000 ml vessels). The test vessels are then incubated for 12 weeks at almost saturated relative humidity and a temperature of 29°C ± 1°. It is recommended that four replicate test specimens each be used, and that a material which is easily attacked by moulds should be included to check the fungal activity.

Evaluation. After 4 and 8 weeks the fungal development on the specimens is examined by the naked eye and, upon completion of the test, with the help of a microscope also (magnification 30 ×). Textiles which exhibit fungal growth on either side after a period of 8 weeks will hardly be expected to give satisfactory results under practical conditions; whereas specimens which do not show any growth even under the microscope after a period of 12 weeks might give satisfactory results under practical conditions.

3.2. Agar Test (according to ISO R 846, methods A and B)

This method is used to determine the resistance of plastics to mildew. Method A is intended to determine the ability of plastics to serve as a nutritive medium for the fungal spore suspension sprayed on the specimens; in method B a nutritive medium is added and the extent of microbial growth is determined. The fungitoxic properties of plastics can thus be determined and the effectiveness of eventually added fungicides may be evaluated.

Method A. The test specimens (edge length 30–50 mm) are placed on 3–4 mm agar plates in petri dishes. The agar contains mineral salts (Table I(C)) but no carbon source. The agar surface with the test specimen is then sprayed with an aqueous spore suspension of the following fungi:

Aspergillus niger
Chaetomium globosum
Paecilomyces varioti
Penicillium funiculosum
Trichoderma viride

The test fungi should be well-defined species.

The spore suspension is prepared as follows. The shaken spore suspension of each culture is filtered through a thin layer of sterile glass wool in order to remove mycelial fragments. The filtered mixed spore suspension is centrifuged aseptically and the supernatant liquid is discarded. The residue is resuspended in sterile water and centrifuged. This procedure is repeated three times. The final washed residue is suspended in 100 ml of sterile aqueous mineral salts solution (Table I(D)).

The dishes are incubated at 29°C ± 1° and 95–100% relative humidity for 4 weeks. Not less than six replicate test specimens should be used; three of them should be placed on the pile and three on the back.

After 28 days the test specimens are inspected for fungal growth. The examination is visual, by the naked eye or, if necessary, with the help of a microscope (magnification 50 × or 60 ×). The intensity of fungal growth is rated as follows:

0 = No growth is apparent under the microscope.
1 = Fungal growth is hardly visible to the naked eye, but is quite apparent under the microscope.
2 = The growth on the surface is visible to the naked eye but does not cover more than 25% of the test surface.
3 = The growth on the surface is visible to the naked eye and covers more than 25% of the test surface.

Method B. The test specimens are placed in petri dishes on agar which also contains sugar as a carbon source in addition to mineral salts (Table I(E)). Thereafter the procedure is the same as described in method A.

TABLE II

Intensity of fungal growth	*Evaluation of testing material*	
	Method A	*Method B*
0 + zone of inhibition	—	Strong fungitoxic effect due to diffusing substance
0	Plastics are inert or fungitoxic	Strong fungitoxic effect
1	Plastics contain nutritive substances; they permit only a slight growth	Not quite a complete fungitoxic effect
2	Plastics contain nutritive substances and are not	} Decreasing effectiveness to complete absence of fungitoxic effect (for 2–5)
3	resistant to fungal attack	
4	—	
5	—	

After 28 days the following degrees of growth can be determined:

0 = No fungal growth is apparent even under the microscope. Inhibition zone absent or present.
1 = Fungal growth is hardly visible to the naked eye, but is quite apparent under the microscope.
2 = Slight growth covering less than 25% of the surface.
3 = Medium growth covering 25–50% of the test surface.
4 = Considerable growth covering 50% or more of the test surface.
5 = Heavy growth covering the entire test surface.

The evaluation of test results is shown in Table II.

REFERENCES

Becker, H. and Gross, H. (1974). Über die Widerstandsfähigkeit makromolekularer Werkstoffe gegen mikrobiellen Angriff (Ein Übersichtsbericht). *Material u. Organismen*, **9**, 81–131.

Fachnormenausschuss Elektrotechnik im Deutschen Normenausschuss (DNA) (1972). *Umweltprüfungen für die Elektrotechnik. Prüfung J: Schimmelwachstum.* DIN 40 046, part 10.

Fachnormenausschuss Materialprüfung im Deutschen Normenausschuss (DNA) (1969). *Prüfung von Textilien: Bestimmung der Widerstandsfähigkeit von Textilien gegen Schimmelpilze, Bewuchsversuch.* DIN 53 931.

Haldenwanger, H. H. M. (1970). *Biologische Zerstörung der makromolekularen Werkstoffe.* Springer Vlg, Berlin–Heidelberg–New York.

International Organisation for Standardisation (1968). *Recommended Practice for the Evaulation of the Resistance of Plastics to Fungi by Visual Examination.* ISO R 846.

Kerner-Gang, W. (1969). Schäden an Kunststoff-Fussbodenbelägen durch Schimmelpilze. *Boden–Wand–Decke*, **15**, 173–80.

Kerner-Gang, W. and Meckel, L. (1972). Beständigkeit textiler Fussbodenbeläge für Nassräume gegen Mikroorganismen. *Melliand Textilber.*, **53**, 1295–8.

Kühne, H. (1975). Beständigkeit von Textilien aus natürlichen und synthetischen Fasern gegen Organismen. *Textil praxis Int.*, 598–602, 718–20.

Mebes, B. (1973). Die Beurteilung antimikrobieller Ausrüstungen von Gebrauchsgegenständen im praktischen Einsatz. *Melliand Textilber.*, **54**, 1232–6.

Pantke, M. (1976). Test methods for evaluation of susceptibility of plasticised PVC and its components to microbial attack. This volume, pp. 51–76.

7

Evaluation Techniques for Resistance of Optical Lenses to Fungal Attack

WALTRAUT KERNER-GANG

Fachgruppe 'Biologische Materialprüfung', Bundesanstalt für Materialprüfung (BAM), 1 Berlin 45, Unter den Eichen 87, Federal Republic of Germany

It is well known that the glass surfaces of optical instruments (*e.g.* binoculars, microscopes, cameras) may be subject to deterioration by mould fungi in areas with high relative humidities. Optical instruments and photographic equipment so affected may soon be rendered largely or completely unserviceable. The damage can often not be repaired by cleaning the fungal mycelium from the glass surface because, as the deterioration persists, the fungal hyphae often produce typical etching patterns, sometimes just called patterns.

This type of deterioration is often encountered in the tropics. Ohtsuki (1943), Magee *et al.* (1944), Hutchinson (1947), Richards (1949), Teitell and Berk (1952), Dade (1958), Kaller (1960), Nagamuttu (1967) and other authors have reported on such deterioration.

Apart from certain temperature and moisture conditions necessary to support fungal development, nutrient substances must be present for the process of deterioration. Even small amounts of organic substances—tiny dead insects, small dust particles, even fingerprints and traces of lubricants or putties on the glass surface—may be utilised by the fungi. Spores of some fungal species also germinate on completely clean glass surfaces and develop hyphae until the nutritive substances are exhausted. Both types of fungal deterioration will be considered in the test methods described below.

The specific types of fungal species which may affect optical lenses are particularly important relative to the selection of suitable test organisms. While Ohtsuki is of the opinion that certain xerophilic fungi (*e.g. Aspergillus vitricolae* Ohtsuki) are the original deteriorating agents, other authors feel that the glass deteriorating agents are widely distributed, no specialised fungi being fully responsible.

In our laboratory, 66 fungal species were isolated from microscopes which had been returned to the manufacturers from all parts of the world (Kerner-Gang and Schneider, 1969). The majority of the isolates were *Aspergillus* and *Penicillium* species. With the exception of two *Aspergillus* and five *Penicillium* species they were known to deteriorate glass. Of the xerophilic fungi only *Aspergillus amstelodami* and *A. versicolor* were isolated.

Several authors have discussed possible factors producing the etching patterns on mildewed glass. Kerner-Gang (1968) presents a review of the chemical behaviour of optical lenses and the production of such patterns in relation to certain factors. In general, the following causes and interactions are responsible.

The etching patterns are due to organic acids produced by fungi; their metabolic products, acid or alkaline in nature, accelerate the etching of glass; the water film surrounding the fungi dissolves some components from the glass; the growing ends of the fungal hyphae are responsible for the etching patterns, and since they comprise more than 90% water they are strongly hygroscopic; etching patterns are attributable to physical factors; chelate-producing microorganisms form water-soluble complexes with the metals present in the silicate, thus accelerating the attack rate of glass.

Kaller (1960) is of the opinion that the intensity of attack is decided primarily by the acidity class of the glass, as well as the time of exposure and the relative humidity. Nagamuttu (1967), who had the opportunity of examining more than 200 optical instruments from different manufacturers, firmly believes that the intensity of fungal growth depends on the material of the lenses. We have also found that fungal growth on glass does not always imply irreversible damage but depends on the acid and weather resistance of the glass.

As shown in Fig. 1, different fungal species produce different patterns on the same type of glass. The appearance of these patterns is influenced equally by temperature and relative humidity (Kerner-Gang and Theden, 1964) (Fig. 2).

Method of test. A method of testing the resistance or a treatment of optical lenses to fungal attack has been developed which may be modified to meet special requirements (Theden and Kerner-Gang, 1964).

The tests are carried out in high-humidity test chambers. Petri dishes of a diameter of 9 cm are generally used unless the size and form of the glass specimens require other test vessels. 10 ml of water is added to the dish to give a relative vapour pressure of 100%, or a certain concentration of a

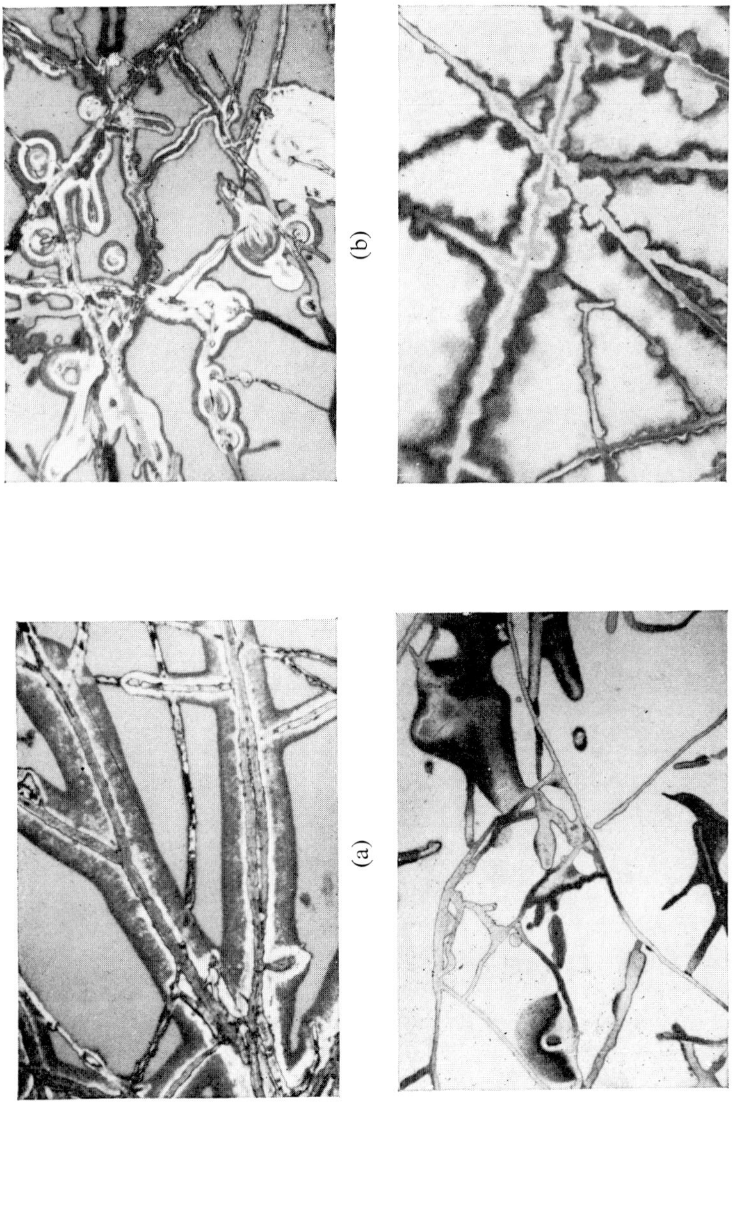

FIG. 1. Etching patterns on SF4 glass after 4 weeks of exposure: (a) *Penicillium funiculosum*; (b) *Dactylium fusarioides*; (c) *Alternaria tenuis*; (d) *Aspergillus fischeri*.

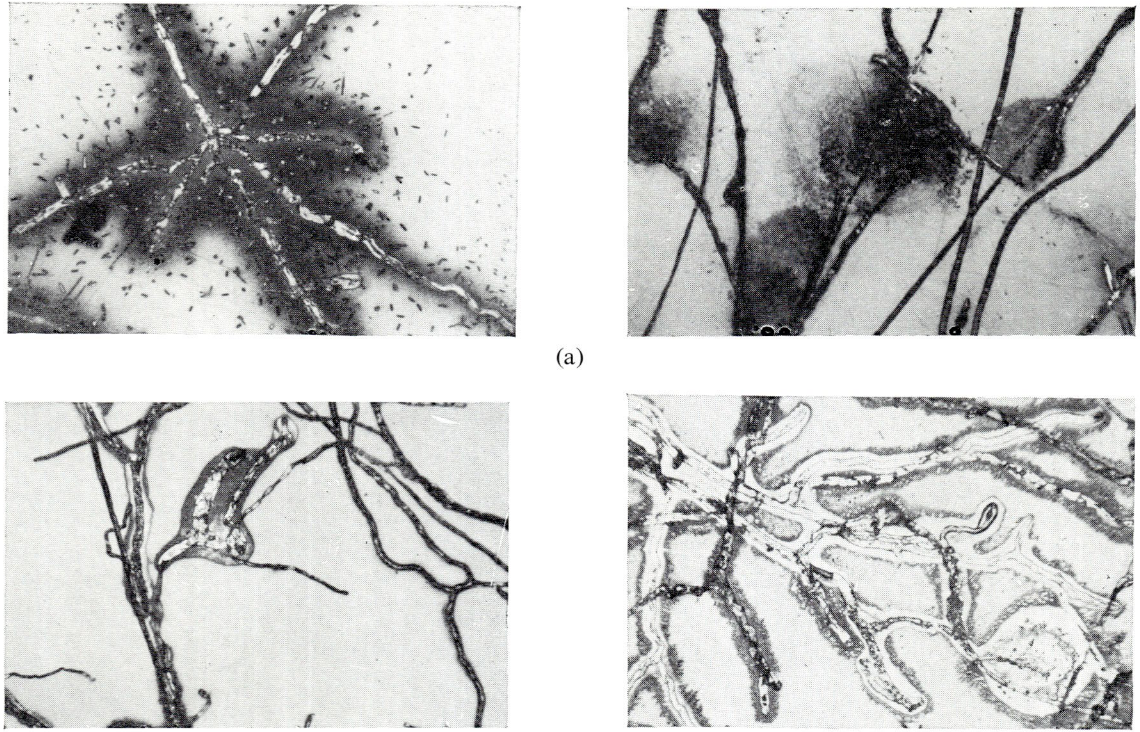

(a)

(b)

FIG. 2. (a) Growth of *Aspergillus niger* on optical glass at 30°C: (*left*) at 97%, (*right*) at 100% relative humidity. (b) Growth of *Aspergillus amstelodami* on optical glass at 95% relative humidity: (*left*) at 14°C, (*right*) at 30°C.

saline solution is used when the test fungus or the type of glass necessitates a lower vapour pressure (*see* Table I). The glass specimens to be tested are placed on 1 cm glass tube sections of a diameter of 2–3 cm which are sterilised before being placed in position.

The glass specimens are cleaned by wiping them with a cloth soaked with acetone or alcohol immediately prior to placing them on the carrier ring. It is recommended that round or square specimens be used, the diameter or edge length of which should be about 4 cm and the thickness about 3 mm. One side of the specimens must be polished or have received the preservative treatment to be tested.

TABLE I

SODIUM CHLORIDE SOLUTIONS TO PRODUCE CERTAIN RELATIVE VAPOUR PRESSURES (according to Walter, 1931)

% *NaCl*	% *RVP*
0	100
5·0	97
7·5	95
10·0	93
15·0	89
20·0	83

The number of specimens should allow for at least three parallel tests with each fungal species. In addition, one type of glass which is not susceptible to fungal attack (normal glass, *e.g.* window glass or coverglass for slides) and one optical glass which is known to be susceptible should be included. The first specimen is used to check the germinating and spreading abilities of the fungi under the chosen test conditions, and the latter specimen is included to check the ability of the test fungi to attack glass.

There are two different methods of applying the fungi to the polished glass surface and the controls:

1. Without adding any nutrient substances: dry spores of the test fungus are applied to the glass surface by means of cottonwool tips (spore germination method).
2. With nutrient substances added: an agar inoculated with a certain fungus is applied to the glass surface (agar method) (Fig. 4).

Spore germination method. The fungi are cultured in petri dishes on malt agar (50 g malt extract and 15 g agar-agar per 1000 ml of water). When they are well germinated—in general, after 2 weeks—the spores are transferred by means of sterile cottonwool tips. The spores should be distributed as evenly as possible and not too densely over the whole surface.

The following fungal species proved to be suitable for this method (the relative vapour pressure, at the optimum temperature for development, is indicated in brackets): *Aspergillus amstelodami* (93–95 %), *Asp. versicolor* (93 %), *Asp. vitricolae* (89 %), *Penicillium funiculosum* (97 %), *Penicillium notatum* (97 %).

Depending on the specific test purpose, other cultures that are thought to yield informative results may also be included in the test. The petri dishes are incubated at 28°C ± 1°. At intervals of 4–5 days the specimens are examined under the microscope (Fig. 3) in order to establish the rate of germination compared with that on normal glass. After an adequate test period (3–4 weeks for resistant types of glass) the specimens are

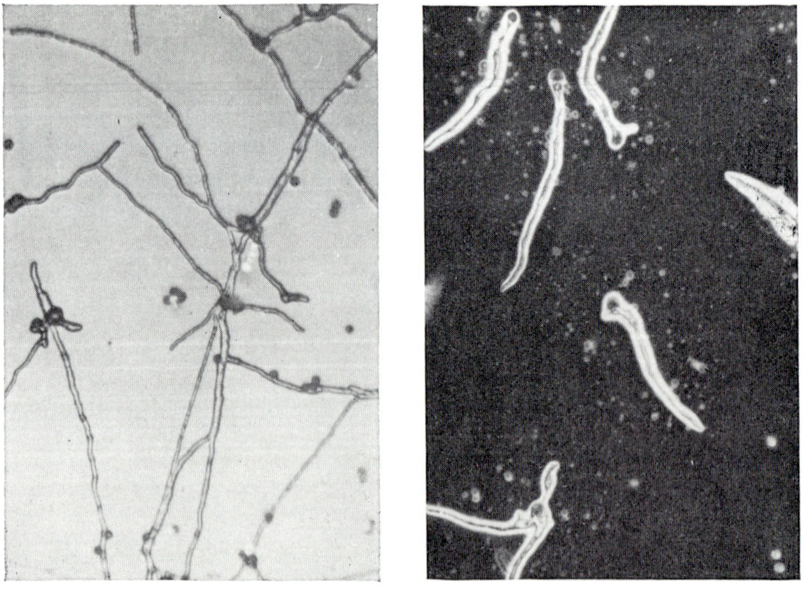

(a) (b)

FIG. 3. Germination of spores after a test period of 5 days. (a) *Penicillium* sp. on untreated optical glass. (b) *Aspergillus versicolor* on optical glass with so-called T-seal.

rinsed with acetone to remove the mycelium, wiped off carefully and examined for traces under the microscope.

Agar method. A thin agar plate (approximately 10 ml per petri dish of a diameter of 9 cm) is inoculated with 0·5 ml of an aqueous spore suspension of the respective fungal species. By means of a Drigalski spatula the suspension is evenly distributed over the agar surface. The dishes are incubated for 12–24 hr at room temperature. By means of sterile glass tube sections (interior diameter 3–4 mm), agar discs are punched out which are placed in the centre of the glass specimen by means of a platinum loop (Fig. 4).

The following fungal species proved to be suitable: *Aspergillus amstelodami* (95%), *Asp. versicolor* (97%), *Dactylium fusarioides* (100%), *Paecilomyces varioti* (100%), *Penicillium funiculosum* (97%). Other cultures which are considered to yield informative results may also be included. The dishes are incubated at 28°C ± 1°. After one week a first statement

FIG. 4. Test arrangement of the agar method. Petri dish with carrier ring and glass specimen on which the inoculated agar plate is placed (lid of the petri dish removed).

may be made on the development of the fungi. The agar surface is either overgrown and the fungus spreads to the glass (Fig. 5) or a possible preservative treatment of the glass surface is already inhibiting fungal development on the agar. All phases of development are possible. The size of the fungal colony is compared with that grown on normal glass. The

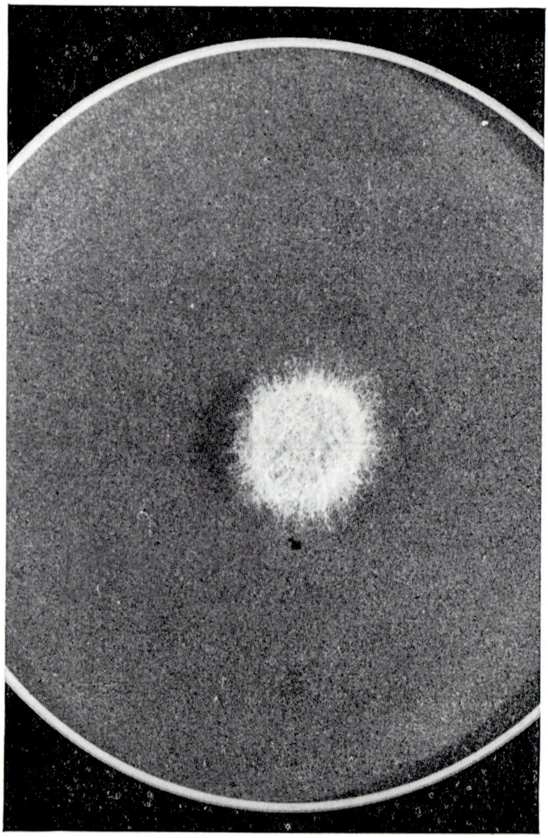

FIG. 5. Glass specimen with agar plate; fungal overgrowth to the glass.

diameter of the colonies is measured at certain intervals. The spreading rate is influenced by the fungal species and the type of glass. The time when the specimens should be inspected for fungal etching cannot definitely be stated. Our own tests have revealed that the period between overgrowing and production of patterns may be less than 24 hr. With less susceptible species the tests will, of course, take longer. When the resistance of a certain type of glass is not known, 2–4 weeks after the test is started the agar discs and the fungal mycelium should be carefully rinsed with acetone, possibly with the help of a soft brush. The specimen is then wiped off and examined under the microscope and compared with the controls.

Figure 6 shows fungal mycelium on type SK16 glass after a test period of 18 days (a) and the same piece after the mycelium was wiped off (b).

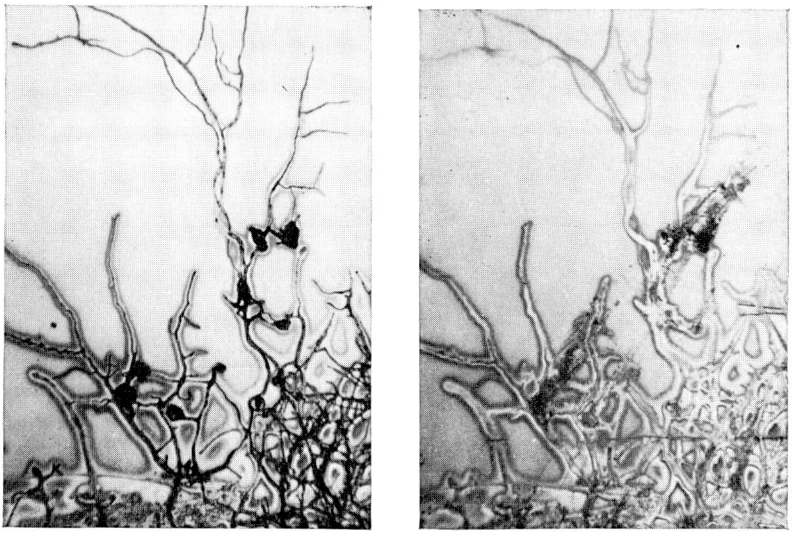

(a) (b)

FIG. 6. Fungal mycelium and etching patterns. (a) Glass with fungal mycelium. (b) The same piece after the mycelium was wiped off.

REFERENCES

Dade, H. A. (1958). Moulding of lenses. *C.P. News*, **4** (2), 17–19.

Hutchinson, W. G. (1947). The fouling of optical glass by microorganisms. *J. Bacteriol.*, **54**, 45–6.

Kaller, A. (1960). Fungusbildung auf Optik. *Feingerätetechn.*, **9** (1), 20–5.

Kerner-Gang, W. (1968). Zur Frage der Entstehung von Schimmelpilzspuren auf optischen Gläsern. *Material u. Organismen*, **3** (1), 1–17.

Kerner-Gang, W. and Schneider, R. (1969). Von optischen Gläsern isolierte Schimmelpilze. *Material u. Organismen*, **4** (4), 291–6.

Kerner-Gang, W. and Theden, G. (1964). Durch Pilze verursachte Schäden an optischen Gläsern. Ein Bildbericht über Laboratoriumsversuche. *Glastechn. Ber.*, **37** (5), 256–9.

Magee, C. J., Hansen, C. T., Grant, C. K. and Chapman, W. D. (1944). *Report on Condition of Service Material under Tropical Conditions in New Guinea*, pp. 56–60. Sci. Liaison Bureau, Melbourne, Australia.

Nagamuttu, S. (1967). Moulds on optical glass and control measures. *Int. Biodetn. Bull.*, **3** (1), 1–3.

Ohtsuki, T. (1943). Über das Verschimmeln der gläsernen Gegenstände. *Proc. Imp. Acad. Tokyo*, **19,** 688–93.

Richards, O. W. (1949). Some fungous contaminants of optical instruments. *J. Bacteriol.*, **58,** 453–5.

Teitell, L. and Berk, S. (1952). Prevention of mold growth in optical instruments: Panama Canal Zone exposure. *Ind. Eng. Chem.*, **44** (5), 1088–95.

Theden, G. and Kerner-Gang, W. (1964). Ergebnisse von Untersuchungen über Schäden an optischen Gläsern durch Pilze. *Glastechn. Ber.*, **37** (4), 200–5.

Walter, H. (1931). *Die Hydratur der Pflanze*. Vlg G. Fischer, Jena.

8

Evaluation Techniques for Corrosion of Ferrous Metals by Sulphate-Reducing Bacteria

D. D. MARA

Department of Civil Engineering, University of Dundee, Dundee DD1 4HN, Scotland, UK

and

D. J. A. WILLIAMS

Department of Chemical Engineering, University College of Swansea, Swansea SA2 8PP, Wales, UK

1. INTRODUCTION

Sulphate-reducing bacteria are obligately anaerobic organisms belonging to the genera *Desulfovibrio* and *Desulfotomaculum*. They are chemolithotrophic heterotrophs (Rittenberg, 1969) which commonly use the sulphate ion as their terminal electron acceptor and thus produce the diagnostically distinctive sulphide ion. Sulphate-reducing bacteria are ubiquitous organisms, occurring everywhere except in the most highly oxygenated environments and often reaching densities of 10^6 or more cells per millilitre or gramme. Anaerobic clay soils, estuarine muds, swamps, industrial cooling waters, oil wells and the water bottoms in gas and petrol storage tanks are examples of their habitats. Their physiology has been extensively reviewed by Le Gall and Postgate (1973).

Sulphate-reducing bacteria are principally responsible for the microbiological corrosion of ferrous metals. This type of corrosion (Fig. 1) is extremely common and its cost to the United Kingdom economy is some £50 million annually. This corrosion occurs extensively in anaerobic soils, and also in 'closed' environments such as gas holding tanks and recirculating cooling systems. In the latter cases control is usually readily achieved by adding a suitable biocide to the water (Purkiss, 1971). However, in an 'open' environment such as the soil, protection against corrosion is more

FIG. 1. A typical example of corrosion by sulphate-reducing bacteria. The specimen is from a 4 in (100 mm) diameter spun iron pipe, $\frac{1}{4}$ in (6 mm) thick, which had been buried in London clay soil for 12 years. Penetration within 4 years is not unknown, but 10–15 years is more usual. (Specimen by courtesy of the National Physical Laboratory.)

difficult; lime is a useful inhibitor of bacterial activity, but frequently cathodic protection is necessary (Booth, 1964).

Sulphate-reducing bacteria corrode ferrous metals by two distinct mechanisms: (1) cathodic depolarisation by means of the enzyme hydrogenase and (2) the production of ferrous sulphides which are corrosive agents in their own right. Reference should be made to Iverson (1974) and Mara and Williams (1972) for detailed discussion of these mechanisms.

2. DETECTION AND ENUMERATION OF SULPHATE-REDUCING BACTERIA

It is frequently necessary to examine a water or soil sample for the presence of sulphate-reducing bacteria and also to determine their numbers, especially in samples from 'closed' environments where this knowledge is useful in determining the required biocide dose.

2.1. Detection

Baars' medium (Postgate, 1966, 1969) is the best medium for detecting sulphate-reducing bacteria. It contains 0·05% KH_2PO_4, 0·1% NH_4Cl, 0·1% $CaSO_4$, 0·2% $MgSO_4 \cdot 7H_2O$, 0·5% Na lactate (70% solution), 0·05% $FeSO_4 \cdot 7H_2O$, 0·1% yeast extract, 0·1% ascorbic acid and 0·1% Na thioglycollate. Add 2·5% NaCl for marine or brackish water samples. Adjust to pH 7·4 (ignore any purple coloration). Distribute into screw-capped bottles, taking care to ensure an even distribution of the precipitate to each bottle, and sterilise at 121°C for 15 min. Wide-necked universal containers (1 oz, 28 ml capacity) are useful for soil samples. For liquid samples use double-strength medium.

Collect a sample of soil (1–5 g) or water (10–50 ml) from the environment under investigation. For soil samples it is usually necessary to use an auger to obtain samples from the desired depth; it is essential to use a clean (but not necessarily sterile) auger. The sample should be collected in as aseptic a manner as is possible. Add the sample to a suitable bottle containing Baars' medium. Fill up the bottle with additional medium to exclude all air bubbles. Incubate at 30°C for up to 10 days, examining daily for the formation of the black precipitate of iron sulphide which indicates that sulphate reduction has occurred. One species, *Desulfotomaculum nigrificans*, is thermophilic and an incubation temperature of 55°C is required for its detection (this organism occurs in heating systems and also causes spoilage of canned foods).

2.2. Enumeration

The simplest method of counting populations of sulphate-reducing bacteria is to use the 'shake-tube' technique with modified iron sulphite agar (Mara and Williams, 1970). This medium is Oxoid iron sulphite agar supplemented with 0·5% Na lactate (70% solution), 0·2% $MgSO_4 \cdot 7H_2O$ and 0·05% $FeSO_4 \cdot 7H_2O$ (add 2·5% NaCl for marine and brackish water samples). After the pH has been adjusted to 7·5, sterilise at 121°C for 15 min, then add 1% (v/v) of a sterile solution containing 7·5% Na thioglycollate and 7·5% ascorbic acid (adjusted to pH 7) to the medium immediately prior to use. The medium cannot be remelted after the reducing agents have been added.

Ten grammes (wet weight) of soil are added aseptically to 90 ml of sterile diluent (quarter-strength Ringer's solution supplemented with the reducing agent mixture above) and thoroughly mixed. With water samples 1 ml is added to 9 ml of diluent. A series of decimal dilutions is made in the usual way (up to a maximum dilution of 1 in 10^6). One ml of each dilution is added in duplicate to separate 16 × 150 mm test tubes. Modified iron sulphite agar at 45°C is poured into the tubes to within 3–4 cm of the top, each one being immediately closed by a sterile No. 13 rubber stopper and 'shaken' by inversion several times. When the medium has solidified, the tubes are filled to the top with additional medium and then incubated at 30°C for 2 days. The black colonies are then counted (Fig. 2). This procedure is suitable only for *Desulfovibrio* spp.; there is no medium presently available for the enumeration of *Desulfotomaculum* spp.

It is essential to work quickly in order to minimise the deleterious effects of atmospheric oxygen on these organisms.

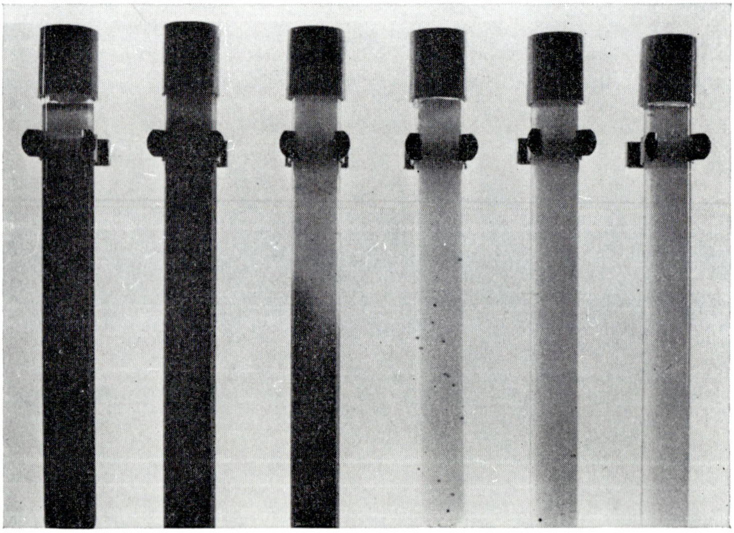

FIG. 2. A typical 'shake-tube' count of sulphate-reducing bacteria in modified iron sulphite agar (decimal dilutions to 10^{-5}, from left to right).

3. PROGNOSIS OF FIELD CORROSION

It is clearly important to be able to predict whether a steel pipe (for example) will corrode in a particular soil or not, for if the soil can be shown to be aggressive or likely to be so, steps can be taken to protect the pipe in an appropriate way.

Booth *et al.* (1967) have shown that the aggressiveness or corrosivity of a soil towards buried ferrous and non-ferrous metals depends on the soil resistivity and its redox potential (on the normal hydrogen scale, corrected to pH 7), according to the following schema:

Positive corrosivity—if *either* resistivity $< 2000\ \Omega$ cm
or redox potential < 400 mV
Negative corrosivity—if *both* resistivity $> 2000\ \Omega$ cm
and redox potential > 400 mV

For predominantly clay soils the critical redox potential is 430 mV rather than 400 mV. For 'borderline' soils (resistivity near 2000 Ω and redox potential 400–430 mV) the soil moisture should be measured; only soils with more than 25% moisture (on a *dry* weight basis) are aggressive.

Thus a soil can be classified as aggressive or non-aggressive on the basis of three or four easily measured soil parameters: resistivity, redox potential, pH (required to correct the field redox value) and moisture content.

3.1. Resistivity

A soil resistivity meter operated on the Wenner four-electrode principle (Fig. 3) is required (for example, model ER2 from Geophysical Specialties Division, Minnetech Laboratories Inc., Minneapolis). The four electrodes are driven into the ground equally spaced along a straight line. In reasonably uniform ground the electrode separation determines the depth of soil over which the resistivity is measured. Small separations lead to estimates

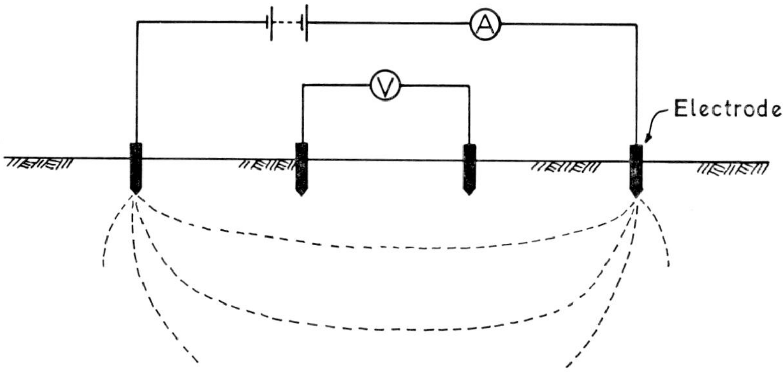

FIG. 3. Principles of soil resistivity measurement. The broken lines represent the paths of equal current flow.

of the resistivity of the surface layers, while as the separation increases the deeper materials have progressively more influence on the meter reading, although there is no precise relationship between electrode separation and depth. The minimum separation is about 1 m and the maximum about 3–10 times the depth to which measurements are required. The bulk resistivity is measured as the product of the meter reading (ohms) and electrode separation (cm).

3.2. Redox Potential

Soil redox is measured by the method of Booth *et al.* (1967). A platinum electrode probe (Fig. 4) is firmly inserted into the soil and connected via a portable pH/millivoltmeter to a reference electrode (normally a saturated calomel electrode) which is also inserted in the soil some 30 cm distant from the redox probe. As the redox probe is inserted, it should be firmly pressed down and rotated through 360°, in order to avoid meter drift and erratic readings. The readings obtained from each of the two platinum electrodes should agree to within 10 mV; if they do not, the electrodes require cleaning by gentle abrasion with a scouring powder and then washing with methylated spirits and distilled water.

The field reading E (mV) is referred to the normal hydrogen electrode and corrected to pH 7 as follows:

$$\text{Eh(mV NHE, pH 7)} = E + 250 + 60(\text{pH} - 7)$$

The figure 250 is the potential in mV of the saturated calomel electrode with respect to the normal hydrogen electrode.

3.3. pH

If the soil is moist, its pH is best measured *in situ* with a portable pH meter fitted with a combination toughened spear-headed electrode of the type normally used to measure the pH of cheese and butter.

If the soil is dry and crumbly the procedure in British Standard 1377 (British Standards Institution, 1975) must be followed. Thirty grammes of soil are placed in a beaker, 75 ml distilled water added and the contents thoroughly stirred. The beaker is left to stand for several hours, preferably overnight, when the pH is measured electrometrically.

3.4. Moisture Content

Approximately 30 g of soil should be placed in a tared container and weighed to the nearest 0·01 g. The soil should be dried to constant weight (usually for 24 hr) at 103°C and then reweighed (British Standards

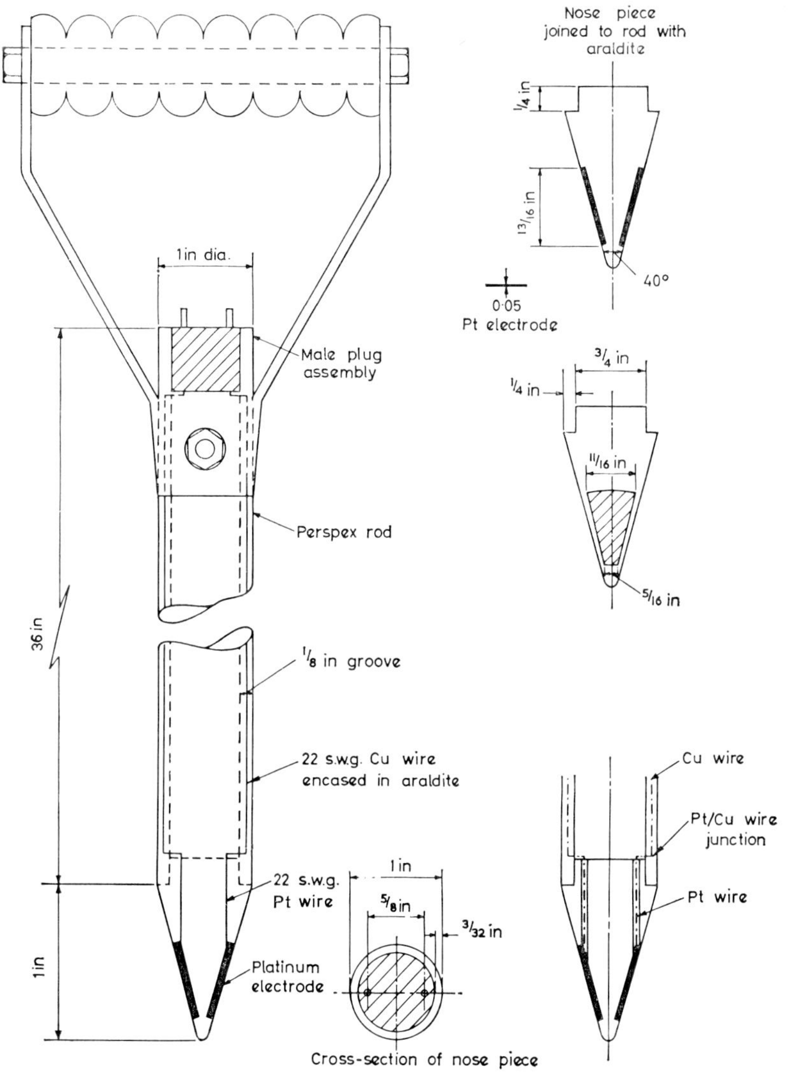

FIG. 4. The National Physical Laboratory field redox probe. (Crown copyright reserved.)

Institution, 1975). The moisture content should be calculated as grammes of water per gramme of dry soil solids × 100.

REFERENCES

Booth, G. H. (1964). *J. Appl. Bacteriol.*, **27,** 174.

Booth, G. H., Cooper, A. W., Cooper, P. M. and Wakerley, D. S. (1967). *Brit. Corros. J.*, **2,** 104.

British Standards Institution (1975). BS 1377: *Methods of Test for Soils for Civil Engineering Purposes*. British Standards Institution, London.

Iverson, W. P. (1974). In: *Microbial Iron Metabolism* (ed. J. B. Neilands), p. 476. Academic Press, New York.

Le Gall, J. and Postgate, J. R. (1973). *Adv. Microb. Physiol.*, **10,** 81.

Mara, D. D. and Williams, D. J. A. (1970). *J. Appl. Bacteriol.*, **33,** 543.

Mara, D. D. and Williams, D. J. A. (1972). In: *Biodeterioration of Materials*, vol. 2 (ed. A. H. Walters and E. H. Hueck-van der Plas), p. 103. Applied Science Publishers, London.

Postgate, J. R. (1966). *Lab. Pract.*, **15,** 1239.

Postgate, J. R. (1969). *Lab. Pract.*, **18,** 286.

Purkiss, B. E. (1971). In: *Microbial Aspects of Metallurgy* (ed. J. D. A. Miller), p. 107. Medical and Technical Press, Aylesbury.

Rittenberg, S. C. (1969). *Adv. Microb. Physiol.*, **3,** 159.

9

Isolation Techniques and Identification of Fungal Biodeteriogens from Soil

D. G. PARBERY

School of Agriculture and Forestry, University of Melbourne, Parkville, Victoria 3052, *Australia*

1. INTRODUCTION

Many thousands of species of insects, protozoans, nematodes, bacteria, etc., live in soil in addition to fungi. Most have the capacity to cause deterioration of materials important to man, so that each has the capacity to earn for itself the title of 'biodeteriogen', or 'deteriogen'. Consequently it is likely that, in time, the student of biodeterioration will need to isolate and study organisms belonging to various phylogenetic groups within the plant and animal kingdoms. A very wide range of techniques would be required to cope with the full range of possibilities. However, because of the large numbers of species and the diversity within each phylogenetic group, requiring a wide range of selective isolation methods within each group, the scope of this chapter is limited to the isolation and identification of fungi from soil.

2. SUBSTRATE RELATIONSHIPS

In the natural biological environment it is known that all organic molecules are produced by organisms and that whatever one organism can build up, other organisms can break down. There are enzyme systems available in a range of organisms which can decompose any natural organic material. With the exception of plastics and some synthetic fibres, the organic materials used by man are composed of organic molecules which have exactly, or almost, the same form as they do in nature. It therefore follows that with the possible exceptions mentioned, there will be a range of organisms with the capacity to attack any organic material used by man.

We should therefore not be surprised to find some sophisticated material attacked by a fungus and we should not assume that the attacking organism is in some way special or unique. The sophisticated material may be no different chemically from its natural raw material. For example, the molecules in various fractions of refined petroleum are the same as molecules being produced today by plants (Jones and Williams, 1967; ZoBell, 1963) and being degraded daily in soil by bacteria, fungi and yeasts. At the same time it is apparent that the number of organisms with the enzymes necessary to degrade organic molecules decreases as the chemical complexity of the molecules increases. For instance, ability to degrade glucose and other simple sugars is widespread among fungi, while the capacity to degrade cellulose and lignin is much more restricted. Again, the number of fungi which attack resin, suberin or cutin appears small by comparison.

Consequently it is regarded as an important step when devising a technique for isolating a specific fungus from soil to know something of its substrate relationships and/or to find out, as far as possible, the composition of the substrate being attacked or the component or components of the substrate being attacked.

Another important consideration regarding substrates is the presence or absence of toxins and inhibitors. These are fairly widespread in nature, but it is usual to find some species of fungi that can tolerate, if not degrade, toxic or inhibitory compounds. For example, a range of phenols, and terpenes such as isoprene, as well as short-chained alkanes, are commonly inhibitory to fungi. A knowledge of an organism's ability to tolerate an inhibitory or toxic substance is valuable information to the researcher attempting to devise a selective isolation technique.

Finally, it may be important to know whether the material being attacked constitutes a nutritional or a physical substrate. In the case of nutritive substrates, it is exclusiveness or specificity through nutritive or toxic characteristics which needs to be known. In the case of physical substrates, it may be structural considerations, as well as associated physical, chemical and nutritive characteristics, which need to be known.

As a prelude to devising a selective isolation technique, it is necessary to know as much as possible about the fungus launching the attack, the substrate being attacked, and the environmental conditions under which the attack and damage are most severe.

3. THE SOIL AS THE PRIMARY HABITAT

It is a convenient starting-point for the exercise of isolating biodeteriogens from their natural habitat to assume that the primary habitat of most non-parasitic fungi is soil. Many saprophytic fungi are found on certain living or dead plants or animals. Various species are commonly found in plant phyllospheres, litter, resinous exudates, insects, etc., and they are also commonly isolated from soil. Indeed, soil is a vast repository of most types of plant and animal remains as well as the fungi and other heterotrophic organisms that degrade them.

In accepting that for every organic molecule built up during the life of an organism there are enzyme systems available to degrade them, we accept that all organic materials are potentially liable to attack by micro-, or other, organisms, We also accept that all saprophytic fungi are potentially biodeteriogens. Since we have also suggested that the bulk of saprophytes are to be found in soil, it is a logical beginning to all research programmes to adopt the hypothesis that the soil is the primary source of the biodeteriogen, unless there are sound reasons for supposing otherwise.

The problem then becomes one of finding a suitable isolation procedure. If the problem is a common one, or if the biodeteriogen is a well-identified, commonly isolated saprophyte, a selective procedure for its isolation may be known. Commonly, however, the biodeteriogen is unknown and it is highly unlikely that any specific procedure would be available for isolating it. Consequently, unless there are some guidelines to follow in setting up procedures, the search for a method is likely to be reduced to an empirical 'trial and discard' procedure involving much waste of time and effort. It becomes a gamble as to whether or not a suitable procedure will be hit upon.

We are fortunate that there is now considerable knowledge of the ecology of microorganisms in soil. Techniques for studying the soil fungi (Barron, 1971; Watling, 1971; Booth, 1971*a*; Warcup, 1960) and conceptual ideas for interpreting observations (Garrett, 1956, 1970; Toussoun *et al.*, 1970; Griffin, 1972) are available from which guidelines can be drawn for developing reliable techniques for isolating specific fungi from soil. In order to do this, however, we need to examine some basic concepts and principles in the ecology of soil microbes.

4. ECOLOGICAL CONSIDERATIONS

The understanding and solving of most problems of biodeterioration come through studying the ecology of the situation in which the problem has occurred. Eggins (1968) says essentially this when he points out that: 'Ecology is primarily concerned with the study of the interaction of organisms with their habitats and with each other. Thus substrates (or materials), organisms and environment, all intimately concerned in biodeterioration, are also the essentials of ecology.' In devising methods for isolating biodeteriogens from soil it is necessary to consider all these elements mentioned by Eggins because some information relative to any one element may provide the main lead into how best to proceed. Consequently there are certain ecological concepts which are important for this purpose and which are briefly described here. More detailed accounts of these concepts can be obtained from Garrett (1956, 1970), Park (1968) or Griffin (1972).

4.1. Competitive Saprophytic Ability (CSA)

This concept, defined by Garrett (1956), is an expression of the ability of a fungus to compete for substrates in soil. In general, fungi which are specialised parasites of plants or other living organisms have a low competitive saprophytic ability and will not grow through soil unless it contains no competitors—this is, unless it is sterile. The concept, however, has a wider application than this. Certain saprophytic fungi do not compete successfully in soil for widely available substrates. For example, *Amorphotheca* (*Cladosporium*) *resinae* will not grow through non-sterile soil, nor can it be isolated by non-selective methods. Yet there seems little doubt that it is a soil-inhabiting organism (Parbery, 1972). It is probable, however, that it colonises substrates not widely available to other microorganisms, thereby compensating for its lack of competition for commonly available substrates. Fungi which might be described as having a high CSA can compete for a fairly wide range of substrates such as sugars, starch, cellulose, etc., which are generally available. Fungi which colonise resin or waxes may well have a low CSA in terms of the above energy sources.

4.2. Habitats

It is unlikely that most soil-inhabiting fungi live only in soil. It seems more likely that they live in specific plant or animal remains in soil from which they either release spores or hyphae which are moved, or grow, through soil or remain dormant until confronted with additional substrate. It is also likely that a fungus which has colonised one organic fragment produces hyphae which grow short distances through soil, colonising other fragments as they go. Perhaps many fungi do so when there are enough organic residues to allow it. The fungi which are most successful may be those which can colonise most organic debris. Such fungi, already defined as having high competitive saprophytic ability, might now be described as colonising 'general habitats'. Park (1968) drew a distinction between two main groups of fungi in soil: those just described, and fungi which cannot compete in this way but nevertheless occur in soil. Park suggested that the latter group succeeds because it has the ability to colonise select habitats not widely available to other fungi. He described the habitats in the following way (Park, 1968):

General habitat. One in which the basic substrate is available to a wide range of fungi which must then compete for a place to colonise it. They may succeed by having a fast germination and growth rate, by producing antibiotics to ward off their potential competitors, and by tolerating any antibiotics or other antagonistic reactions in their competitors. Cultural methods using agar media may not be generally useful in isolating such fungi.

Select habitat. This is one composed of a substrate which is highly selective, either because it is so chemically complex that very few organisms have the necessary enzyme systems to degrade it, or because it contains a toxin which few fungi have the capacity to withstand. Some are both chemically complex and toxic, *e.g.* resins. Selective isolation based on the use of agar culture media are potentially useful for isolating fungi from such habitats.

If possible, some insight should be gained as to whether a biodeteriogen is likely to be found in the general habitat or in the select habitat, and this can give a very important lead into the best approach to setting up a selective isolation procedure.

5. APPROACHES TO ISOLATION

Until something is known of the biology of an organism, it is difficult to see the best approach to isolating it, while at the same time, until that organism is successfully isolated, there is much still to learn about its biology. This is the most likely reason for so many workers taking an empirical approach to the problem.

Frequently, something can be learned about the biology of the organism by observing the problem itself. It is often possible to identify the deteriogen and to find out something about the chemical composition of the substrate. Such knowledge of the substrate can indicate whether or not the organism is likely to be selective in ways which may be exploited in its isolation. For most practical purposes there is a choice between two basic approaches to isolation. The first is to bait soil with small pieces of the substrate in question. The second is to suspend small portions of soil in an agar medium and to incubate it. The choice will depend, as previously stated, on whether the organism is a highly competitive saprophyte occupying a general habitat or whether it is a selective organism. Selective media may be more difficult to devise for organisms in the first group, which would therefore appear to be more readily isolated using baiting procedures. However, if the basis for the selectivity of the species in the group is known, then selective media may be readily devised.

Any method for isolating fungi from soil, or any other habitat, will need to be specifically selective for some species, possibly more highly selective for one species or, at most, a few species. In doing this, however, selectivity may be approached in two quite different ways. This may be important even if the difference is largely philosophical. Either positive selection is made for a particular species by providing an essential ingredient in the medium which only the one species can use, or negative selection is attempted by adding an ingredient which positively discourages everything else. For example, by adding a particular purified keratin as the only carbon source to a medium containing all other essential ingredients for growth, conditions would be provided which, in general, only certain species in the Gymnoascaceae could utilise. This would negatively discriminate against other fungi lacking the enzyme systems required to metabolise keratin. On the other hand, a medium containing glucose as the carbon source provides nutritional conditions favourable to a very wide range of fungal species; however, by adding creosote, which is biocidal to most organisms, most potential competitors of *Amorphotheca resinae* will be removed, thus allowing this organism to grow notwithstanding its low competitive saprophytic ability. In this case, by positively selecting against all the fungi which would exclude *A. resinae* through competition with it, a negative selection for *A. resinae* is achieved.

6. SELECTIVE AGAR MEDIA

6.1. Basic Ingredients

In devising selective media it is necessary to provide the basic nutritional requirements of the fungus being sought in such a way that other fungi cannot compete. In order to culture fungi there are certain basic nutrients which must be supplied. These include water, carbon, nitrogen, potassium, phosphorus, magnesium and sulphur. Selectivity may be altered by changes in the source, availability or concentration of any one of these. It is most common, however, to modify the carbon source or sometimes the nitrogen source. Selectivity can also be altered by including a range of toxic materials or changing physical conditions in the agar, such as pH or osmotic pressure, or in the incubation environment, such as temperature, relative humidity, light or relative concentrations of oxygen and carbon dioxide. Not all fungi will grow on the simple media above. Some fungi, *e.g.* Basidiomycetes, have additional requirements (Booth, 1971*a*). Basidiomycetes, for example, often require sources of protein, amino-acids, base compounds and vitamins in addition to trace elements. Consequently no one basic set of minerals and nitrogen will be universally satisfactory. However, if the species is known or if the taxonomic group to which it belongs can be identified, then it is often possible to select a suitable basic medium with which to work. Hence, Eggins and Pugh (1962) devised a mineral solution which they found more satisfactory than others for use in media containing cellulose as the carbon source. Parbery and Thistlethwaite (1973) found one mineral solution better than others for culturing *C. resinae*. Czapek-Dox mineral solution (Ainsworth, 1971; Booth, 1971*a*) is generally useful for a wide range of species.

In preparing such media it is preferable not to make them too nutritious. It is also important to preserve certain balances in the ratio of the various components of the medium to each other. A guide for this can be obtained from studying the general composition of widely used media such as Czapek-Dox agar. In general, the carbon source should be no greater than 1–2%, and the nitrogen source should be kept at about one-third of that. Richer media will generally encourage luxuriant vegetative growth but disfavour sporulation. The second chapter of Booth (1971*a*) provides valuable information in this regard.

6.2. Incorporation of Soil

For isolation from soil, either a water suspension can be prepared over a range of dilutions (Barron, 1971) and aliquots incorporated into molten agar at about 45°C, or the soil can be added directly to a petri dish, dispersed in a drop of sterile water and then incorporated into an agar medium in the above manner (Warcup, 1960). Pre-treatment of the soil sometimes favours the isolation of certain groups. Air drying and partial heating, for example, favour the isolation of Ascomycetes (Warcup, 1960).

The use of molten agar at a high temperature may have the same effect. Basidiomycetes are very rarely encountered in soil plates. They may be found, however, if Warcup's (1960) procedure for the manual removal of mycelial fragments from soil is followed.

6.3. Incubation Time

The incubation time required before fungi appear and reproduce varies considerably. Zygomycetous species often appear in 6–12 hr. Many Fungi Imperfecti appear within a few days to a week. Ascomycetes may also appear in the vegetative state within a few days, but the time required for fructification varies from a week or so to several months, depending on the species. In addition, unless both mating types are present, many Ascomycetes will not produce sexual structures at all, although they may be identifiable from their imperfect states. Mating apparently identical cultures isolated from the soil may induce perfect state formation.

6.4. Selective Agents

Isolating fungi from soil is assisted by depressing bacterial growth. The following procedures are commonly followed for this purpose:

(a) The acidification of agar to a pH of approximately 4·5 (Warcup, 1950).
(b) Inclusion in the medium of the dyestuff Rose Bengal (Booth, 1971*a*). Rose Bengal, however, does depress the growth of some Oomycetes (Langdon and Parbery, unpublished).
(c) Inclusion of antibacterials such as penicillin, streptomycin, etc., in the medium.

Tsao (1970) provides a comprehensive list of selective media used predominantly for isolating plant pathogens. Domsch and Schwinn (1965) have also compiled a list of selective isolation methods and they, in common with Tsao, discuss the principles behind the success of such methods and consider how new selective methods may be devised.

Unfortunately the lists of selective media are short because few, relative to the number of fungi to be isolated, have been developed (Tsao, 1970). Lack of suitable media often results in failure to isolate the appropriate organisms and may lead to erroneous conclusions about the causes of problems.

7. BAITING PROCEDURES

Isolation procedures which select only one organism are comparatively rare. This is also true of baiting procedures. Therefore, the initial baiting technique needs to be supplemented by the transfer of each isolate into pure culture. Each can then be tested, individually or in various combinations, for ability to degrade the original material.

In setting up a baiting chamber it is necessary to pay attention to several factors which may be important either singly or together. Successful baiting procedures can be devised provided initial attempts are tried and refined, using soil known to contain the deteriogen. To ensure this, it may be necessary to use soil deliberately seeded with a culture of the deteriogen, or with small fragments of the degraded material containing it. Then, when it is known that the organism is present in the soil, failure to isolate it must be due to inadequate technique which can then be modified until it will work. Complete failure to obtain isolation should raise the question as to whether the organism did come from soil in the first place.

Consideration of several points is required when developing a baiting procedure. These will be discussed in relation to the direct baiting procedure which is more or less standard, but there are many modifications, some of which will also be discussed. The various modifications of the basic procedure may have advantages in particular cases.

7.1. Direct Baiting from Soil

This is the most commonly used method of isolating particular fungi from soil, whether they are plant pathogens or deteriogens. Points to be considered are:

7.1.1. *Collecting Soil*

Soil should be collected as near to the site of the problem as possible when attempting to establish the soil-inhabiting nature of the deteriogen. Samples can be taken further from this point when it is desirable to establish the distribution of the deteriogen. Samples should be collected as near to the time of isolation as possible, and if the isolation procedure cannot begin soon after collection, the samples should be stored under cool conditions until required.

In collecting soil, the litter layer should be cleared away first, or collected and tested separately. Soil samples should be collected down to a depth of 10–20 cm, especially in warm, dry climates.

7.1.2. *Setting Up Baiting Chambers*

In many cases baiting chambers may need to be incubated for several weeks. Unless care is taken, soil and baits within them will dry out. It is therefore useful to select petri dishes with close-fitting lids so as to reduce the rate of desiccation. Petri dishes, or other suitable containers, should be loaded with as much soil as possible but so as to leave sufficient head room to accommodate the bait without its contacting the upper half of the dish. Since there is usually some condensation of water inside the lid, any bait in contact with it is frequently drawn away from contact with the soil by surface tension, thereby lessening the chances of colonisation by fungi. Conveniently sized pieces of material, as baits, should be either firmly pressed into close contact with the soil, or partly buried in it. Room should be provided for the curling of fabrics (Figs. 1 and 2).

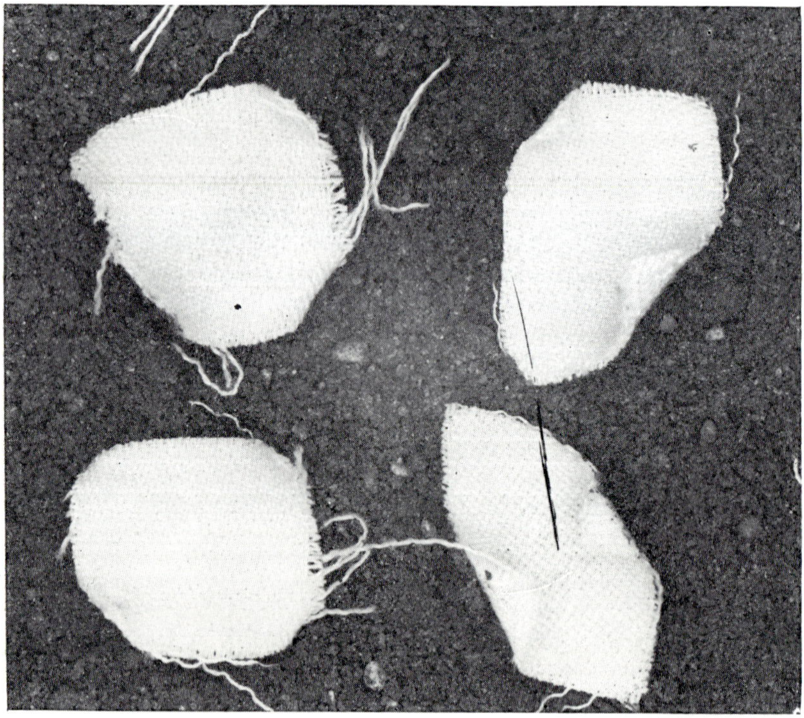

FIG. 1. Pieces of undyed wool placed on moist soil in a baiting chamber. A drop of water would then be placed on each and would be replaced about every 2–3 days, as required.

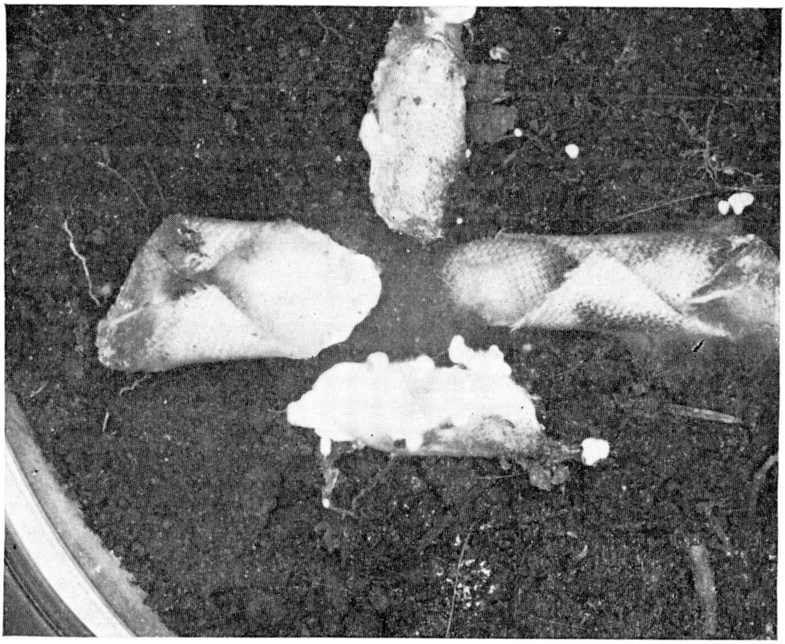

(a)

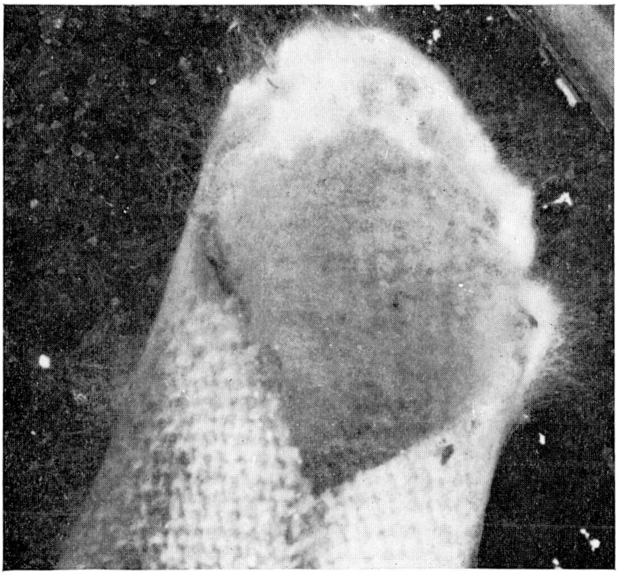

(b)

FIG. 2. (a) Wool baits after one month showing growth of *Arthroderma torryei*. (b) A colonised bait (magnification ×4·5).

7.1.3. *Moisture*

Fungi require water for growth; however, while soil fungi are generally favoured by moist conditions in the soil, *i.e.* soil wet to the point where free water is evident (Fig. 3), filling in all air spaces discourages most fungi other than water moulds. Waterlogged soil favours the growth of bacteria and so generally discourages fungi. Soil which is moist and friable without being wet and sticky is usually favourable to fungi (Fig. 4). Baiting procedures will fail equally commonly when soil is too wet or too dry. It is

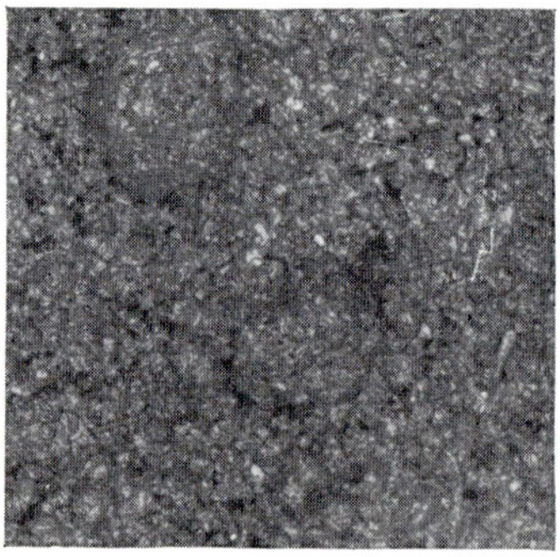

FIG. 3. Soil which is much too wet for isolation of fungi.

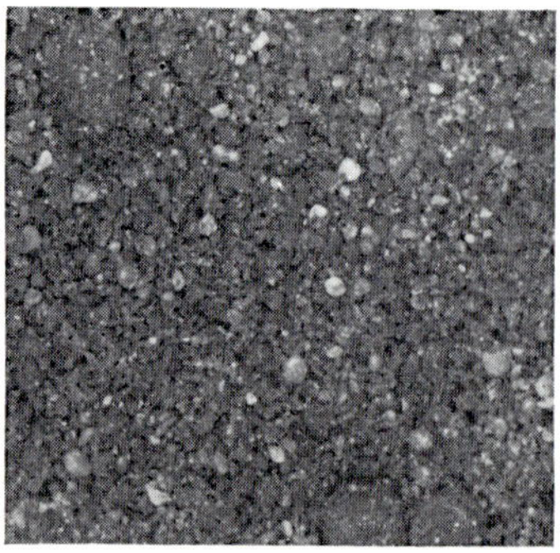

FIG. 4. Soil ideally moist for use in baiting chambers, when wet baits are used.

generally most satisfactory to place the already suitably moist soil into the petri dish and then to place the moistened bait in position. It is important during the incubation period to make regular checks to ensure that the soil and bait remain suitably moist. The incubation of sets of baiting chambers in closed polythene bags is helpful in preventing desiccation. When required, a few drops of water should be applied to the bait and surrounding soil. Some baits, *e.g.* paper, require daily wetting.

7.1.4. *Temperature*

Usually the best guide to the temperature required is to study the range of temperatures registered in the environment where the problem developed. This may not be the optimum temperature for growth of the organism, but may be nearer its minimum or maximum temperature. In some cases it may be that the deteriogen has a greater tolerance of an extreme temperature (in mycological terms approaching 0° and 50°C) than its competitors. The higher the incubation temperature, the greater the need to guard against desiccation.

7.1.5. *Period of Incubation*

The incubation time required for the colonisation of baits can vary considerably according to substrate and colonising species and is often underestimated. Colonisation may occur some time before it is apparent. Vegetative growth is usually apparent well before fruiting bodies are seen in fungi such as Ascomycetes. When conditions are favourable, most baits will be colonised in 3 weeks or less. If fruiting structures are required, however, some baits may have to be incubated for many weeks. As a general guide, Fungi Imperfecti usually fruit within a few days to 2 weeks, Ascomycetes from about 2 weeks to several weeks. Basidiomycetes do not usually fruit readily on baits, but where they do it usually takes at least 1 month. In some cases it may be necessary to maintain baits for 3 months or longer to obtain fructifications. During that time it is often advisable to allow some degree of alternate wetting and drying, and to incubate the chambers according to how temperatures fluctuate more or less in accordance with the season. However, there are many other factors which can be considered to induce fruiting, such as the application of black light, raising CO_2 levels, and transfer to special agar media. Much time can be lost in attempting to induce reproduction unless (and even when) an experienced mycologist is consulted.

7.1.6. *Aeration*

Griffin (1972) has studied the influence of various gas tensions on the growth of soil fungi. It seems that there is a strong case for giving much more serious consideration to carbon dioxide levels particularly, but also to oxygen levels, when attempting to isolate soil fungi. The behaviour of some fungi is altered by changing gas levels. Rhizomorph production in *Armillariella elegans*, for example, is limited by partial pressures of oxygen exceeding 0·04 (Smith and Griffin, 1971). Similarly, *Gibberella zeae* grows best when carbon dioxide levels are higher than in the aerial environment (Macauley and Griffin, 1969). The aquatic fungus which frequents stagnant ponds, *Aqualinderella fermentans*, can be isolated and grown only in an atmosphere of greatly reduced oxygen and increased carbon dioxide levels (Emerson and Held, 1969). Similarly, Hintikka and Korhonen (1970) have demonstrated that several wood-decomposing fungi grow best at higher CO_2 levels than occur in the normal atmosphere. They noted CO_2 levels of up to 11·4% in wood.

Soil carbon dioxide levels vary between 0·3% (10 times atmospheric) and 5% (Leeper, 1964). Thus it is probable that some soil fungi complete their life cycles under relatively high levels of CO_2 compared to atmospheric levels and it is probable that some compete best under these conditions. It is suggested, therefore, that some deteriogens would be easier to isolate if the gaseous environment in the isolation chamber were richer in CO_2, *i.e.* if it were nearer the levels pertaining in soil. The best guide here might be the aeration conditions under which the problem developed.

Although, to the best of the present writer's knowledge, this idea has not been tested in relation to deteriogens, a further example supports the suggestion of attempting to isolate some species in a CO_2-enriched atmosphere. Park (1961) found that *Fusarium oxysporum* could be selectively isolated from soil by increasing the level of CO_2 in the atmosphere of the isolation chamber.

7.1.7. *Mycophagous Arthropods*

A common problem associated with the isolation of identifiable cultures by baiting soil is the prevention of fruiting by the continual grazing of colonies by mycophagous insects, mites and other arthropods. At times this can wreck experiments. In recent years, however, observations at Melbourne University (Morgan and Parbery, unpublished) on the control of red-spider mite on angiosperms under glasshouse conditions have offered a solution to the problem of mycophagous insects. It was found that the hanging of 'Shelltox Pest Strips' in glasshouses gave complete control of red-spider. However, as it was found that some plant species were sensitive to the active ingredient, Vapona, it was considered that the use of this compound in baiting chambers might inhibit the appearance of some fungi sensitive to it. So far, tests conducted at Melbourne University have not identified any local fungal species other than possibly some Gymnoascomycetes that are depressed by the use of pieces of Shelltox Pest Strips used to inhibit arthropods in baiting chambers. It has been found that a 1 cm^2 piece of Pest Strip will prevent any arthropod problem in a petri dish of 9 cm diameter. The possibility that some species of fungi will be sensitive to Vapona should not be overlooked, however, as their growth might be inhibited.

7.1.8. *Transfer to Pure Culture*

There are various ways of transferring fungi from baits into pure culture. Preference will be given to the least troublesome and most reliable. This, however, frequently depends on the operator. Three commonly used methods which are useful are:

(i) *Direct transfer*. Fungi that sporulate or produce hyphae free of the bait surface can usually be transferred direct from the bait to a suitable agar medium by collecting an inoculum of spores or hyphae with forceps or a needle. Fine forceps are required and care should be taken to avoid picking up soil or portions of the bait which may be contaminated with other organisms, including bacteria. It is advantageous to moisten a needle before attempting to pick up spores or hyphae. This is simply done by stabbing a flamed needle into agar.

(ii) *Plating out.* Fungi sometimes grow and sporulate in the fabric of the bait in such a way as to make it difficult to transfer inoculum directly. This is the case with species of *Humicola* growing in linen or cotton. It is then necessary to transfer a piece of the infected material on to a culture medium. If the bait is strong enough, a thorough wash in sterile water to remove soil and surface contaminants reduces problems of contamination. The use of an antibacterial agent in the agar is also advisable. Subcultures of resultant colonies should be made as soon as possible to avoid problems of mixed cultures.

(iii) *Dilution plate method.* When plating out is not practicable because of the large number of fungi or the fragile nature of the substrate, it may be necessary to use this method of obtaining a pure culture. A portion of the colonised bait is dispersed in sterile water to provide inoculum. Then there is a choice of procedures. First, a streak plate may be made to separate the various propagules in the inoculum sufficiently to allow selection of the individual fungal colonies which result. A loopful of inoculum is smeared in a straight line across the agar surface of the medium close to one side of a petri dish. Without collecting fresh inoculum, the loop is then streaked across the surface a second time in close parallel to the first. The loop is then sterilised and, once cool, is used to make a second series of streaks at a right-angle to the first, starting on the second line of the first series of streaks. This procedure is repeated four or five times. Eventually the discrete colonies which appear can be transferred to a suitable medium in slope cultures. The second procedure is to prepare serial dilutions of the original suspension. Aliquots of the dilutions, usually 1 ml each, are transferred to petri dishes to which is added 6–8 ml of molten antibacterial agar at about 45°C. Again, when discrete colonies appear they are transferred to suitable agar slopes. Barron (1971) offers valuable comment on how to achieve the best results using this method.

7.1.9. *Culture Media*

While many fungi sporulate well on baits, some even better than on agar media, sporulation and growth on selective media are frequently limited or atypical. Consequently it is usually necessary to transfer fungi from baits and selective media on to more general media for maintenance, experimentation and identification. It is important to select suitable media for this purpose. Fungi such as Basidiomycetes are, at best, difficult to grow so that the selection of the most suitable medium is critically important (Booth, 1971*a*; Watling, 1971). Other species grow and sporulate readily on a wide range of media. Certain groups of fungi which will grow on a range of media will sporulate only on specific media. Consequently anyone wishing to grow fungi for identification purposes should refer to the relevant chapters in Booth (1971*a*) or consult an expert. There are, however, some general comments which may be of guidance to those entering this field:

(i) *Fungi Imperfecti*. In general, these are the most commonly encountered fungi. If the fungus to be identified is *not* a species of *Aspergillus*, *Penicillium* or *Fusarium*, then cultures on tomato decoction agar and on oatmeal agar are usually satisfactory for maintenance and identification. For maintenance purposes it is advisable when subculturing to alternate isolates from one medium to another. Not all species, however, will sporulate on these or other agar media. It is then useful to try other natural substrates such as the bait from which the fungus was obtained. Growth on boiled oats grains set into water agar is useful for inducing sporulation. Pieces of carnation leaf, picked green, dried and stored, then soaked in water and sterilised prior to use when they too are set in water agar, is valuable for inducing sporulation in Fusaria and other fungi. With many natural substrates of this kind, sterilisation by boiling or exposure to gases such as propylene oxide (Bruch, 1965; Klarman and Jemens, 1960) is frequently preferable to autoclaving, which can change the nature of the substrate to an extent that is detrimental to the growth and reproduction of the fungus.

Most species of *Aspergillus* and *Penicillium* need to be grown on Czapek-Dox agar (Ainsworth, 1971; Booth, 1971*a*) for identification purposes. A few species, however, do not grow satisfactorily on Czapek-Dox agar and need to be grown on malt agar. Several species of *Aspergillus* are osmophilic and will grow best on a medium with increased osmotic pressure. This can be achieved by increasing the glucose or sucrose level of the medium or by adding 15–20% (dry weight) of sodium chloride to the medium. Indeed, salt malt agar (Booth, 1971*a*) is a selective medium for the osmophilic Aspergilli. Species of *Fusarium* can be selectively isolated using PCNB agar but require culturing on potato sucrose agar for identification (Booth, 1971*b*).

(ii) *Ascomycetes*. Many different media are used to culture specific

Ascomycetes. Growth of many species can be obtained on a wide range of media, but reproduction is not always as readily achieved. In addition, Ascomycetes not uncommonly lose their ability to sporulate in culture. One of the most successful agars for inducing reproduction in a wide range of Ascomycetes is cherry agar, for which stocks of cherry decoction need to be prepared and stored during the cherry season. The maintenance of baits on soil or on agar is also a valuable means of obtaining fruiting bodies. Dickinson and Pugh (1965) advocate this method for inducing fructification in fungi which are cellulose destroyers. Sometimes it may be necessary to maintain cultures under more or less seasonal conditions for several months before fruiting occurs.

(iii) *Basidiomycetes.* As a group the Basidiomycetes are generally difficult and often impossible to culture. Long-term baiting in the field at selected sites is probably necessary in many cases (Wälchli, 1970). Watling (1971), however, refers to several techniques which have been used successfully to isolate certain Basidiomycetes from soil. A method for obtaining fructifications from and thereby identifying Basidiomycetes was developed by Tamblyn and DaCosta (1958). This method would appear to offer great potential for developing selective isolation procedures for Basidiomycetes. The method involves the growing of mycelium in a mixture of sawdust plus nutrients in a jar on the top of which a block of wood is then attached. The mycelium permeates the enriched sawdust and eventually colonises the block of wood. Approximately 33% of all isolates grown in this way can be identified from the fruits they produce. It is possible that by adding an antibacterial agent and soil to the jar, and by varying the type of sawdust, the composition of nutrients and the type of wood block on top, this basic technique might be adapted to isolate a wide range of Basidiomycetes direct from soil.

7.1.10. *Identification*

Precise identification of isolates may be difficult. It is, however, essential for the retrieval of information and for valid comparison of research results that identifications are accurate. Consequently, once it has been determined which isolates are of interest to the problem, they should be sent to a reliable taxonomist or institution for identification. There is generally a charge made for this service, but the cost is small compared with the confusion caused by, and the depreciation of effort through, incorrect identification. Three centres where fungi will be identified are the Centraalbureau voor Schimmelcultures (CBS), Oosterstraat 1, Baarn, The Netherlands; the Commonwealth Mycological Institute (CMI),

Ferry Lane, Kew, Surrey, England; and the American Type Culture Collection (ATCC), Beltsville, Maryland, USA.

Useful references for identifying fungi in the laboratory are:

General: von Arx (1974); Ainsworth *et al.* (1973).
Ascomycetes: Dennis (1968); von Arx and Muller (1954); Muller and von Arx (1962); von Arx and Muller (1975).
Fungi Imperfecti: Ellis (1971); Barron (1968); Kendrick (1971).
Penicillium: Raper and Thom (1949).
Aspergillus: Raper and Fennell (1965).
Fusarium: Booth (1971*b*).
Thermophilic fungi: Cooney and Emerson (1965).

The direct baiting procedure has been used at the University of Melbourne over the last five years and has consistently yielded the fungi listed in Table I. Each time a new bait is introduced, additional isolates have been obtained from the soils used, which have come from southern New South Wales and Victoria.

7.2. Soil Enrichment

The incorporation of finely ground materials into soil prior to baiting may increase the efficiency of this method in detecting fungi present in soil in low numbers. This may be useful for survey purposes. Our experience with *Arthroderma curreyi* is an example of how soil enrichment can improve isolation frequencies. Usually it takes up to 3 weeks for *A. curreyi* to appear on wool baits on soil. Even then, not all baits attract the mould, suggesting that when the fungus is present in soil at a low frequency it is a matter of chance as to whether it is detected and recorded or overlooked and assumed absent. We find, however, that when an infected bait has disintegrated and been returned to the soil sample which is baited a second time, the colonisation of the fresh baits is apparent in about a week. Similarly, the isolation of *Nannizzia gypsea* from Melbourne soils, while common, is not consistent enough to be relied on for teaching purposes. However, by regularly (twice yearly) incorporating human hair into a specific area of garden soil, *N. gypsea* can be isolated at will in about 10 days by baiting soil taken from the amended area with human (child's) or dog hair.

The principle underlying the success of soil amendment is that of increasing inoculum potential of the deteriogen. It is regarded as a valid

TABLE I

A LIST OF FUNGI RECOVERED FROM AUSTRALIAN SOILS BY DIRECT BAITING, AT TWO DIFFERENT RANGES OF TEMPERATURE

Fungus	Bait							
	Wool	*Hair*	*Feather*	*Cotton*	*Linen*	*Wood*	*Paper*	*Silk*
At 25–30°C:								
Ctenomyces serratus			+					
Chrysosporium sp. A.			+					
Chrysosporium sp. B	+							
Microsporum sp. A	+	+						
Microsporum sp. B			+					
Arthroderma curreyi	+		+					
Arthroderma tuberculata		+						
Nannizzia gypsea	+	+						
Trichophyton sp.	+							
Aspergillus fumigatus			+				+	
Acremonium sp.						+		
Chaetomium globosum					+		+	
Curvularia harveyi					+	+		
Ceratocystis sp.						+		
Fusarium solani							+	
Graphium putredinis					+	+		
Gonytrichum macrocladium						+		
Humicola grisea				+	+	+	+	
Humicola sp.					+	+		+
Melanospora anoeta						+		
Penicillium sp.					+	+	+	
'*Pseudo-humicola*' sp.*					+	+		
Staphylotrichon sp.				+			+	
Trichoderma sp.				+		+	+	
At 45–50°C:								
Streptomyces sp.	+	+	+					
Aspergillus fumigatus				+	+	+	+	
'*Acremonium*-like' sp.†				+		+		
Chaetomium thermophile				+		+		
Sporotrichum thermophile				+	+		+	+
Thermomyces lanuginosa				+	+	+	+	+
Malbranchea pulchella							+	+

* A fungus which exhibits features of the sporing structures of *Gilmaniella* and a catenate *Humicola* is common in Melbourne soils. Satisfactory pure cultures have not been obtained and described.

† This thermophilic species is difficult to retain in culture at temperatures below 40°C. It produces single-celled hyaline conidia from simple or occasionally branched condiophores. The conidia may be produced in short chains or in small heads. The simple conidiophore or its one or two branches are phialides. This species is common but as yet has not been assigned to a genus.

procedure in that unless the organism is present, its inoculum cannot be increased, so that unless a quantitative survey was required, soil enrichment with whole or component parts of materials is regarded as a useful aid to direct baiting procedures.

7.3. Perfusion Techniques

Under natural conditions materials are colonised by progressions of organisms rather than a single species. The progression is governed in part by the changes wrought in the material by successive species' metabolism. Thus the material is continually changing together with the species colonising it. A technique has been developed which, to some extent, can delay the progression by at least keeping nutrient levels and pH in the material more or less constant (Eggins *et al.*, 1968). This involves the perfusing of the material, which is in contact with soil in a petri dish, via two pieces of 'wick', one leading from a reservoir of nutrient, the other leading into the atmosphere. The entire wick and material, apart from where it is in contact with the soil, is encased in an impermeable sheath so that only the ends are exposed. This allows the nutrient solution to pass from the reservoir, through the wick and material by capillarity, and to evaporate into the air.

The technique has been modified in various ways for different purposes —mostly, however, for ecological studies. However, it has potential for use as a selective means of isolating fungi from soil. Since 1968, Eggins and co-workers have published at least ten papers based on the use and modification of perfusion methods. Several other papers have been written by other authors. Further information on the use of the technique is available from the Biodeterioration Information Centre, The University of Aston in Birmingham, England.

7.4. The Creosoted Matchstick Method for Isolating *Amorphotheca resinae* from Soil

This method (Parbery, 1967) is highly selective for one species, *Amorphotheca resinae*. During the first 3–10 days, only *A. resinae* usually appears in plates. However, after that time species such as *Paecilomyces varioti*, *Absidia glauca*, *Penicillium* sp. and sometimes *Fusarium solani* may appear. Strictly speaking, this is not a baiting technique in that *A. resinae* most frequently grows on the soil surface for some days before it colonises the matchsticks, which it does usually eventually do.

Since the publication of this method, the writer has received several complaints that it has not worked for others. In most cases, however, the likely cause is not that the fungus is absent, but that either the quantity of creosote has been inadequate, the soil has been too wet, or the fungus has been looked for only on the matchsticks. When the following procedure is followed strictly, *A. resinae* is usually isolated from at least some soil samples taken from any area examined (Parbery, 1969; Sheridan and Steel, 1971). The procedure is as follows:

1. An even layer of soil is placed in a 9 cm diameter petri dish to one-third to one-half of its depth. The soil should be firmed without being compacted so as to allow good aeration.
2. Two beheaded matchsticks, freshly soaked in creosote for from 5 to 10 min and allowed to drain of the free liquid, are placed centrally in the petri dish approximately 2 cm apart (Fig. 5).
3. The petri dish lid is then replaced and the dishes are incubated at 25–30°C until the fungus appears. If this takes several days, water may need to be added to the dish, but care should be exercised to ensure that the soil is not flooded.
4. When examining the plates for the presence of the fungus, the entire soil surface needs to be scanned. *A. resinae* appears first on the soil surface, or colonised plant debris (Fig. 6), appearing only later on the matchstick. There are several points to which attention must be paid to ensure success:

   (a) The soil should be no wetter than necessary to bind together when squeezed in the hand. This is especially so with sandy to loam soils, but clay soils may need the addition of about 1 ml of water to the dish.

   (b) Creosoted matchsticks must be freshly prepared. This is due to the evaporation of certain volatiles which soon disappear and remove with them the general suppressiveness of the creosote as far as other fungal species are concerned. Similarly, the quantity of creosote is important. For a 9 cm diameter petri dish, the amount on two matchsticks is adequate. Too much more will inhibit *A. resinae*. Even so, some isolates of *A. resinae* do not have a high tolerance to creosote and take longer to emerge than more tolerant isolates (Parbery, 1971).

   (c) It is common after a few days to find sparse sterile mycelial wefts in some plates. Transfer of these to agar medium occasionally yields a colony of *A. resinae*; however, more frequently it proves

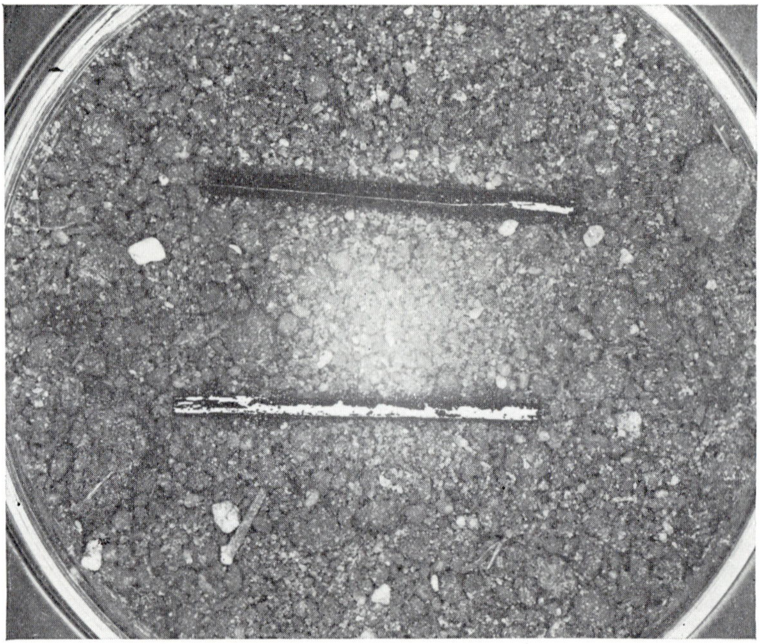

FIG. 5. Creosoted matchsticks in position on moist soil in a petri dish. Note the low degree of wetness of the soil, which is ideal.

to be *Paecilomyces varioti*, *Penicillium* sp. (which usually fructifies on the matchstick in 14–21 days) and, very occasionally, *Fusarium solani*.

(d) In hot or dry regions the chances of isolation are improved if soil samples are taken from a depth of 10–20 cm rather than the more usual 5–10 cm. The isolation of *A. resinae* from the litter layer is extremely rare.

The writer is confident that *A. resinae* is ubiquitous in soil, since it has been isolated in Australia, Britain, France, Sweden (Parbery, 1969), New Zealand (Sheridan and Steel, 1971) and almost certainly from North and South Africa and America (Parbery, 1971). Consequently, although it may not always occur in high frequency, it is highly probable that procedures were not being strictly followed if it were not obtained from at least 1 in 10 soil samples.

FIG. 6. The cladosporial state of *Amorphotheca resinae* growing on soil and plant debris 7 days after the setting-up of the baiting chamber. Note that no colonisation of the creosoted matchstick is yet apparent.

ACKNOWLEDGEMENT

I wish to acknowledge the contribution to this chapter of Dr H. J. Swart of the Department of Botany, University of Melbourne. He shared in many of the isolation experiments reported in Table I and has been largely responsible for the identification of isolates. I am also grateful to him for critically reading the manuscript and for suggesting modifications.

REFERENCES

Ainsworth, G. C. (1971). *Dictionary of the Fungi*. Commonwealth Mycological Institute, Kew. 663 pp.

Ainsworth, G. C., Sparrow, F. K. and Sussman, A. S. (1973). *The Fungi: An Advanced Treatise*, vol. IV A: *A Taxonomic Review with Keys: Ascomycetes and Fungi Imperfecti;* vol. IV B: *A Taxonomic Review with Keys: Basidiomycetes and Lower Fungi*. Academic Press, New York and London. 621 and 504 pp.

Barron, G. L. (1968). *The Genera of Hyphomycetes from Soil*. Williams & Wilkins, Baltimore. 364 pp.

Barron, G. L. (1971). Soil fungi. In: *Methods in Microbiology*, vol. 4 (ed. C. Booth), Chap. XIV, pp. 405–27. Academic Press, New York and London. 795 pp.

Booth, C. (1971*a*). Introduction to general methods (Chap. I); Fungal culture media (Chap. II). *Ibid*., pp. 1–94.

Booth, C. (1971*b*). *The Genus Fusarium*. Commonwealth Mycological Institute, Kew. 237 pp.

Bruch, C. W. (1965). Gaseous sterilisation. *Ann. Rev. Microbiol*., **15**, 245–62.

Cooney, D. G. and Emerson, R. (1965). *Thermophilic Fungi: An Account of Their Biology, Activities and Classification*. W. H. Freeman & Co., San Francisco. 188 pp.

Dennis, R. W. G. (1968). *British Ascomycetes*. J. Cramer, Stuttgart. 453 pp.

Dickinson, C. H. and Pugh, C. J. F. (1965). Use of selective cellulose agar for isolating soil fungi. *Nature* (*London*), **207**, 440–1.

Domsch, K. H. and Schwinn, F. J. (1965). Nachweis und Isolierung von Pflanzen-pathogenen Bodenpilzen mit selektiven Verfahren. *Anreicherungskultur und Mutantenauslese*. Suppl. 1, pp. 461–85.

Eggins, H. O. W. (1968). Ecological aspects of biodeterioration. In: *Biodeterioration of Materials*, vol. 1 (ed. A. H. Walters and J. J. Elphick), p. 22. Applied Science Publishers, London.

Eggins, H. O. W. and Pugh, C. J. F. (1962). Isolation of cellulose-decomposing fungi from soil. *Nature* (*London*), **193**, 94–5.

Eggins, H. O. W., Malik, K. A. and Sharp, R. F. (1968). Some techniques to investigate the colonisation of cellulose and wood substrates. In: *Biodeterioration of Materials*, vol. 1 (ed. A. H. Walters and J. J. Elphick), pp. 120–30. Applied Science Publishers, London.

Ellis, M. B. (1971). *Dematiaceous Hyphomycetes*. Commonwealth Mycological Institute, Kew. 608 pp.

Emerson, R. and Held, A. A. (1969). *Aqualinderella fermentans* gen. nov., spec. nov., a Phycomycete adapted to stagnant water. II. Isolation and growth. *Amer. J. Bot.*, **56,** 1103–20.

Garrett, S. D. (1956). *Biology of Root-Infecting Fungi*. Cambridge Univ. Press. 293 pp.

Garrett, S. D. (1970). *Pathogenic Root-Infecting Fungi*. Cambridge Univ. Press. 294 pp.

Griffin, D. M. (1972). *The Ecology of Soil Fungi*. Chapman & Hall, London. 193 pp.

Hintikka, V. and Korhonen, K. (1970). Effects of CO_2 on the growth of lignicolous and soil-inhabiting hypomycetes. *Commun. Inst. for Fenn.*, **69,** 5–29.

Jones, G. and Williams, A. (1967). Hydrocarbon biogenesis. Symposium on Microbiology, London, 19–20 Sept., pp. 1–7.

Kendrick, B. (1971). *Taxonomy of Fungi Imperfecti*. Univ. of Toronto Press. 309 pp.

Klarman, W. C. and Jemens, C. (1960). Sterilisation of agar medium with propylene oxide. *Phytopathol.*, **50,** 868.

Leeper, G. W. (1964). *Introduction to Soil Science*, 4th ed. Melbourne Univ. Press. 222 pp.

Macauley, B. J. and Griffin, D. M. (1969). Effects of carbon dioxide and oxygen on the activity of some soil fungi. *Trans. Brit. Mycol. Soc.*, **53,** 53–62.

Muller, E. and von Arx, J. A. (1962). Die Gattungen der didymosporeen Pyrenomyceten. *Beitr. Krypt. Fl. Schweiz*, **11** (2), 922 pp.

Parbery, D. G. (1967). Isolation of the kerosene fungus, *Cladosporium resinae*, from Australian soil. *Trans. Brit. Mycol. Soc.*, **50,** 682–5.

Parbery, D. G. (1969). The natural occurrence of *Cladosporium resinae*. *Trans. Brit. Mycol. Soc.*, **53,** 15–23.

Parbery, D. G. (1971). Biological problems in jet aviation fuel and the biology of *Amorphotheca resinae*. *Material u. Organismen*, **6,** 161–207.

Parbery, D. G. (1972). Ecological and genetic factors causing variation in *Cladosporium resinae*. In: *Biodeterioration of Materials*, vol. 2 (ed. A. H. Walters and E. H. Hueck-van der Plas), pp. 19–26. Applied Science Publishers, London.

Parbery, D. G. and Thistlethwaite, P. J. (1973). Improved mineral salts solution for the growth of *Amorphotheca resinae*. *Int. Biodetn. Bull.*, **9,** 11–16.

Park, D. (1961). Isolation of *Fusarium oxysporum* from soils. *Trans. Brit. Mycol. Soc.*, **44,** 119–22.

Park, D. (1968). The ecology of terrestrial fungi. In: *The Fungi*, vol. III (ed. G. C. Ainsworth, F. K. Sparrow and A. S. Sussman), pp. 5–39. Academic Press, New York and London.

Raper, K. B. and Fennell, D. I. (1965). *The Genus Aspergillus*. Williams & Wilkins, Baltimore. 686 pp.

Raper, K. B. and Thom, C. (1949). *A Manual of the Penicillia*. Williams & Wilkins, Baltimore. 875 pp.

Sheridan, J. E. and Steel, J. (1971). The natural occurrence of the 'kerosene fungus' *Amorphotheca resinae* in New Zealand soils. *N.Z. J. Sci.*, **14,** 147–60.

Smith, A. M. and Griffin, D. M. (1971). Oxygen and the ecology of *Armillariella elegans*. *Aust. J. Biol. Sci.*, **24,** 231–64.

Tamblyn, N. and DaCosta, E. W. B. (1958). A simple technique for producing fruit bodies of wood-destroying basidiomycetes. *Nature* (*London*), **181,** 578–9.

Toussoun, T. A., Bega, R. V. and Nelson, P. E. (1970). *Root Diseases and Soil-Borne Plant Pathogens*. Univ. of California Press, Berkeley and Los Angeles.

Tsao, P. H. (1970). Selective media for isolation of pathogenic fungi. *Ann. Rev. Phytopathol.*, **8,** 157–86.

Wälchli, O. (1970). Notes on the problem of the distribution of wood-destroying fungi at different altitudes. *Int. Biodetn. Bull.*, **6,** 43–52.

Warcup, J. H. (1950). The soil plate method for isolation of fungi from soil. *Nature* (*London*), **166,** 117.

Warcup, J. H. (1960). Methods for isolation and estimation of fungi in soil. In: *The Ecology of Soil Fungi* (ed. D. Parkinson and J. S. Waid), pp. 3–21. Liverpool Univ. Press.

Watling, R. (1971). Basidiomycetes: Homobasidiomycetidae. In: *Methods in Microbiology*, vol. 4 (ed. C. Booth), pp. 219–36. Academic Press, New York and London.

von Arx, J. A. (1974). *The Genera of Fungi Sporulating in Pure Culture*. 2nd ed. J. Cramer, Stuttgart. 288 pp.

von Arx, J. A. and Muller, E. (1954). Die Gattungen der amerosporen Pyrenomyceten. *Beitr. Krypt. Fl. Schweiz*, **11** (1), 434 pp.

von Arx, J. A. and Muller, E. (1975). A re-evaluation of the bitunicate Ascomycetes with keys to families and genera. *Stud. Mycol.*, **9,** 159 pp.

ZoBell, C. E. (1963). The origin of oil. *Int. Sci. & Technol.*, (Aug.) pp. 42–8.

10

Methods for Studying Water Vapour Sorption and Water Activity of Materials

J. L. MULTON and H. BIZOT

Ministère de l'Agriculture, Institut National de la Recherche Agronomique, Laboratoire de Biophysique des Aliments, Chemin de la Géraudière, 44072—Nantes Cedex, France

1. INTRODUCTION

The study of water vapour sorption properties is a very useful investigation technique for ascertaining the distribution and degree of water-binding in biochemical and biological products, in foods, and in any other substances of low and intermediate moisture content. Sorption studies may be linked with other research methods, such as solvent water determination (Guilbot and Lindenberg, 1960; Duprat and Guilbot, 1975), non-freezing water measurement, NMR and Hertzian spectroscopy (Guilbot *et al.*, 1960).

The study of sorption curves is particularly useful for investigating all reaction mechanisms with low or intermediate moisture content involving water activity: enzymatic reactions (Acker, 1962; Drapron and Guilbot, 1962), Maillard's reactions (Eichner and Karel, 1972), protein denaturation (Multon and Guilbot, 1970, 1975) and microorganism development (Scott, 1957; Leistner and Röer, 1975). Such phenomena govern food technological processes like drying, cooling, sterilisation and the storing of all substances.

From a fundamental point of view, each point of a sorption isothermal curve characterises the equilibrium state between the moisture content of a product and the water vapour pressure (WVP) of the surrounding atmosphere, the equilibrium being reached by gaseous exchange. Very commonly, these relations are graphically expressed by plotting the moisture content of the product (wet or dry basis) versus the relative humidity (RH), definable as the ratio:

$$\mathrm{RH}\ (\%) = \frac{p}{p_s} \times 100$$

where p = water vapour pressure in air in equilibrium with the sample and p_s = pure water vapour pressure at the same temperature.

'Water activity' (a_w) and RH do not have exactly the same physical meaning, although both are expressed by the same number; a_w results express a modified Raoult's Law generalisation (Loncin, 1975):

$$p = p_s X\gamma$$

where X = mole fraction of water in the sample and γ = activity coefficient of water in the sample.

When the mole fraction of water in a solid product is unknown, one can only measure the expression $X\gamma$, or 'water activity', which is, of course, equal to p/p_s, *i.e.* to RH. Hence a product with $a_w = 0{\cdot}9$ is in moisture equilibrium with an atmosphere of 90% RH. Sorption curves may offer different shapes, but generally speaking, in macromolecular components, for example in foodstuffs, a sigmoid appearance (Fig. 1) is observed.

At the same RH, when temperature increases, the water content usually

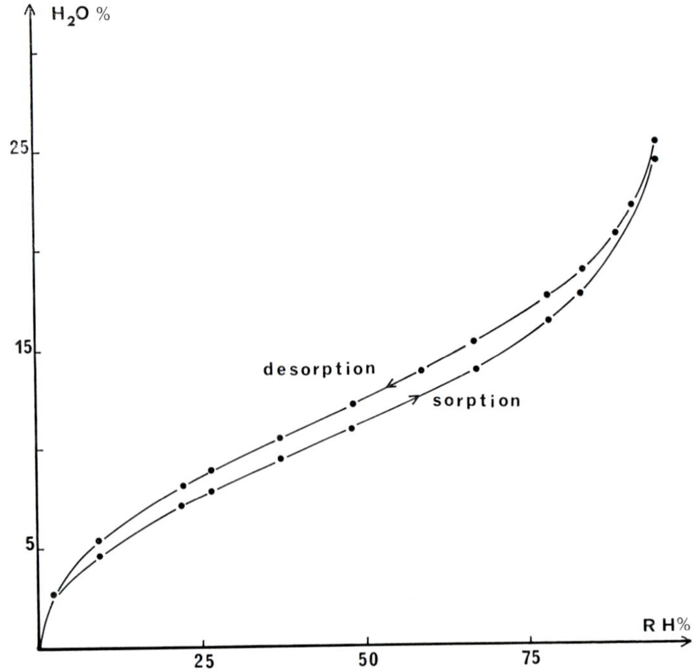

FIG. 1. Sorption and desorption curves of maize at 20°C.

decreases slightly. For example, in a foodstuff, for each 10°C rise in temperature the water content may be expected to fall by 1 %.

Curves show a hysteresis: at the same RH, the adsorption isotherm point (the product starting dry and becoming increasingly wet) is usually lower than the desorption isothermal curve point (the product starting very wet and drying progressively).

Because of this hysteresis effect, it is not possible to deduce from the sorption curve the equilibrium RH or water activity of a product when only its moisture content is known. In this chapter, our purpose is (i) to introduce the experimental methods for sorption isothermal curve determination and (ii) to describe the techniques for precise water activity measurement.

2. DETERMINATION OF SORPTION EQUILIBRIUM CURVES

It is usual (Multon *et al.*, 1971; Gal, 1975) to classify the sorption determination methods in two classes: (i) gravimetric methods following weight changes, and (ii) manometric methods using pressure measurements. But this classification makes an insufficient distinction between the sorption equilibrium study and the ERH measurements.

The determination of an adsorption or desorption curve consists of allowing the dry (adsorption) or wet (desorption) sample to reach the water vapour pressure equilibrium with a RH regulated atmosphere (in a special conditioned cabinet), and then:

(i) to measure the water content of the sample;
(ii) to measure or to check the RH of the atmosphere. The RH may be, according to the situation, directly determined in the conditioned cabinet (manometric methods), or measured on a sample of air taken in the cabinet, or directly connected with the chemical or physical regulation process (gravimetric methods).

The sorption therefore brings into play three different techniques: water content measurements, RH regulation and RH determination in small cabinets.

The experimental procedure involved is in three steps:

(i) preparation of samples (milling and conditioning);
(ii) achievement of water vapour equilibrium;
(iii) measurement of the two parameters characterising the equilibrium (moisture content and RH).

In order to draw a correct sorption curve, 9–10 experimental points are necessary—sometimes even more when better precision is required around a particular zone of the curve, such as the solvent water zone (Guilbot and Lindenberg, 1960). The determination of these 10 points may be obtained according to one of the following experimental procedures (Fig. 2):

(i) Several samples are prepared and each of them is directly brought from its initial state (dry or wet) to one of the required RH equilibria (Fig. 2(a)).
(ii) One trial sample is prepared and successively brought to each required RH equilibrium; the sample weight is then obtained at each level for computation of the water content (Fig. 2(b)).

These two procedures do not have exactly the same meaning even if, generally speaking, the equilibrium moisture contents obtained are the same. The second method is more hazardous (the alteration risk is increased because of the longer time) but is more suitable than the first one when only a small amount of product is available. It is also more simple because only one moisture determination is necessary.

(iii) In other respects, when it is necessary to plot a sorption isotherm network at different temperatures (in order to compute the adsorption heat, for example), it is more accurate (Fig. 2(c)) to prepare a sufficient amount of trial sample, at a given RH, to bring each trial sample to its RH equilibrium at the lowest temperature (adsorption) or the highest temperature (desorption), and then bring the sample to another temperature equilibrium (increasing for adsorption, decreasing for desorption), keeping the RH of the sample at the same value during this change.

2.1. Sample Preparation

The sample used for a sorption test must be representative of the original from a statistical point of view, even if this is sometimes difficult, particularly when the product has a very heterogeneous structure. Preparation consists generally of grinding and tempering the samples for adsorption or desorption.

2.1.1. *Grinding*

If the product is not naturally pulverulent, the experiment may be done with or without grinding. Generally, in food, biochemical and biological

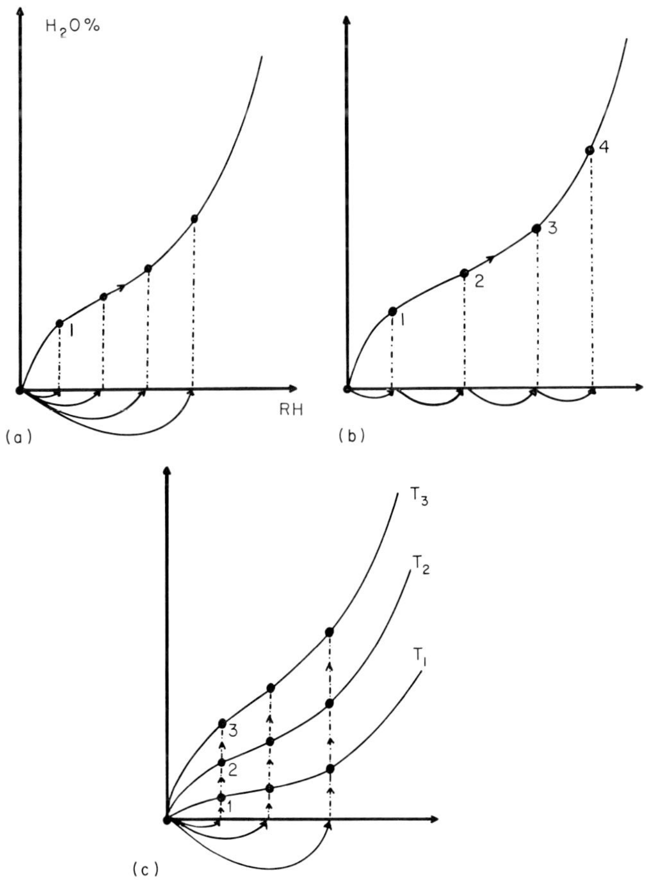

FIG. 2. Possible experimental procedures for studying sorption or desorption curves. (a) One sample for each point. (b) One sample for all the experimental points. (c) A network of isothermal curves, at different temperatures.

products, the natural porosity allows a free circulation of water molecules inside the molecular structure. When this is so, it is best to grind the sample: this operation increases the 'exchange' speed, without displacing the final equilibrium point, *e.g.* equilibrium is reached in 1–4 days with ground maize instead of 3–4 weeks, or more, with the whole kernel. On the contrary, when the natural porosity is too small for a free diffusion of water molecules, the quantity of water adsorbed is a function of the external area of the product. Thus the grinding process, by increasing this surface, increases the adsorption and thus changes the equilibrium, as for example in the case of sucrose.

The grinding difficulties are the same as those encountered when grinding before a moisture determination (Guilbot *et al.*, 1973; Multon and Martin, 1975). To obtain a good grinding, the water content of the substance must be approximately between 9 and 15%. When too dry, the product grinds badly, giving a heterogeneous granulometry, with too large particles. If too wet, the product may give a kind of paste. Sometimes such difficulties may be solved by grinding the product previously frozen, and mixed with liquid nitrogen or with solid carbonic acid. The grinding must be done without noticeable heating, by using a refrigerated mill. After grinding, the particle sizes must be the same as those required by the technique used for the water content determination; if not, some serious errors may arise.

Another very important point is to homogenise the ground product as well as possible so that every sample used for assays will be statistically similar and representative of the product.

2.1.2. *Conditioning*

Before equilibrating to a particular relative humidity, the sample must be conditioned. The method of doing this is different for an adsorption than for a desorption procedure.

(i) *For adsorption measurement*, the sample must be dried below a water content of 1%. The drying must be done gently and with care, in order to avoid any chemical or structural alteration. Heating, especially of the wet product, is therefore impossible. Freeze-drying is also to be avoided, because of a possible change in surface structure. Generally, the best solution is to dehydrate the sample at 20°C in a desiccator with P_2O_5 (residual partial water vapour pressure 10^{-5} mm Hg), with the desiccant renewed as often as necessary. In these conditions, the water content of the sample is lowered to below 1%. This threshold ensures that sorption isotherms made with identical samples do not show any difference, provided the initial water content is between 0 and 1%.

(ii) *For desorption determination*, the sample must be moisturised until the water content is as high as possible, about 50%. Direct soaking in an excess of water must be avoided because the extraction of soluble components (glucides, proteins) that will result will change the dry matter composition. Hydration must be performed by mixing the ground sample with the right quantity of water, following which a period of several hours is necessary to homogenise the water repartition in the sample. A little trituration may be helpful, but must be avoided when inducing a physical change (*e.g.* with wheat flour the trituration induces the formation of dough, the sorption properties of which are quite different from those of flour).

In the case of a crystalline and soluble product, it is recommended that a saturated solution with a minimum of liquid phase be prepared as a starting sample. When tempered samples have been prepared, they must be set out in thin layers in a tared dish and weighed.

2.2. Equilibrium of Water Vapour Pressure: Regulation of Relative Humidity in Small Cabinets

Prepared samples are allowed to adsorb or desorb in special cabinets, with regulated RH and temperature, until equilibrium weight is reached. Several chemical or physical systems are used to regulate the water vapour pressure (and consequently RH) in such small closed cabinets. In fact, the same systems may also be used for other purposes such as a study of reactions in low hydrated media, storage studies and calibration of hygrometric probes. Whichever system is chosen, the temperature must be kept very constant, with an accuracy of $\pm 0{\cdot}1°C$, in order to regulate RH with a reproducibility of about 1% (absolute value).

2.2.1. *Regulation by Aqueous Solution*

It is well known that when any component goes into solution in pure water, the result is a tonometric decrease of water vapour pressure (WVP) which is a function of the solute concentration. At infinite dilution the solution tends to ideality and the WVP can be calculated according to Raoult's Law:

$$p = p_s \,.\, X$$

where p is the water vapour pressure in equilibrium with the solution, p_s is the pure water vapour pressure and X is the molar fraction of solute in water. Even if the solution is not ideal, if we know the activity coefficient γ it is still possible to calculate the WVP:

$$p = p_s \gamma X$$

Timmermans (1960), American Society for Testing and Materials (1963) and Boublik *et al.* (1975) have reviewed the vapour pressure of numerous substances. When the solution is more concentrated, *a fortiori*, if a solubility threshold exists (saturated solution), calculation becomes impossible or very inaccurate, and only experimental measurement can give the WVP in equilibrium with the solution. Incidentally, an aqueous solution, the

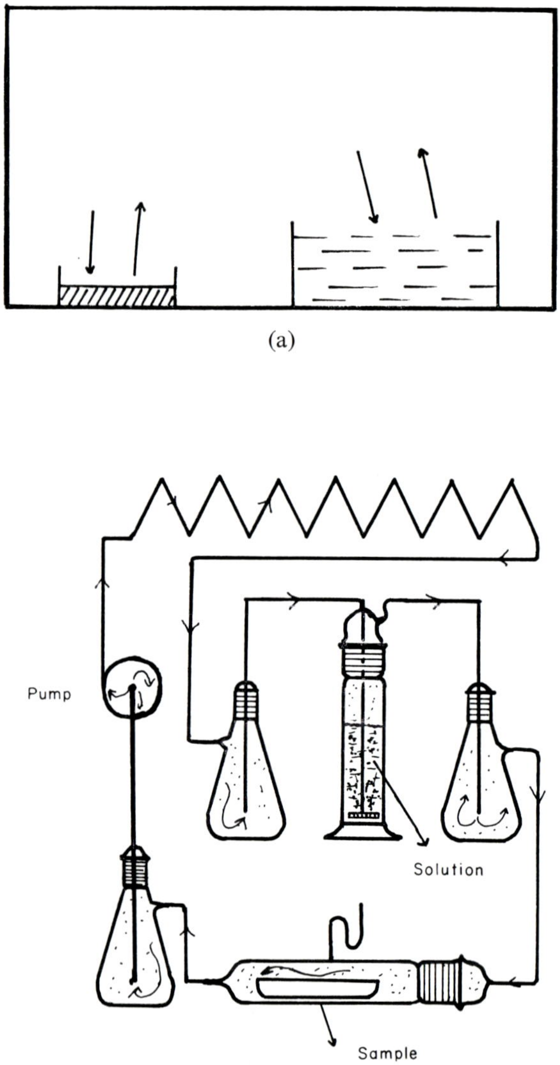

(a)

(b)

FIG. 3. Experimental device for RH regulation with solution. (a) Static atmosphere. (b) With recirculation of air.

WVP of which is known, may be used to keep the RH of the small air volume of a closed cabinet constant (Fig. 3). In fact, insofar as the volume of water contained in solution may reasonably be supposed infinite compared with the quantity of water contained in the cabinet atmosphere and in the samples themselves, the solution will exert a true 'buffer effect' on the WVP, *i.e.* on the RH of the cabinet.

Generally speaking, two types of aqueous solutions are used, the saturated solutions and the diluted solutions of liquids miscible with water.

(i) *Aqueous saturated solutions.* The particular properties of sorption isothermal curves of some crystalline salts of small ion size (*e.g.* oligosides) are of advantage in this case. On increasing RH with initially dry crystals, the following equilibration steps may be observed (Fig. 4):

(a) Formation of one or several hydrates. In its stability zone the unique water content of hydrate corresponds to an indeterminate RH (plateaux of the curve).
(b) Formation of a liquid phase in equilibrium with crystals. This is the saturated solution, the unique RH equilibrium of which corresponds to large changes of water content (vertical line of the curve).
(c) Disappearance of the solid phase, the WVP increasing as a function of dilution.

The saturated solution can therefore be used for the RH regulation in a closed space, with an excellent 'buffer effect'. No concentration measurement is required after the solution has become saturated. This technique is very attractive as the materials required are cheap. However, there are some disadvantages, the most serious of which is probably the relative inaccuracy of ERH. In this regard, the data given in the literature show important differences; for example, in the case of NaCl, at 25°C, Rockland (1960) gives 75%, O'Brien (1948) 76·5% and Sakai (1940) 74·06%. These authors, and Winston and Bates (1960), Stokes and Robinson (1949), Wexler and Hasegawa (1954), Martin (1965), Acheson (1965), Hedlin and Trofimenkoff (1965) and Wylie (1965), give the ERH corresponding to numerous saturated solutions. Finally, it must be emphasised that the salts and the water used must be very pure, and that it is necessary to keep the saturated solution perfectly homogeneous without a non-saturated zone. The best method is to heat a mixture of water and salt with a minimum of water and an excess of salt and let this mixture cool to give a residue which must be stirred up frequently.

(ii) *Aqueous solutions of water-miscible liquids.* The solutions mainly used

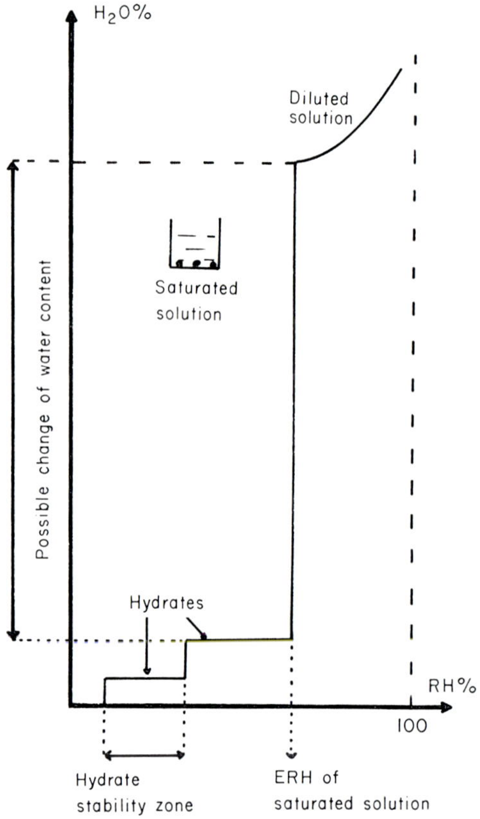

FIG. 4. Sorption curve of a crystallised system (salt, sugar, etc.).

are sulphuric acid, calcium chloride (Breese, 1955), or glycerol, the partial pressures of which are negligible, the WVP being only a function of the dilution.

Perry (1941), Stokes (1947), Stokes and Robinson (1949), Spencer-Gregory and Rourke (1957) and American Society for Testing and Materials (1963) give the data concerning the water vapour pressure of these solutions. These data are known with a better accuracy than those of saturated solutions, and it is easier to avoid a concentration gradient by stirring up a diluted solution than a saturated one. Therefore we prefer the diluted solutions of miscible liquids, even if they need, at the end of each experiment, a precise measurement of concentration in order to obtain the

ERH. However, the 'buffer effect' is less important than in the case of saturated solutions, the WVP changing continuously as a function of dilution (Fig. 5). Consequently it is advisable to use a volume of solution large enough to ensure that its water content changes are comparatively negligible.

(iii) *Practical points.* Before using the saturated or diluted solutions, it is first necessary to solve the problem of exchange and quick WVP equilibrium between the solution and the air. There are several possibilities.

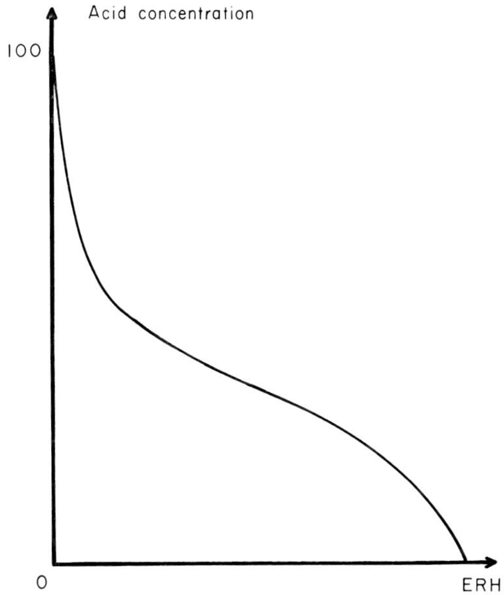

FIG. 5. Sorption curve of an acid solution.

The first is to evacuate the assay chamber, in order to quicken the diffusion of water vapour and to increase the free average run of water molecules. This is the principle used by Guilbot (Drapron and Guilbot, 1962). A glass tube is employed composed of two units separated by a ground joint; the sample is placed on one side, and in the other is placed a solution of sulphuric acid (Fig. 6). A nozzle with a stopcock is used for evacuation (vacuum corresponding to a water pump). Under these conditions, the ultimate equilibrium between air–solution and sample is reached in 48 hr, instead of 15 days or more under atmospheric pressure.

FIG. 6. RH regulation with solutions. Pyrex glass tube for sorption equilibrium: (*left*) acid solution (*right*) sample studied.

Several authors have recommended similar systems (Kapsalis, 1967; Kapsalis *et al.*, 1970; MacKenzie and Rasmussen, 1972; Hamm *et al.*, 1974; Spiess *et al.*, 1969; Wolf *et al.*, 1973; Bushuk and Hlynka, 1960).

Under atmospheric pressure, it is convenient to agitate the air of the cabinet with a ventilator blowing on the surface of the solution (Young and Nelson, 1967; Bosin and Easthouse, 1970; Gal, 1975). It is also possible to use a pump to bubble the air directly through the solution. This procedure allows simultaneous agitation of the solution and the air, and gives the same equilibrium WVP as those observed in a static process (Timofeev, 1960). We have used such a system in our laboratory. The recirculating air is swept along with a 'peristaltic' pump, and bubbles in a flask containing sulphuric acid. Duprat and Charbonnière (1974) use the same system to temper the samples before studying their X-ray diffraction spectrum, and Drexler (1972) describes a similar apparatus but using salt-saturated solutions.

It is sometimes desirable to renew the air. In this case it is necessary to increase the contact time between air and solution, for example by arranging several flasks, one after the other, containing only saturated solutions. Care must be taken to avoid aerosols of the solute being carried towards the zone where the sample has been placed.

2.2.2. *Regulation by Physical Processing*

Several physical processes can be used to regulate the WVP and RH in small cabinets. Especially interesting are the systems involving saturation of air under particular conditions of temperature and pressure; they allow study of the simultaneous sorption of water vapour mixed with other volatile liquids, such as alcohol (Duprat and Guilbot, 1975).

(i) *Regulation involving a two-temperature procedure*. The principle of this regulation is easy to understand on a psychrometric chart (Fig. 7). Air, swept along by a pump (Fig. 8), goes through a saturator, the temperature of which is equal to the dew-point temperature (θ_r) expected in the cabinet (let AB be the course of the representative air state point on the chart). The saturated air stream is then warmed up again until its

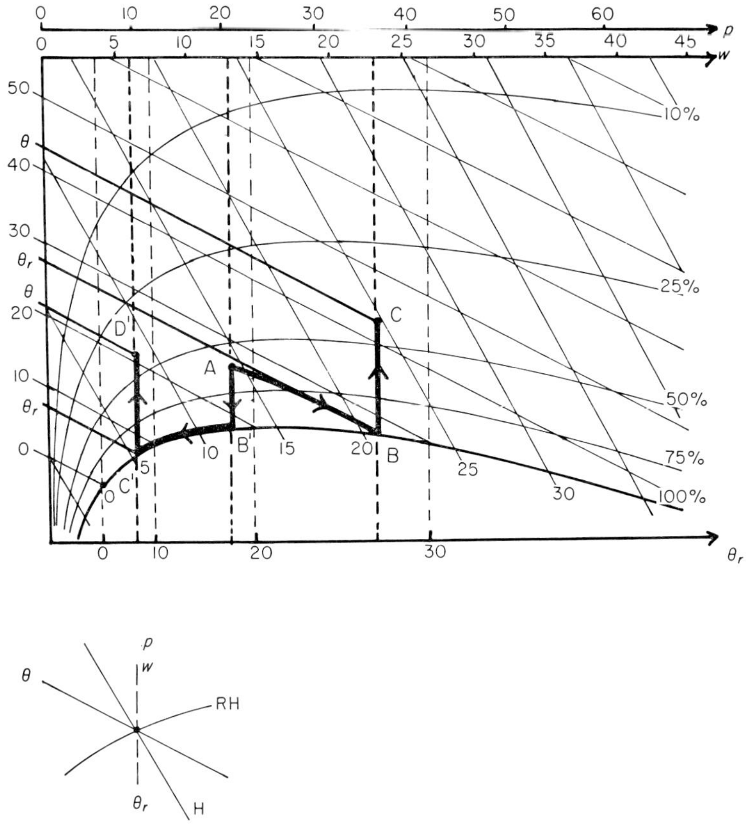

FIG. 7. Psychrometric chart.

temperature is equal to that of the cabinet (θ), where it is entering. On the chart the representative point is moving along a vertical line (specific humidity and partial water vapour pressure are constant) from B to C. The relative humidity is then:

$$\mathrm{RH}\ (\%) = \frac{p_{\theta_r}}{p_\theta} \times 100$$

where p_θ and p_{θ_r} are the pure WVP at the temperatures θ and θ_r. This system may work either with recirculation of air (closed system) or not (open system).

Several authors have used this general technique, with many minor variations of saturation and temperature regulation devices (Wentzel,

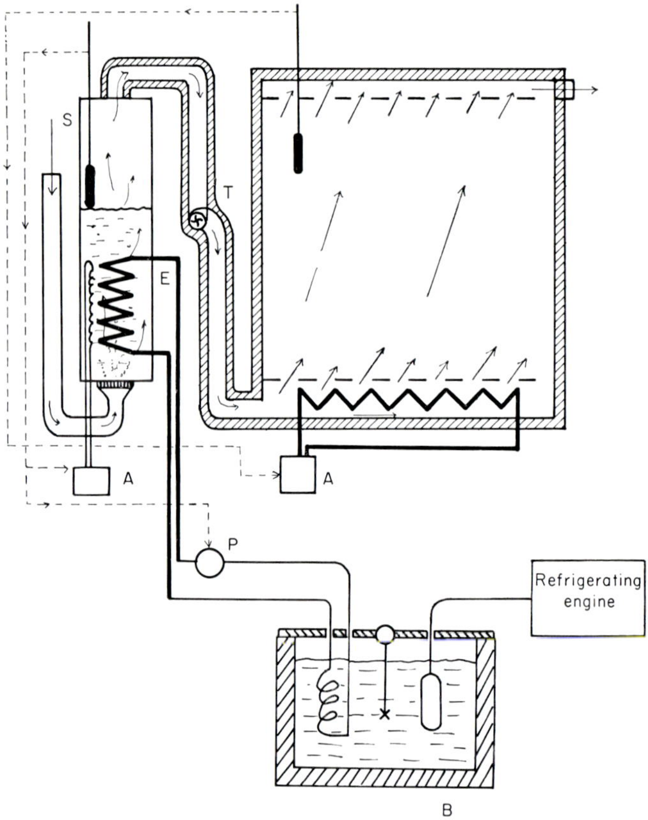

FIG. 8. Principle of RH regulation with a two-temperature system. A. Switches commanding heating of saturator and test chamber. E. Heat-exchanger cooling saturator. P. Pump for circulation of cooling fluid. B. Cooled bath. S. Saturator. T. Air-circulating pump.

1965; Handegord and Hedlin, 1965; Le Maguer, 1972). More recently, in our laboratory, Duprat (1975) has applied the same basic principle in order to obtain a RH controlled cabinet which makes possible a programme using several successive RH cycles. Some cabinets built according to this principle are also commercially available (particular by Weiss). The more interesting advantages of this method are the complete independence of air speed, RH and temperature, and the stability of RH; there is no cyclic fluctuation, any hygrometric probe being used for its regulation. Temperatures also are easy to regulate with great precision.

On the other hand, the air produced by the saturator is not completely saturated and this may lead to a RH error, which can be avoided using a special device, as described below. When the saturator temperature is higher than the room temperature, some water condensation may occur in pipes; it is therefore necessary to lag the whole system, or to heat the pipe in which the saturated air leaves the saturator. Loncin *et al.* (1965) and Krueger (1965) avoid this difficulty by using a saturator filled with a NaCl-saturated solution.

The RH corresponding to dew-points lower than room temperature may be obtained only with a refrigerating device, accurately regulated. Dew-point lower than 0°C may be obtained by cooling the air in a small chamber, the temperature of which is kept equal to that of dew-point, and reheating to cabinet temperature (let AB′C′D′ be the course of the representative air point on the chart (Fig. 7)) (Chopin and Chopin, 1958).

The same principle (two-temperature procedure) may be applied to a high vacuum system. Figure 9 shows such a system; the sample and the water vapour source (distilled water) are frozen (about −80°C), then all the non-condensable gas is evacuated until a high vacuum has been obtained (10^{-6} mm Hg). The apparatus is isolated from the vacuum pump, and the water vapour source reheated until the temperature θ_A is equal to the dew-point wanted (let p_{θ_A} be the saturated WVP created in A).

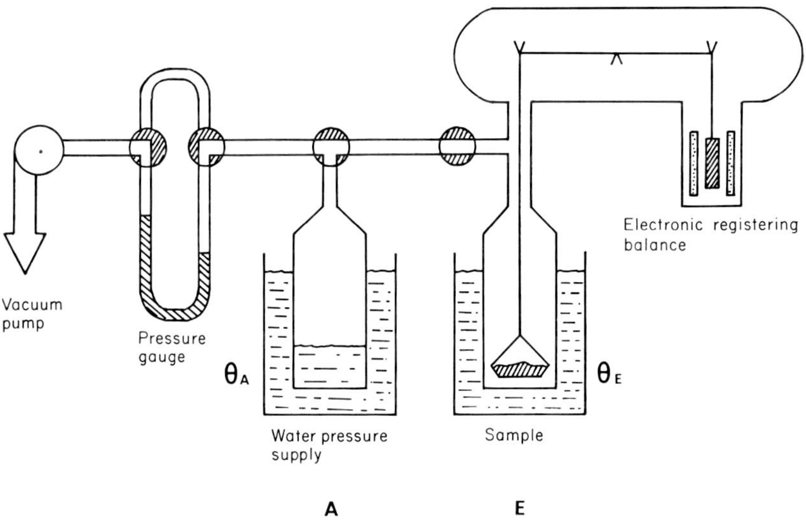

FIG. 9. Evacuated registering balance with RH regulation (two-temperature system).

Since the chamber (E) is being kept at a temperature $\theta_E > \theta_A$, when the link between A and E is open, the sample adsorbs water until it is in equilibrium with the water vapour pressure p_{θ_A} (easily measured with a pressure gauge); RH in the apparatus is equal to $p_{\theta_A}/p_{\theta_E}$, if p_{θ_E} is saturated pure WVP at the sample temperature. This high vacuum system is expensive because of the balance part of the system working under vacuum. One of the more important and interesting aspects of this method is the ability it gives to study sorption phenomena easily at any temperature, with very small samples (10–100 mg).

Several authors have described different kinds of apparatus using this vacuum system (Gast, 1966; Berlin *et al.*, 1968; Cramer *et al.*, 1969; Williams, 1969; Herrmann, 1971; Palnitkar and Heldman, 1971; Dickel and Fiederer, 1972; MacKenzie and Rasmussen, 1972; Elvanides, 1973).

It is also interesting to note that the two-temperature system may be used with or without vacuum to regulate RH in a very small volume testing tube (Corry, 1975).

(ii) *Regulation involving a two-pressure procedure*. The principle described by Weaver and Riley (1948) for measuring hygrometry has been used by Laine and Gac (1951), Jones and Petersen (1965), Amdur and White (1965) and Multon (1969) in order to regulate the RH in a small cabinet. Air is saturated (Fig. 10) by bubbling in water under an absolute total pressure P, regulated by the valve (c). In order to improve the saturation process, the saturator is fitted with 'Raschig' rings and the air entering is divided into very small bubbles with sintered glass. The pure WVP in air is independent of absolute total pressure P, and is only a function of the temperature t. With the valve (h), the total air pressure is expanded to atmospheric pressure (H), therefore the saturated WVP (p_s) is also expanded in the same ratio, to a value p, according to Dalton's Law on gas mixture. Consequently one has:

$$\frac{H}{P} = \frac{p}{p_s}$$

which is equal to the RH according to its definition.

RH in the chamber is independent of temperature and is a function of absolute air pressure (P) only during the saturation process. During the water transfer in the saturator from the water towards the air, the temperature of the water decreases slowly until a value equal to the 'humid temperature' of the air injected is reached. However, the air leaving the saturator is not entirely saturated, for the same reasons as given in the two-temperature system. Both disadvantages can be eliminated as follows.

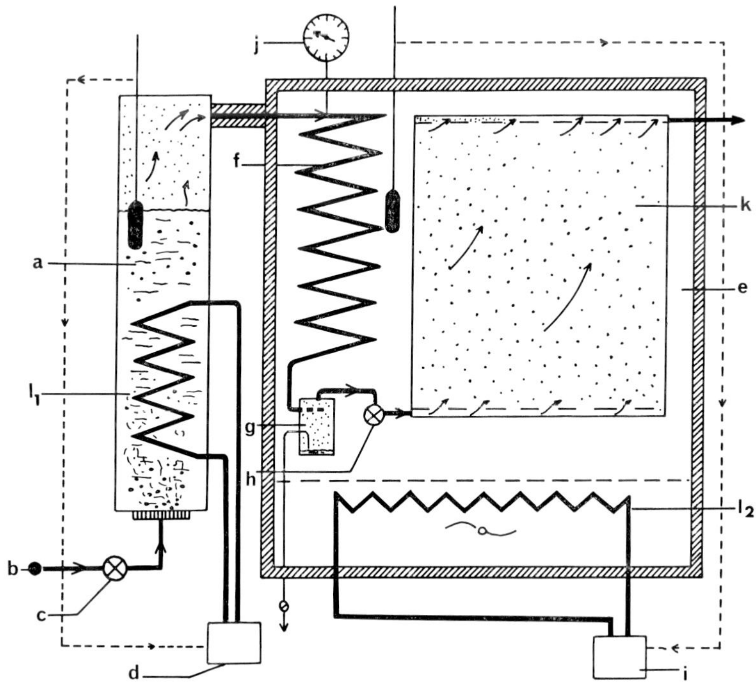

FIG. 10. Principle of RH regulation with a two-pressure system. (a) Saturator; (b) compressed air inlet; (c) first valve; (d) and (i) switches; (e) constant temperature chamber; (f) condensor; (g) flask for condensed water; (h) second valve; (j) pressure gauge; (k) test chamber. I_1, I_2, heating radiators.

The saturator is heated and kept at a temperature (t_s) a little higher (0·5–1°C) than that of the cabinet itself. When the air enters the cabinet, its temperature decreases slightly in a heat-exchanger made of a copper coil, and a small excess of condensed water is collected in the flask (g). Then, just before expanding, the air is exactly saturated at the cabinet temperature.

RH and temperature are completely independent entities. A hygrometer is unnecessary for RH regulation since any changes may be simply and quickly obtained through a simple adjustment of pressure. RH may be easily regulated from 15% up to 99%, with excellent accuracy (about 1% absolute value). A great advantage of this principle is that a cheap cabinet can be made with several chambers at different RH for the same temperature.

The two-pressure apparatus may also work as a two-temperature one.

It is only necessary to open completely valve (h) (Fig. 10), to inject valve (c) with expanded air and to regulate the saturator temperature so that it is equal to the dew-point temperature. It is, of course, advisable to provide the apparatus with clean air, without oil aerosol, and with a constant pressure.

In our laboratory we have used this principle to convert a multicellular oven (Chopin) into an apparatus for sorption studies (Multon *et al.*, 1971). We have also used the same principle to build cabinets for studying the storage of grain and for breadmaking tests (fermentation test).

2.2.3. *Regulation by Blending Wet and Dry Air*

RH may be adjusted at any value by blending two air streams, one completely dried, the other water-vapour saturated (Fig. 11). It is easy to compute or to determine on a psychrometric chart the correct ratio of mixture corresponding to a certain RH.

The greatest difficulty is to obtain a mixture chamber giving a perfectly homogeneous blend. On the other hand, this system can give any RH between 0 and 100 %, and is suitable for a cabinet with several chambers at different RH (Spencer-Gregory and Rourke, 1957; Cherry, 1965; Humphreys and Henschel, 1965).

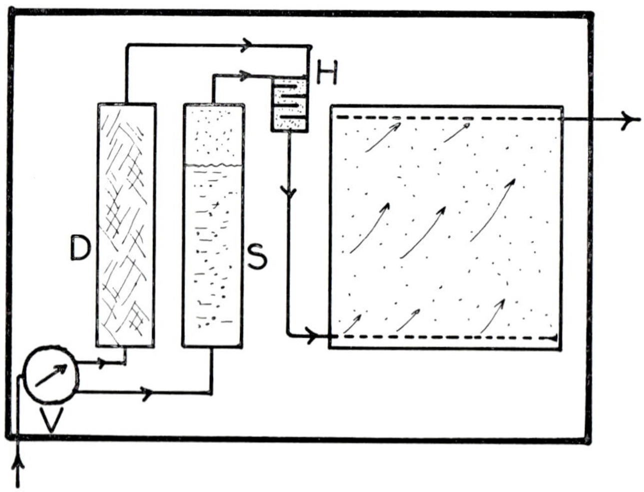

FIG. 11. Principle of RH regulation with mixture of wet and dry air. D. Drying column. H. Wet and dry air mixing chamber. S. Saturator. V. Stopcock dividing air stream between D and S.

2.2.4. *Regulation by Direct Water Injection*

In this kind of regulation (Fig. 12) the air in the cabinet is recirculating, being swept along by a pump with, eventually, a partial renewal of the used air. The air-conditioning process is generally made in a special chamber contiguous with the test chamber. Temperature regulation is obtained with a heat-exchanger, and RH adjusted by steam or water aerosol injection, directly into the air stream, with a hygrometer set in the test chamber.

When a low RH is needed, air must be dried when leaving the chamber and remoisturised again in order to reach the right RH. This principle of regulation, which is obtained in a different way, is that most often used in commercial climatic cabinets; but this system does not fit well for small volumes and is not very accurate (from ±3 to ±5% RH), particularly because of the hygrometer-driving. This is why we cannot recommend this system for precise sorption research.

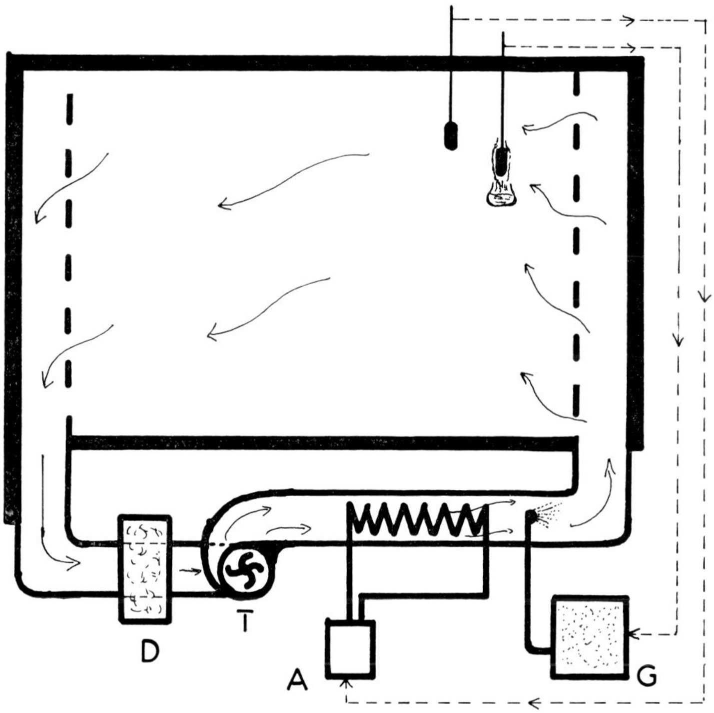

FIG. 12. Principle of RH regulation with injection of water vapour. A. Heating system. D. Drying system. G. Water-vapour generator. T. Ventilator.

2.2.5. *Regulation by Capillary Suction Pressure*

A special device (Gur-Arieh *et al.*, 1965*a*,*b*) allows sorption equilibrium to be obtained in very high RH areas (95–99·99 %). This method equilibrates the capillary suction pressure of water in the product with an external pressure exerted on the product placed on a porous plate (Fig. 13).

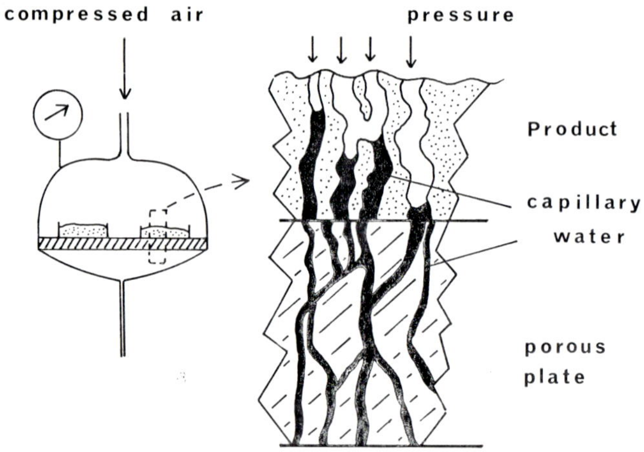

FIG. 13. Principle of very high ERH measurement, by equilibrating capillary suction pressure.

When equilibrium is reached, the RH is given by the Thomson equation:

$$\mathrm{RH} = \frac{p}{p_s} = \exp\left(-P/\rho RT\right)$$

where p is the WVP of the product, p_s is the pure WVP, P is the external pressure exerted on the product, ρ is the specific gravity of water, R is the perfect gas constant and T is the absolute temperature.

The pressure may be exerted by a piston (Kneule, 1959) using compressed air or by centrifugation. Here RH is an indirect relation which may be computed from thermodynamic relations, but is impossible to measure experimentally.

2.3. Detection and Measurement of the Equilibrium Parameters

2.3.1. *Detection of the Equilibrium*

When the sample is placed in the cabinet where the relative humidity is maintained at a constant level by one of the previously described processes, the moment when the water vapour exchange stops must be determined and the pressure equilibrium obtained (end of the reaction).

The end of the reaction is mostly detected by the observation of constant-mass of sample, within the limitations of the weight method precision and as far as other reactions do not interfere (Section 2.4). Since to obtain the constant weight is often a very long and tedious process, Seidl and Sczigel (1966) proposed that the level of the plateau be calculated from several experimental points, obtained before the complete equilibrium is reached, postulating the last phase of the sorption kinetic as an exponential function. The weighing can be made either continuously *in situ*, or discontinuously.

(i) *Weighing of the sample in situ*. This technique needs either an insulated balance fixed as part of the relative humidity regulation device, or an outside balance allowing the weighing to be done in the cabinet. The first system must be used for the high vacuum regulation.

Usually, the insulated balances (Cahn, Beckmann, Sartorius, Mettler) have an automatic electric compensation system, with a recording readout. A quartz spring balance (MacBain) may also be used, or a balance with a special metallic spring (Kapsalis *et al.*, 1970). A good method is to use an automatic electric balance outside the cabinet weighing under the pan. The only difficulty is in passing the sample suspension wire through the cabinet without exchange between the inside RH regulated air and the outside laboratory air. Drexler (1972) and Bolliger *et al.* (1972) have described different means of reducing the weighing error, according to the fact that the sample is suspended in a dynamic air stream.

It is preferable, when possible, to make a continuous recording of the sample weight, which is necessary for a precise study of the sorption kinetic. Meanwhile, the inconvenience of using a balance fixed as part of the apparatus is that only one sample can be studied at the same time. This device must therefore be used only for high-precision measurements. Several authors propose automatic handling systems where the samples are transferred successively on to the pan of the recording balance without being taken out of the regulated cabinet (Best and Spingler, 1972; Ferguson *et al.*, 1972). It is important to notice that automatic weight recording does not dispense with making, at the end of the experiment, a water content determination of each sample. Indeed, it is necessary to know the exact dry weight in order to compute the water content of a sample at each successive equilibrium. Moreover, using the balance part of the system becomes very difficult when high temperatures are reached, and then a thermo-analytical balance must be used.

(ii) *Weighing of the sample outside the conditioned cabinet*. This procedure is naturally less precise than weighing *in situ* and needs several precautions. When taking back the pan containing the sample, the cover must be replaced immediately (the pan and its cover must be previously tared) to avoid, as far as possible, water exchange between the product and the laboratory atmosphere.

If the sorption is made at a very low or a very high temperature, the pan containing the sample with its cover must be put inside a glass tube closed by a ground glass stopper, previously tared. In this way, it is certain that all the water of the sample has been weighed, even if there is some water condensation on the sample, or some water evaporation from the sample. All these operations must be done quickly, and the balance has to be put very close to the sorption apparatus.

Generally speaking, the precautions taken during the weighing are very similar to those recommended for the water content determinations (Multon and Guilbot, 1975; Guilbot *et al.*, 1973).

2.3.2. *Measurement of the Equilibrium Parameters: Water Content and RH*

When the sorption equilibrium is reached, first the water content of the sample must be determined, and secondly the RH, in order to get the corresponding point of the sorption isothermal curve. We can only give a reminder here of some important aspects of these measures, the details of which are extensively published for each particular product.

(i) *Measurement of the water content.* The determination of the water content is simplified because the sample is ground. The choice of an appropriate measurement technique must be made. Generally, for food products a standard method may be used. A large effort towards standardisation has been made for cereals and cereal products (Multon and Martin, 1975). For other materials there are also standard methods published by international bodies such as the International Standardisation Organisation. Otherwise, any method described by a regulating body or in the literature is satisfactory. However, it is necessary, before beginning the sorption study, to develop an appropriate water content determination technique.

(ii) *Measurement of RH.* Usually it is not necessary to measure the RH in the sorption cabinet, because RH is known either according to the principle of the regulation process, or relative to the aqueous solution, from the tables.

The concentration of the miscible liquid solution (H_2SO_4, alcohol, glycerol) is changing slightly during the exchanges between water vapour, air and product. Therefore it is necessary, when the equilibrium is reached, to measure the final concentration of the solution to determine the exact RH. This concentration can be obtained by measuring the density (picnometry), the acidity (in the case of H_2SO_4), the refractive index or any other appropriate method.

Although the RH is known, it can be useful to measure it in the cabinet, especially in order to check the regulation system. This measurement is very hard to do because of the small volume of the cabinet and because it must be more precise than RH regulation itself. Apparatus using an electric probe may be used (Gröninger, 1968), but these are subject to important errors. Such apparatus, whatever the type, is calibrated and subject to hysteresis. The best of them indicate the RH with a precision of ± 3 to $\pm 5\%$ (absolute value), but this is not sufficient for a sorption study. Also, the reproducibility of these probes is poor; hence it is necessary to check their calibration frequently.

All the indirect methods (variations of the length, the colour, electric sensitivity of an hygrometric element) have to be excluded because they have the same disadvantages. Lithium chloride probes, measuring the temperature at which the WVP of the LiCl-saturated solution is equal to the atmospheric WVP, cannot be used in small volume because of the heat they produce. A detailed discussion of RH measurement can be found in Wexler (1965).

More appropriate methods for measuring RH in a small cabinet are the following:

(a) Psychrometric method. The psychrometric method measures the air under test for the dry and the wet temperature (temperature given by a thermometer, the bulb of which is surrounded by a pure liquid water film, and ventilated). For small volumes it is advisable to build the psychrometer with thermocouples or resistance probes; the wet bulb must be ventilated with adequate speed (minimum speed 2 m/sec).

(b) Measurement of the dry and dew-point temperatures. Much progress has been made since the first apparatus of Regnault (1845), and today automatic dew-point hygrometers are available (model HCP of Schlumberger, model 880 of EG & G, Cambridge Systems). The mirror of the instrument is cooled by an electronic thermoelectric cooler (Peltier element), and when the condensate appears on the mirror there is a difference of reflectance registered by a photo-cell. The cell develops a proportional signal which regulates the cooling supply. The temperature of the mirror, when equilibrium is obtained, is measured with a thermocouple.

Such an apparatus needs a very slow air stream near the mirror, about 0·5 litre/min, when sampled in the conditioned cabinet with a small pump and eventually recirculated. Dew-point hygrometers are therefore convenient for all systems of dynamic regulation where the air sampling is possible without changing the cabinet RH.

The dew-point measurement can be made at every temperature and RH adjustment.

(c) Gas chromatography. Gas chromatography allows determination of water content in a gas sample taken in a conditioned cabinet with a special syringe. However, besides the difficulty of sampling, the system needs calibration.

Le Maguer (1972) describes an original sorption apparatus where the sample is used as packing for the gas chromatographic column. The air injected in the column has a known RH (two-temperature process) and he deduces from the chromatogram analysis the quantity of water fixed by the sample for each RH change. This technique allows the study of simultaneous sorption of water and volatile substances.

(d) Phosphorous pentachloride electrolysis. A very precise apparatus, specially suitable for low RH (dew-point between $-20°$ and $-80°C$), has been developed by Beckmann. A known air flow, taken in the cabinet, passes through the measurement cell of the apparatus; this cell is formed by a capillary pipe with two rhodium coils passing through it, one being isolated from the other by a P_2O_5 packing. The water vapour is fixed by P_2O_5 and electrolysed; the intensity of the electrolysis current is a function of the water content of the air, and of the air flow, through the cell.

(e) Gravimetric methods. These are based on forcing a known and constant flow of the air to circulate through a series of bubbling flasks or through tubes filled with hygroscopic products; thus, knowing the absorbent weight increase and the air flow, the RH can be calculated.

(f) Manometric methods. Manometric methods are based on measuring the partial WVP in the air. A reference method, slow, discontinuous but very precise, has been developed by Laine. An air sample is picked up in the cabinet with a glass bottle previously dried and evacuated. The water vapour is condensed on the inner side by very slow cooling, down to $-80°C$, and the incondensable gas (CO_2, rare gases) is evacuated (10^{-6} mm Hg). The bottle is connected to an accurate manometer, the other end of which is connected to the vacuum pump, and reheated up to the initial air temperature. Then the water, which was ice-condensed, is again vaporised and the pressure produced can be read on the manometer.

Another manometric method consists of measuring the difference in pressure in the sampling bottle caused by fixation of the water vapour with a powerful desiccant.

Taylor (1961), Block *et al.* (1965) and Labuza (1972) describe for sorption studies some apparatus involving direct manometric measurement of the product. These systems consist of measuring the WVP in equilibrium with the sample, previously equilibrated in a separate chamber where RH is regulated according to one of the methods previously described. The sample is then placed in the measurement apparatus, frozen at a very low temperature (−80°C) in order to fix the water, and the apparatus is evacuated with a mercury or oil duffusion pump. As in the method described by Laine, it is necessary to heat the sample again and measure the WVP in the apparatus with the manometer (Fig. 14). It is notable that this system allows the direct measurement of the water activity of a product (cf. Section 3).

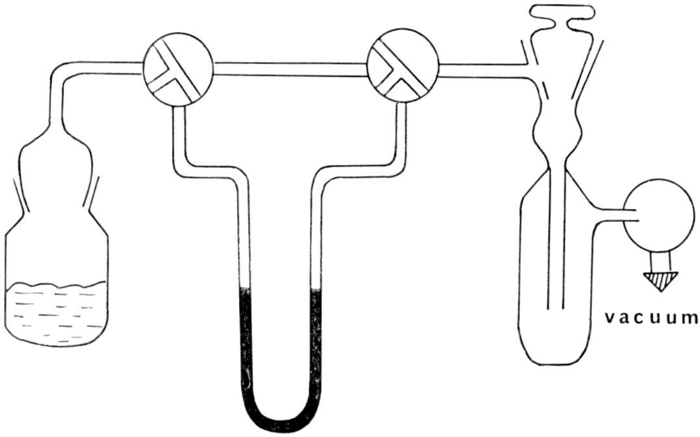

FIG. 14. Principle of manometric device for measuring water activity.

2.4. Accuracy of Sorption Measurements

The accuracy of sorption isotherms depends on the apparatus, the adsorbing product and the manipulations. After the study of the most critical aspects has been presented, an idea of the accuracy obtainable will be given.

2.4.1. *Analysis of the Main Sources of Error*

(i) *Errors proper to the RH-regulating apparatus.* Sources of errors examined here are those which are general and common to several appliances.

(a) Temperature regulation. Temperature regulations must be carried out with great care. In practice, a small variation within the limit of ±1% RH (absolute value) is obtained by a temperature regulation of ±0·1°C. This stability of temperature must be obtained also in the regulation working (heating cycle) as well as in the different points of the cabinet (homogeneous thermal geography). Moreover, a large thermal inertia of the apparatus is very desirable in order to reduce, as much as possible, the duration of the temperature recovery after each loading. Accurate cold regulation being very difficult to achieve, the best regulation will therefore always be obtained by heating. The apparatus is placed in a climatised room at a temperature slightly lower than that defined for the inside of the cabinet. Thus the heating periods are far apart, with short duration and little intensity. If a cold regulation is positively necessary, then operate as previously described.

(b) Saturator. Different regulation systems (cf. Section 2.2) work with a device saturating the air by bubbling either in water or in solutions. When making these saturators, several important points need remembering. First, a total saturation requires an infinite time of contact between the liquid and gaseous phases, and as this time is limited, the output air is not 100% saturated, even with several saturators in series. Secondly, the transfers are improved when the air dispersion in the water is made of smaller bubbles, and a piece of sintered material (glass, bronze, steel) of very thin porosity can be used to disperse the air flow. Thirdly, the mixing of the bubbles with the water must be made turbulently; the air flow will then reach a minimum value depending on the geometrical characteristics of the apparatus. Also, the transfers are improved by a flow deflected and split on Raschig rings, or some similar process as in a distilling tower.

Finally, as the vaporisation of water for the saturation of the air leads to a certain cooling of the saturator, the temperature stabilises at a value lower than that of the cabinet which varies from the wet bulb temperature of the injected air.

From these different phenomena it follows that the air coming out of the saturator is slightly unsaturated and at a temperature lower than required. A device which corrects the two defects at the same time has already been described (Section 2.2.2(ii)).

(c) Adsorption by the constitutive materials of the cabinet. The whole apparatus, pipes included, must be built with low hygroscopic materials (metal, glass, Teflon), avoiding wood and most plastics. Without this precaution, an important delay will be observed after each changing of the RH until its regulation is stabilised, due to sorption or desorption on the materials.

(d) *Case of extreme conditions.* Conditions lying outside the 5–95 % RH range and 0–80°C temperature range are difficult to achieve and need much care when manipulating. Low values of RH (below 5–10 %) or high ones (over 97–98 %) cannot be easily regulated, and, as the sorption isotherms show steeper slopes in these areas, the moisture content determinations are less accurate.

Temperatures far from the ambient lead to difficulties in weighing; these can be solved by using a tared tube, as described. Temperatures over 100°C cannot be reached in most regulating systems because of the boiling that may happen inside the saturator.

The highest RH combined with high temperatures need special saturating systems because of the great demand for water vapour. At the opposite end, many difficult problems in regulation are encountered with cold temperatures, since, owing to the corresponding small specific humidity, the slightest alteration leads to huge differences in the RH value.

(ii) *Errors due to the products themselves.* Biochemical products, mixtures of lipids, glucids and proteins, are very complex products allowing many types of errors. Among them is the ability of these products to change with time, mainly during the rather long period necessary to reach the sorption equilibrium. Thus the final value is no longer representative of the initial product, but is related to the transformed product after different biological and biochemical changes. Many deteriorating mechanisms, such as enzymatic reactions, glucid and protein denaturation, Maillard reactions, non-enzymatic oxidation and microorganism development, particularly of moulds, are greatly dependent on the water activity. With variations in the temperature and RH fields, one or other of these reactions may develop, and lead to a modification of the composition of the product studied. Moreover, *the constant alteration of the dry portion makes the constant weight very difficult to ascertain* and then the notion of reaching equilibrium becomes almost meaningless.

Among all these reactions, two are of particular interest:

The enzymatic reactions caused by some enzymes existing in the products, namely hydrolases which produce some short and soluble molecules from the glucids or protein macromolecules. Because of this, there is a risk of limited zones of saturated solutions in the high RH area. Under such conditions, the moisture content of the sample will increase considerably and such a risk is intensified by the osmosis phenomenon which occurs when some cellular structures are present.

Other important alterations appear when the product contains some living cells, because in the high RH, respiration speeds up and fermenting processes may develop. But undoubtedly the greatest inconvenience in the high RH range is mould growth, the enzymes of which add their effects to those of the product itself. The development of moulds is extremely difficult to avoid, since all sterilisation processes alter the substrate; possibly the whole experimental system would need to be sterilised.

All the side reactions which occur at high RH add to the difficulties of regulation, and explain the poor results for sorption isotherms in the higher RH (over 90%) and the lack of reproducibility for experiments performed in such zones.

2.4.2. *Reproducibility of Sorption Results*

Accuracy depends on:

(a) the product (risks of deterioration);
(b) the RH regulating system;
(c) the moisture determination method;
(d) the range of RH variations; and
(e) the temperature.

With so many parameters, the accuracy quantification can only be achieved for each particular case with the help of an error calculation.

However, and for information purposes only, we give here the results obtained in our laboratory in the simple case of tapioca starch with a regulating system working on pressure differences. The repeatability of the values was:

For the absolute value of the moisture content: ±0·03 to ±0·5%.
For the absolute value of RH: from ±1 to ±2% (according to the different zones).

Comparing the results obtained on the same product by different laboratories using their own methods appears difficult. This is shown by the following figures, from an international ring analysis performed on the same tapioca starch:

Moisture content: differences up to ±3% (absolute value).
RH: differences up to ±5% at RH of 50%.

Only an elaborate standardisation of methods may improve these results and help to simplify the comparison between results from different laboratories.

3. MEASUREMENT OF WATER ACTIVITY

Measuring the ERH (or 'water activity', a_w, when referring to the availability of water as a solvent or as a diffusion medium) is achieved by the determination of the air RH in equilibrium with the product as presented, at its own temperature, and without taking into account the method of reaching equilibrium, either by sorption or desorption, or by addition of a component which alters the original state. However, in this measurement we assume that all transfers have stopped inside the product, otherwise the ERH changes until the WVP equilibrium settles.

a_w determination is very significant with regard to the stability of a low-moisture product, as long as other parameters remain constant. Thus there can be a good estimate of the spoilage risks, or of the efficiency of technological processing performed in order to reduce the a_w (*i.e.* the use of additives). Hence, such a measure is essential for intermediate moisture food technology.

The next section outlines the main methods of ERH measurement, some having already been described in the study of sorption isotherms.

3.1. Direct ERH Measurement of the Air Surrounding the Product

In these techniques the RH inside a small air- and water-tight cell containing a sample of the product is measured after the necessary time has been allowed to reach the vapour pressure equilibrium. During the water vapour transfer to reach equilibrium inside the cell, the air volume should be as limited as possible. Much care must be taken over temperature variations. Measurements need to be done at the same temperature as that of the original product. The effects of temperature variations may be different according to the conditioning method, *i.e.* the sample may be sealed in a tight container with or without a RH regulator, or kept in a non-hermetic device where many kinds of water content variations may occur. When such temperature variations take place, the conditions relative to them must be defined with precision, otherwise the meaning of any measurement would be compromised.

The reaching of equilibrium between the sample and its surrounding atmosphere is confirmed by successive measurements showing constant ERH value. The abovementioned conditions, among which is the small volume of gas available, lead to a choice of the following methods to determine the RH right inside the container. The manometric techniques (Taylor, 1961; Block *et al.*, 1965; Labuza, 1972; Sood and Heldman, 1974), which are very similar to Laine's method previously described, are very suitable for the low range.

Manometric techniques require freezing the cell, evacuation of non-condensable gases, reheating the cell and then directly measuring the WVP. The determination may be performed directly with the product in the vacuum cell; but when the freezing–defreezing process may cause a change in the a_w of a sample, the best way is to equilibrate the product with air in a special chamber, then to take a sample of air in a special glass flask for measuring the WVP by manometric methods.

Dew-point measurement may be performed directly in the chamber where the product is, or in a closed cell with continuous recirculation of air between cell and chamber by means of a small pump (Leistner and Röer, 1975). The two methods—the manometric and the dew-point—may be considered as reference methods, which may be used for the calibration of others.

On the other hand, measurement of a_w with hygrometric electric probes is suspect, because of their lack of reliability and their general behaviour, characterised by a hysteresis cycle. However, for routine tests in which only little precision is required, such probes, or other calibrated systems such as hair hygrometers, may be sufficient provided they are always used in sorption and recalibrated very frequently on standard solutions or by the above methods (Leistner and Röer, 1975).

3.2. a_w Determination by Successive Approximation

In this simple method the product a_w is considered to be equal to the RH of an atmosphere which does not induce any change in the sample weight. Obviously, as soon as a weight variation is noticed, a new test must be made on another sample. In order to limit the length of these tests, a series of containers with constant RH regulators will be of great help. The container corresponding to no weight change of the respective sample has the same RH as the product.

The graphical interpolation method is quicker. One defines a series of RH, lower and higher than the supposed a_w of the product; different samples are exposed to these RH and their weight changes are plotted as a function of time; then one finds by interpolation the RH value equal to the actual a_w, for which the weight does not change (Landrock and Proctor, 1951).

3.3. Conditioning of a Standard Test Substance

A small quantity of a dry standard substance, the sorption curve of which is well known, is equilibrated in the range of a large amount of the product; all this is placed in a small gas-tight container. With respect to the mass of the tested product and its moisture content, the amount of water adsorbed by the standard must be negligible. (Here the product acts as a RH regulator in the container.) When the equilibrium is reached, that is to say when the mass remains constant, the ERH is determined by referring to the sorption curve, just by measuring the moisture content of the standard. Standards may be products such as starch, microcrystalline cellulose (Vos and Labuza, 1974), or even salt solution (Robinson and Stokes, 1955). In a similar method, the standard substance is replaced by a series of samples (some crystals) of different compounds for which the ERH-saturated solution is well known. The ERH in the container is then indicated as the value between that of the saturated solution, with the highest ERH (deliquescence of the crystal), and that of the dry crystals, with the lowest ERH (for its saturated solution value).

These techniques are very time-consuming but this disadvantage is balanced by their simplicity in equipment.

3.4. Solvent Extraction

Hasegawa and Sekine (1965) and Mallet *et al.* (1974) have proposed measuring the a_w by solvent extraction. The sample is ground, then immersed and stirred in benzene. The water is partly extracted in liquid phase in proportion to its bound state and the a_w of the product. A moisture content determination by the Karl Fischer method on the liquid phase then indicates the a_w of the product when compared to a pure liquid water equilibrated standard. From the author's findings, this method seems adapted to the high a_w range, but its accuracy was comparable only to electric probe technique results.

4. CONCLUSION

We have reviewed the main techniques and apparatus available for drawing the sorption isotherms of biochemical material. One of the main problems lies in the nature of the product itself, since it can change during the measurements, particularly because of mould growth. Many of the methodological problems discussed overlap fundamental questions concerning the sorption phenomenon, and in many instances we are far from having all the answers (*i.e.* hysteresis, the role of temperature). As our knowledge improves, methods of studying sorption isotherms may become more precise, especially in the high RH range. With ERH measurement methods we have tried to show how they differ from the sorption methods, although the techniques are often very similar.

Undoubtedly the control and standardisation of all these measurements will contribute to the knowledge of the role of water which nowadays is of ever greater technological interest.

REFERENCES AND BIBLIOGRAPHY

Acheson, D. T. (1965). In: *Humidity and Moisture* (ed. A. Wexler), vol. III, p. 521. Reinhold, New York.

Acker, L. (1962). *Adv. Food Res.*, **11**.

Amdur, J. E. and White, R. W. (1965). In: *Humidity and Moisture* (ed. A. Wexler), vol. III, p. 445. Reinhold, New York.

American Society for Testing and Materials (1963). Recommended Practice E-104–51.

Ayerst, G. (1965). *J. Sci. Food Agric.*, **2,** 71–8.

Bem, Z. and Leistner, L. (1970). *Fleischwirtschaft*, **50,** 1412.

Berlin, E., Anderson, B. A. and Pallansch, M. J. (1968). *J. Dairy Sci.*, **51,** 1339.

Best, R. and Spingler, E. (1972). *Chem.-Ing.-Tech.*, **44,** 1222.

Block, A., Ellison, F. E. and Hewitt, E. (1965). *US Army Natick Laboratories Tech. Rep.* FD-19. Contract No. DA 19-129-AMC-55(N).

Bolliger, W., Gal, S. and Signer, R. (1972). *Helv. Chim. Acta*, **55,** 2659.

Bosin, W. A. and Easthouse, H. D. (1970). *Food Technol.* (Champaign), **24,** 1155.

Boublik, T., Fried, V. and Hala, E. (1975). *The Vapour Pressures of Pure Substances*. Elsevier, Amsterdam.

Breese, M. H. (1955). *Cereal Chem.*, **32,** 481–7.

Bushuk, W. and Hlynka, I. (1960). *Cereal Chem.*, **37,** 391–8.

Cherry, R. H. (1965). In: *Humidity and Moisture* (ed. A. Wexler), vol. I, p. 539. Reinhold, New York.

Chopin, M. and Chopin, G. (1958). *C.R. Acad. Agr.*, séance du 5 Février 1958.

Chordash, R. A. and Potter, N. N. (1972). *J. Milk Food Technol.*, **35,** 395.

Chou, H. E. and Labuza, Th. P. (1972). *J. Agr. Food Chem.*, **20,** 1288.

Cole, K. M. and Reger, J. A. (1969). Paper presented at 24th Annual ISA Conference, Pittsburgh, Pa.

Corry, J. E. L. (1975). In: *Water Relations of Foods* (ed. R. B. Duckworth). Academic Press, London.

Cramer, W. D., Hanson, T. P., Tsao, G. T. and Lancaster, E. B. (1969). *J. Macromol. Sci.—Phys.*, **B3,** 611.

Dickel, G. and Fiederer, D. (1972). In: *Progress in Vacuum Microbalance Techniques* (ed. Th. Gast and E. Robens), vol. I, p. 119. Heyden & Son, London.

Drapron, R. and Guilbot, A. (1962). *Ann. Technol. Agr.*, **11,** 3.

Drexler, H. (1972). Untersuchungen zur Technologie und Qualitätsbeurteilung zerstäubungsgetrockneter Fruchtpulver. Ph.D. Thesis, University of Hohenheim.

Duprat, F. (1975). Personal communication.
Duprat, F. and Charbonnière, R. (1974). Personal communication.
Duprat, F. and Guilbot, A. (1975). In: *Water Relations of Foods* (ed. R. B. Duckworth). Academic Press, London.
Dyer, D. F., Carpenter, D. K. and Sunderland, J. E. (1966). *J. Food Sci.*, **31**, 196.
Eichner, K. and Karel, M. (1972). *J. Agr. Food Chem.*, **20**, 218.
Elvanides, S. N. (1973). An analysis and interpretation of thermodynamic properties of foods as determined from moisture sorption data. Ph.D. Thesis, Michigan State University.
Ferguson, J. M., Livesey, P. M. and Mortimer, D. (1972). In: *Progress in Vacuum Microbalance Techniques* (ed. Th. Gast and E. Robens), vol. I, p. 87. Heyden & Son, London.
Fett, H. M. (1973). *J. Food Sci.*, **38**, 1097.
Gal, S. (1967). *Die Methodik der Wasserdampf-Sorptionsmessungen.* Springer Vlg, Berlin–Heidelberg–New York.
Gal, S. (1972*a*). *Alimenta*, **6**, 213.
Gal, S. (1972*b*). *Helv. Chim. Acta*, **55**, 1752.
Gal, S. (1975). In: *Water Relations of Foods* (ed. R. B. Duckworth). Academic Press, London.
Gast, Th. (1966). *Arch. Tech. Messen*, Lieferung 361, R. 31.
Gröninger, K. G. (1968). *Mitt. Geb. Lebensm. Hyg.*, **59**, 460.
Guilbot, A. and Lindenberg, A. B. (1960). *Biochim. Biophys. Acta*, **39**, 389.
Guilbot, A., Charbonnière, R., Abadie, P. and Girard, P. (1960). *Die Stärke*, **11**, 328.
Guilbot, A., Multon, J. L. and Martin, G. (1973). *Seed Sci. Technol.*, **1**, 711–33.
Gur-Arieh, C., Nelson, A. I. and Steinberg, M. P. (1965*a*). *J. Food Sci.*, **30**, 188–91.
Gur-Arieh, C., Nelson, A. I., Steinberg, M. P. and Wei, L. S. (1965*b*). *J. Food Sci.*, **30**, 105.
Hägerdal, B. and Löfqvist, B. (1973). *J. Agr. Food Chem.*, **21**, 445.
Hamm, R., Potthast, K. and Acker, L. (1974). *Z. Lebensm. Untersuch.-Forsch.*, **154**, 73.
Handegord, G. C. and Hedlin, C. P. (1965). In: *Humidity and Moisture* (ed. A. Wexler), vol. I, p. 265. Reinhold, New York.
Hasegawa, Y. and Sekine, T. (1965). *Bull. Chem. Soc. Japan*, **38**, 1713.
Hedlin, C. P. and Trofimenkoff, F. N. (1965). In: *Humidity and Moisture* (ed. A. Wexler), vol. III, p. 519. Reinhold, New York.
Heiss, R. and Eichner, K. (1971*a*). *Food Manuf.* (May), 53.
Heiss, R. and Eichner, K. (1971*b*). *Food Manuf.* (June), 37.
Heldman, D. R., Reidy, G. A. and Palnitkar, M. P. (1973). *J. Food Sci.*, **38**, 282.
Henderson, S. M. (1970). *Trans. Am. Soc. Agr. Engrs.*, **13**, 762.
Henderson, S. M. (1973). *Trans. Agr. Eng. Res.*, **18**, 55.
Herrmann, W. (1971). Die Adsorption von Wasserdampf in Schwerpat Presslingen und ihr Einfluss auf deren Festigkeit. Ph.D. Thesis, University of Karlsruhe.
Hill, J. E. and Sunderland, J. E. (1967). *Food Technol.* (Champaign), **21**, 2176.
Humphreys, C. M. and Henschel, A. F. (1965). In: *Humidity and Moisture* (ed. A. Wexler), vol. II, p. 293. Reinhold, New York.

Jones, R. H. and Petersen, A. (1965). In: *Humidity and Moisture* (ed. A. Wexler), vol. I, p. 507. Reinhold, New York.

Kapsalis, J. G. (1967). *US Army Natick Laboratories Tech. Rep.* 67–87 FL AD 655 488, Series FL-56.

Kapsalis, J. G., Drake, B. and Johansson, B. (1970). *J. Texture Stud.*, **1,** 464.

Karan-Djurdjic, S. and Leistner, L. (1970). *Fleischwirtschaft*, **50,** 1104.

Kiermeier, F. and Wolfsleder, H. (1972). *Z. Lebensm. Untersuch.-Forsch.*, **149,** 218.

Kneule, F. (1959). *Das Tröcknen*. Sauerländer & Co., Aarau, Switzerland.

Krueger, L. L. (1965). In: *Humidity and Moisture* (ed. A. Wexler), vol. III, p. 495. Reinhold, New York.

Labuza, Th. P. (1972). Sorption phenomena in foods: theoretical and practical aspects. Paper presented at 1st Annual Conference on Physical Properties of Food, University of Massachusetts.

Laine, N. (1968). *US Army Natick Laboratories Tech. Rep.* 68-62-FL (Phase II). Contract No. DA 19-129-AMC-252(N). Project reference: 1C01 450 1A7, 1C Series FL-75.

Laine, P. (1950). Unpublished results.

Laine, P. and Gac, A. (1951). *Proceedings of the 8th International Congress on Refrigeration*, p. 292.

Landrock, A. H. and Proctor, B. E. (1951). *Mod. Packaging*, **24,** 123, 130–80.

Leistner, L. and Röer, W. (1975). In: *Water Relations of Foods* (ed. R. B. Duckworth). Academic Press, London.

Le Maguer, M. (1972). Rétention des matières volatiles sur des substrats à humidité variables. Thèse de Docteur-Ingénieur, Université de Paris.

Loncin, M. (1975). In: *Freeze Drying and Advanced Food Technology* (ed. Goldblith, Rey and Rothmayr). Academic Press, New York.

Loncin, M., Bimbenet, J. J. and Lenges, J. (1968). *J. Food Technol.*, **3,** 131.

Loncin, M., Lenges, J., Bimbenet, J. J. and Jacqmain, D. (1965). *Ind. Aliment. Agr.*, **82,** 923.

MacKenzie, A. P. and Rasmussen, D. H. (1972). In: *Water Structure at the Water–Polymer Interface* (ed. H. H. G. Jellinek), p. 146. Plenum Press, New York and London.

Mallet, D., Kohnen, J. B. and Surles, T. (1974). *J. Food Sci.*, **39,** 847–8.

Martin, S. (1965). In: *Humidity and Moisture* (ed. A. Wexler), vol. III, p. 503. Reinhold, New York.

Mattern, P. J. and Bishop, J. B. (1973). *Cereal Sci. Today*, **18,** 8.

Multon, J. L. (1969). Thèse de Doctorat d'Etat, Faculté des Sciences, Université de Paris.

Multon, J. L. and Guilbot, A. (1970). *Ann. Technol. Végétale*, **19** (3), 239–60.

Multon, J. L. and Guilbot, A. (1975). In: *Water Relations of Food* (ed. R. B. Duckworth). Academic Press, London.

Multon, J. L. and Martin, G. (1975). *Bull. anciens élèves EFM*, **269,** 239–51.

Multon, J. L., Trentesaux, E. and Guilbot, A. (1971). *Lebensm.-Wiss. u. Technol.*, **4** (6), 184–9.

O'Brien, F. E. M. (1948). *J. Sci. Instr.*, **25,** 73–6.

Palnitkar, M. P. and Heldman, D. R. (1971). *J. Food Sci.*, **36,** 1015.

Perry, J. H. (1941). *Chemical Engineer's Handbook*. McGraw-Hill, New York.

Potthast, K., Hamm, R. and Acker, L. (1975). In: *Water Relations of Foods* (ed. R. B. Duckworth). Academic Press, London.

Regnault, H. (1845). *Ann. Chim. Phys.*, **15,** 129.

Robens, E. and Sandstede, G. (1967). *Z. Instrumentenkunde*, **75,** 167.

Robinson, R. A. and Stokes, R. H. (1955). *Electrolytes Solutions.* Academic Press, New York.

Rockland, L. (1960). *Anal. Chem.*, **32** (10), 1375.

Rödel, T. A. W. (1973). *Fleischwirtschaft*, **53,** 27.

Rödel, T. A. W. and Leistner, L. (1972). *Fleischwirtschaft*, **52,** 1461.

Sakai, W. (1940). *J. Soc. Chem. Ind. Japan*, **43,** Suppl. binding 104 B.

Salwin, H. and Slawson, V. (1959). *Food Technol.* (Champaign), **13,** 715.

Saravacos, G. D. and Stinchfield, R. M. (1965). *J. Food Sci.*, **30,** 779.

Schmidhofer, Th. and Egli, H. R. (1972). *Alimenta*, **11,** 169.

Scott, W. I. (1957). *Adv. Food Res.*, **7,** 84.

Seidl, G. and Sczigel, R. (1966). *Luft u. Kältetechn.*, **4,** 114.

Simatos, D. (1971). *Rev. Gén. Froid*, **62,** 183.

Smith, P. R. (1971). *BFMIRA Scientific & Technical Surveys*, No. 70.

Sood, G. and Heldman, D. (1974). *J. Food Sci.*, **39,** 1011.

Spencer-Gregory, H. and Rourke, E. (1957). *Hygrometry.* Crosby Lockwood & Son, London.

Spiess, W. E. L., Sole, P. C. and Pritzwald-Stegmann, B. F. (1969). *Deut. Lebensm. Rdsch.*, **65,** 115.

Stokes, R. H. (1947). *J. Am. Chem. Soc.*, **69,** 1291–2.

Stokes, R. H. and Robinson, R. A. (1949). *Ind. Eng. Chem.*, **41,** 2013–15.

Storey, R. M. and Stainsby, G. (1970). *J. Food Technol.*, **5,** 157.

Taylor, A. A. (1961). *Food Technol.*, **15** (12), 536.

Timmermans, J. (1960). *Physico-Chemical Constants of Binary Systems in Concentrated Solutions.* Interscience, New York.

Timofeev, D. P. (1960). *Ber. Acad. Sci. USSR*, **131** (6), 1390–2.

Vos, P. T. and Labuza, T. P. (1974). *J. Agr. Food Chem.*, **22** (2), 326–7.

Weaver, E. H. and Riley, R. (1948). *J. Res. Nat. Bur. Std.*, **40,** 169–214.

Wentzel, J. D. (1965). In: *Humidity and Moisture* (ed. A. Wexler), vol. III, p. 461. Reinhold, New York.

Wexler, A. (1965). *Humidity and Moisture.* Reinhold, New York.

Wexler, A. and Hasegawa, S. (1954). *J. Res. Nat. Bur. Std.*, **53,** 19.

Williams, C. J. (1969). *Am. Lab.* (June), 40.

Winston, W. and Bates, D. H. (1960). *Ecology*, **41** (1), 232–7.

Wolf, M., Spiess, W. E. L. and Jung, G. (1973). *Lebensm.-Wiss. u. Technol.*, **6,** 94.

Wolf, M., Walker, J. E. and Kapsalis, J. G. (1972). *J. Agr. Food Chem.*, **20,** 1073.

Wylie, R. G. (1965). In: *Humidity and Moisture* (ed. A. Wexler), vol. III, p. 507. Reinhold, New York.

Young, J. H. and Nelson, G. L. (1967). *Trans. Am. Soc. Agric. Engrs.*, **10,** 756.

11

Enumeration of Fungi and Assay for Ability to Degrade Structural and Storage Components of Grain

B. FLANNIGAN

Department of Brewing and Biological Sciences, Heriot-Watt University, Chambers Street, Edinburgh EH1 1HX, Scotland, UK

A. *ENUMERATION OF FUNGI IN GRAIN*

1. INTRODUCTION

The fungi which are found in and on cereal kernels may be classified for convenience as field fungi, which contaminate and colonise kernels before harvest, and storage fungi, some of which may be noted in a small percentage of kernels before harvest, but which are generally considered to contaminate kernels at or after harvest and to colonise them in storage. The field fungi include filamentous fungi of genera such as *Alternaria*, *Aureobasidium*, *Cladosporium*, *Epicoccum*, *Fusarium* and *Verticillium*, and yeasts such as *Cryptococcus*, *Rhodotorula* and *Sporobolomyces*. The storage fungi are predominantly members of two genera, *Aspergillus* (mainly *A. glaucus* and *A. restrictus* groups) and *Penicillium*, and are xerophilic, unlike the field fungi, which are mesophilic or hydrophilic. In commercial storage of grain dried to 12–14% moisture content there should be little or no development of fungi, but above 14% some storage fungi may begin to grow, replacing field fungi and generating heat and moisture which can initiate the development of a succession of microthermophilic and, ultimately, thermotolerant and thermophilic species. Similar successions of species have been observed in cereal bulks stored moist for animal feed (Clarke *et al.*, 1967, 1968), where *Aspergillus flavus*, *A. nidulans* and *A. terreus*, the yeast *Hansenula anomala*, and later *Absidia* spp., *A. fumigatus*, *Thermoascus crustaceus* and *Thermomyces lanuginosus* may succeed the storage fungi.

The most widely used technique for enumeration of fungi in cereals is the dilution plate method, but there are probably as many variations of this technique as there are laboratories using it. Outlined below is one procedure, currently used in this laboratory, in which steps have been taken to reduce variability in application to the minimum. Also described is the direct plating method for assessment of frequency of distribution of species on or in kernels.

2. DILUTION PLATE METHOD

2.1. Homogenisation of Sample

Weigh out 10 g of grain on sterile aluminium foil and add to 100 ml of sterile 0·12% Agar No. 3 (Oxoid), or other agar at one-tenth of the concentration recommended by the manufacturer, in the autoclaved vortex beaker of a MSE Homogeniser and homogenise for 10 min at 14 000 rpm. All but the largest fragments should remain in suspension.

Alternatively, where comparison is to be made at intervals, *e.g.* during an extended period in storage, a fixed number of kernels, say 100, should be homogenised. Since the weight of 100 kernels of barley, oats or wheat is usually 3·5–5·0 g, the volume of agar used should be reduced proportionally.

The shaft and blades of the homogeniser should be sterilised before and after use, either by autoclaving or by immersion in absolute ethanol for 5 min followed by burning off the ethanol.

2.2. Preparation of Dilutions

Using a sterile plugged wide-mouth 10 ml blow-out pipette with the tip held 1–2 cm below the surface, remove 10 ml of the primary suspension, touching the tip against the side of the vortex beaker as the charged pipette is withdrawn.

With the tip of the charged pipette against the neck of the bottle, blow out the contents into a 4 oz (123 ml) medical flat* containing 90 ml of sterile 0·12% agar, allow 6 sec for the residue of the viscous suspension to drain to the tip, then blow out again and discard the pipette.

Shake the bottle 50 times by hand in an up-and-down movement of about 30 cm, the procedure taking about 25 sec.

Prepare further dilutions in the same manner, using a fresh pipette each time.

* 'Medical flat' is a UK standard shape of medicine bottle.

2.3. Inoculation of Agar Medium

2.3.1. *Pour Plate Method*

Using a sterile plugged 1 ml blow-out pipette for each appropriate dilution, withdraw as above 1 ml into each of three sterile disposable non-vented 9·0 cm petri dishes. Holding the tip of the charged pipette about 0·5 cm above the base of the petri dish, blow out, allow 6 sec and blow out again with the tip against a fresh area of the base.

To each petri dish add 15 ml of molten agar medium at 45°C. Immediately mix inoculum and medium by moving the dish rapidly five times to and fro on the bench, then five times in a clockwise circle, then five times from side to side and finally five times in an anticlockwise circle. Allow the medium to gel and then transfer to an incubator.

Although the above procedure is suitable in most cases, it has been found that both the volume and temperature of the molten agar can affect counts of field fungi (Flannigan, 1973, 1974). Because of this, and because agar medium cools more slowly in plastic dishes than in the traditional glass petri dishes, it is advisable to take adequate precautions when samples are likely to contain temperature-sensitive filamentous field fungi and, especially, common field yeasts such as *Sporobolomyces roseus*. In such cases, the volume of medium poured should be reduced to 10 ml and its temperature reduced to 40°C before pouring.

In all cases, it is advisable to pour measured volumes from individual tubes, or McCartney or Universal bottles, rather than estimated volumes from a larger bottle, in order to reduce variability in counts. Medium should also be held at the required temperature ($\pm 1°$) in a thermostatically controlled water bath, the outside of each tube being wiped dry and the mouth flamed before the medium is poured. The temperature of the medium for pouring should never be judged by rule-of-thumb procedures or experience.

2.3.2. *Spread Plate Method*

In addition to the temperature-sensitivity of certain field yeasts, another problem encountered in enumerating yeasts is that in pour plates containing lower dilutions it may be difficult to distinguish between small, submerged, lenticular yeast colonies and coarse particles of grain. It is also virtually impossible to distinguish between submerged colonies of different types of yeasts. Consequently, the spread plate method is to be preferred not only where temperature-sensitive yeasts are likely to be present (Flannigan, 1973), but also where it is intended to enumerate different yeasts.

Agar plates should be opened and dried in a dry-air oven at 50°C for 20–30 min, depending on the medium and the amount of surface moisture, so that they will absorb liquid spread over the surface.

Deliver 0·15–0·20 ml of suspension as a known number of discrete drops from a sterile plugged 1 ml graduated (0·01 ml) pipette. The volume of the drop, and hence the volume delivered, is calculated from the mean of three trials in which ten drops of the suspension are discharged from the vertical pipette at the same rate as on to the plates. Use a fresh pipette for each dilution.

Disperse the inoculum over the surface of the agar medium, as the plate is rotated in the hand, using the shorter limb (4 cm) of a 16 cm glass rod (3 mm in diameter) bent to form a right-angle. Do not continue to rub the surface of the plate after the liquid has been absorbed into the medium as this will result in irregular counts, or the surface of the plate may be torn so that the plate has to be discarded. Store spreaders in absolute ethanol, and burn this off before use. Use a fresh spreader for each plate and discard into a 10% solution of hypochlorite before washing the spreader and returning it to ethanol, otherwise retained particles will be baked on to the glass when the ethanol is burned off and increase (a) the chances of tearing the surface of the plate and (b) irregularity in counts.

2.4. Media

Potato dextrose agar (Oxoid) is a good medium for the enumeration of most fungi, except xerophilic storage species, as colonies of many species are more easily identified on this medium than on, say, malt extract agar. To prevent bacterial growth the medium should contain 50 units/ml benzyl penicillin (Glaxo) and 100 units/ml streptomycin (Glaxo, streptomycin sulphate). It is most convenient to prepare a sterile concentrated solution containing 5000 and 10 000 units/ml, respectively, and add this to medium cooled to 45°C, just before pouring on to plates, at the rate of 1 ml per 100 ml medium.

Xerophilic storage fungi are enumerated on malt salt agar, the formula being:

Malt extract (dehydrated)	30 g
Mycological peptone (Oxoid)	5 g
Sodium chloride	100 g
Agar (Oxoid No. 3)	12 g
Distilled water	1000 ml

The medium should be autoclaved for 10 min at 10 lb/in^2 (70×10^3 N/m^2) at 115°C. Although the high osmotic pressure and low pH (5·4) of the medium ensure that bacteria seldom grow, the antibiotics noted above may be added routinely to the medium. Certain field fungi such as *Alternaria* and *Cladosporium* grow on this medium.

Occasionally, potato dextrose agar plates may be unreadable because of rapidly growing *Rhizopus* spp., the mycelia of which may spread over the plates and inhibit growth of other species. In such cases, Littman oxgall agar (Difco) can be used for enumeration, either with only streptomycin added according to the manufacturer's instructions, or with the antibiotic mixture used in potato dextrose agar. Growth is slow on this medium and the colonies of filamentous fungi are quite different in appearance from those on potato dextrose agar. Consequently, most can only be identified after subculturing on to the latter medium.

2.5. Incubation of Plates

Incubate malt salt agar plates at 25°C and potato dextrose agar plates at 25°, 37° (for microthermophilic and thermotolerant species) and 50°C (for thermotolerant and thermophilic species). Plates should be incubated base upwards in stacks of no more than five, and at 50°C no more than two. Plates at 25°C should be removed after 7 days, and at 37° and 50°C after 4 days.

2.6. Choice of Plates for Counting

Various workers, including Jensen (1962), have drawn attention to the fact that counts of fungi in successive dilutions do not decrease in the proportions expected, *e.g.* there may be only a four- or fivefold reduction instead of the expected tenfold reduction. This discrepancy has been attributed to reduced competition, cumulative volumetric errors and increased dispersion due to additional shaking involved in preparing higher dilutions. There is no apparent solution, but steps should be taken to minimise volumetric errors, *e.g.* by sealing dilution blanks before autoclaving. The main problem arising from this relative increase with dilution is the choice of dilution to be counted. For bacterial counts, the dilution giving a total of 30–300 colonies per plate is normally recommended, and this can reasonably be recommended if only a total for yeasts is required. As far as filamentous fungi are concerned, the choice is complicated by the range of fungi encountered. James and Sutherland (1939) counted plates with 5–50 colonies, and as far as is possible the author counts plates with 20–50 colonies, but it is possible reasonably to count up to 100 slow-growing colonies on malt salt agar or Littman oxgall agar. Jensen (1962) considers that results are not comparable unless the same degree of dilution is used throughout a study. Whilst this appears to be advisable, difficulties are experienced where numbers are changing with time in storage, or where there are large differences between the numbers of colonies of different species so that they must be counted in different dilutions. Counts can only, therefore, be taken as giving relative numbers of viable units, and, where numbers are changing, as indicating only broad trends.

2.7. Counting of Colonies

Plates should be examined under a magnifying lens (2 ×) in incident light, and also on a translucent surface illuminated from behind so that small yeast colonies masked by fungal mycelium can be seen. Where differential counts are to be made, a dissecting microscope with a magnification of 20–30 × is useful when it is necessary to identify filamentous fungi by spore type and arrangement rather than by the gross appearance of the colonies.

To calculate the number of viable units per gramme of grain, multiply the mean of the counts on triplicate plates by the reciprocal of the dilution from which pour plates were prepared or by the reciprocal of the volumes of dilution spread multiplied by the reciprocal of the dilution from which spread plates were prepared, the primary homogenate being taken as a 10^{-1} dilution.

Where a specific number of kernels have been homogenised, to calculate the number of viable units per kernel, multiply the mean in pour plates by

$$\frac{\text{Volume of primary homogenate}}{\text{Number of kernels homogenised}} \times \frac{1}{\begin{array}{c}\text{Dilution of primary homogenate}\\ \text{from which plates prepared}\end{array}}$$

and in spread plates by

$$\frac{\text{Volume of primary homogenate}}{\text{Numbers of kernels homogenised}} \times \frac{1}{\text{Volume spread}} \times \frac{1}{\begin{array}{c}\text{Dilution of primary homogenate}\\ \text{from which plates prepared}\end{array}}$$

Where differential counts are made and some fungi appear on both media, *e.g.* *Alternaria alternata* and *Cladosporium* spp., or at two temperatures, *e.g.* *Aspergillus fumigatus* at 37° and 50°C, the higher count should be accepted. For example, although *Cladosporium* spp. are field fungi, they may appear in larger numbers on malt salt agar, showing some tolerance of high osmotic pressure. In contrast, although *A. alternata* appears on both media, its colonial morphology is quite different on malt salt agar from that on potato dextrose agar and its numbers fewer.

3. DIRECT PLATING METHOD

This method is used by plant pathologists for detection of seed-borne pathogens. The seed is generally pre-treated to eliminate surface-borne saprophytic fungi which might otherwise mask the emergence of deeper-seated pathogens on to the agar surface. It is, however, possible to acquire a better picture of the mycoflora of kernels by plating out both untreated and pre-treated grain.

3.1. Pre-treatment

3.1.1. *Surface Disinfection with Hypochlorite*

Immerse 100–200 kernels from a sample in approximately 200 ml of a sodium hypochlorite dairy disinfectant such as Chloros (ICI), diluted 1 in 10 with water to give a solution containing 1% (w/w) available chlorine. Stir vigorously to disperse air bubbles from the surfaces of the kernels and to ensure that all surfaces are wetted. If the indicator incorporated in the disinfectant changes colour, the available chlorine has been exhausted and the solution should be decanted and replaced by fresh disinfectant. Steep the kernels with occasional stirring for 10 min, decant the disinfectant, rinse in 200 ml of sterile water, and finally decant this water before plating on to agar.

3.1.2. *Exposure to Dry Heat*

Following the method of Malone (1962) for detection of *Pyrenophora avenae* in oats, place an open sterile 9·0 cm glass petri dish containing 100–200 kernels in a dry-air oven at 100°C. After 1 hr, remove, partially cover the dish with a sterile lid and allow to cool before plating out the kernels.

Most common field fungi are inactivated by this treatment, but colonies of *Mucor* and *Chaetomium* and the yeast *Sporobolomyces roseus* are more easily counted in the absence of competing species.

3.2. Plating Out

Plate out untreated or heat-treated kernels using fine forceps sterilised between transfers by dipping in ethanol and burning off. Place eight kernels equidistantly round the circumference of each of ten 9·0 cm agar plates and two kernels towards the centre to give the maximum spacing between kernels. Surface-disinfected kernels should be plated out in the same way, but forceps need only be sterilised after the inoculation of each plate.

3.3. Media

Malt salt agar and potato dextrose agar containing antibiotics should be prepared as previously, and approximately 15 ml poured into each petri dish. It is not necessary to dry plates at 50°C, unless there is excessive moisture on the agar surface.

3.4. Incubation

Incubate petri dishes, lids uppermost, in stacks of five (or two at 50°C) according to the following schedule:

Pre-treatment	*Medium*	*Incubation*	
None	Potato dextrose agar	25°C	7 days
Surface disinfection	Potato dextrose agar	25°C	7 days
Dry heat	Potato dextrose agar	25°C	7 days
None	Potato dextrose agar	37°C	4 days
Surface disinfection	Potato dextrose agar	37°C	4 days
None	Potato dextrose agar	50°C	4 days
None	Malt salt agar	25°C	7 days
Surface disinfection	Malt salt agar	25°C	7 days

3.5. Examination of Plates

Plates should be examined with the naked eye for filamentous fungi which are recognised according to their colonial morphology, and also with the aid of a dissecting microscope at a magnification of 20–30 × for slow-growing species which may be masked by other mycelium, and for sporing structures produced on, and often restricted to, the kernels, *e.g.* those of *Cladosporium* spp. The reverse side of plates should also be examined, since small colonies adjacent to the kernels can often be detected in this way, *e.g.* those of the pink yeasts, such as *Sporobolomyces*, on potato dextrose agar, and the dark colonies of *Cladosporium* spp. which are frequently masked by *Aspergillus* spp. on malt salt agar.

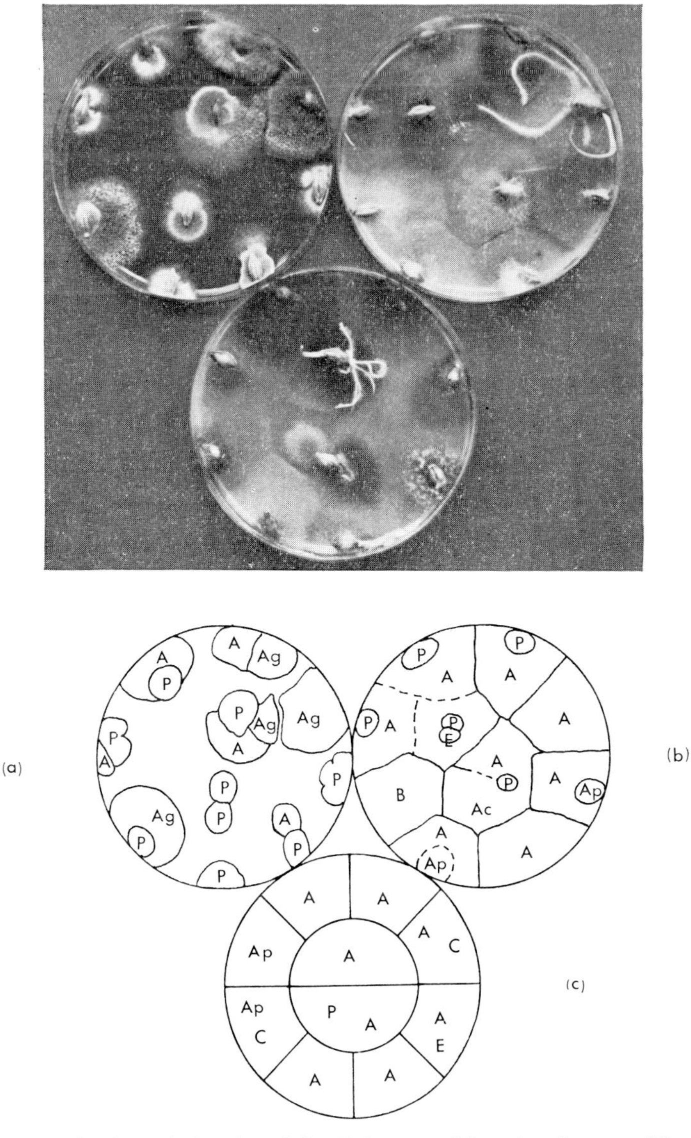

FIG. 1. Barley kernels incubated for 7 days on (a) malt salt agar, (b) potato dextrose agar, and (c) after surface disinfection of kernels, on potato dextrose agar.

Key: A, *Alternaria alternata*; Ac, *Absidia corymbifera*; Ag, *Aspergillus glaucus*; Ap, *Aureobasidium pullulans*; B, *Botrytis cinerea*; C, *Cladosporium* sp.; E, *Epicoccum nigrum*; P, *Penicillium* sp.

Plate (c) is recorded as on a duplicated chart for recording the fungi on ten plates.

Results may be recorded merely by scoring against a species list, or by inserting a specific code in the appropriate compartment on a duplicated sheet (*see* Fig. 1). Comparison of plates bearing untreated and surface-disinfected kernels should give some indication of the relative frequency of distribution of deeper-seated and superficial species.

B. *ASSAY OF FUNGI FOR ABILITY TO DEGRADE STRUCTURAL AND STORAGE COMPONENTS OF CEREALS*

4. INTRODUCTION

Although the principal component of the cereal kernel is the storage polysaccharide starch, which may constitute 50–80% of the total dry weight, cellulose and hemicelluloses are also present as cell-wall components. The amount of cellulose present is highest in hulled cereals, barley, oats and rice, since cellulose, along with lignin and hemicelluloses, is a major component of the cell walls in the hull, as well as being present in the pericarp-testa investing all cereal kernels and to a lesser extent in the embryo. The hemicelluloses vary in composition but are of two generic types, arabinoxylans and mixed-linkage (1,3; 1,4) β-glucans. In barley hull, for example, the main hemicellulose is arabinoxylan (Aspinall and Ferrier, 1957), which is composed of long chains of 1,4-linked β-D-xylopyranose units to which are attached side-chains of three types: single arabinose and 4-*O*-methyl glucuronic acid residues, and xylose-terminated chains linked to the long-chain 'backbone' by arabinose. Similarly, arabinoxylan is present in the outer layers of the wheat kernel (Pomeranz, 1971), as well as in the walls of the starch endosperm of both barley and wheat. However, in wheat endosperm, arabinoxylan is the main cell-wall component, whilst in barley it is present in smaller amounts than water-soluble 1,3; 1,4-β-glucan (Cook, 1962). It is to be expected that fungi actively involved in the degradation of cereal kernels will produce enzymes active against these types of components, and against the lipid and protein of the kernels. Some of these enzymes, such as amylases and 1,3-β-glucanases, are constitutive in many fungi, but others, such as cellulases and xylanases, are adaptive enzymes. It is therefore advisable to grow all fungi on the substrates of interest, whether or not the relevant enzymes are thought to be constitutive.

5. TECHNIQUE

5.1. Growth of Fungi

Fungi can be grown conveniently in 100 ml of liquid medium in 250 ml conical flasks incubated at 25°C (mesophiles) or 37°C (thermotolerant and thermophilic species) in a reciprocal shaker (75 strokes/min) or an orbital incubator (150 cycles/min) for 7 days (14 days if cellulose is the sole carbon source).

5.2. Media

The basic composition of medium, except for the production of proteases, is:

$NaNO_3$	2·0 g
KH_2PO_4	1·0 g
KCl	0·5 g
$MgSO_4 \cdot 7H_2O$	0·5 g
$FeSO_4 \cdot 7H_2O$	0·01 g
Yeast extract (dehydrated)	0·05 g
Carbon source	10·0 g
Distilled water	1000 ml

The carbon source for the production of amylases is soluble starch; for cellulases, crystalline cellulose (Whatman Chromedia CC41, or Koch-Light Avicel SF); and for lipases, corn oil or wheat germ oil. For production of enzymes active against arabinoxylans, barley-hull arabinoxylan prepared by the method of Aspinall and Ferrier (1957) or wheat straw arabinoxylan (Aspinall and Mahomed, 1954) can be used, and for those active against 1,3; 1,4-β-glucans, barley β-glucan (Preece and MacKenzie, 1952). In these cases, where the carbon sources are not available commercially, the concentration in the medium can satisfactorily be reduced to one-tenth of that above to economise. For protease production, the carbon source is glucose (10 g) and casein (2 g) is substituted for $NaNO_3$. All media are autoclaved for 15 min at 15 lb/in^2 (105×10^3 N/m^2) at 121°C.

5.3. Inoculation

With sporing fungi, 10 ml of sterile 0·01% Tween 80 should be poured on to an agar-plate culture and a sterile glass spreader used to detach spores, which should then be filtered through sterile organdie muslin into a sterile container. From this a suspension of 10^6 spores/ml should be prepared and flasks inoculated with 1 ml of this suspension.

Where fungi do not sporulate, two 4 mm plugs cut from the margin of an agar-plate culture may be used as inoculum.

5.4. Preparation of Enzyme Solutions

After the required incubation period, the contents of the flasks should be filtered through sintered glass (porosity 3) and the culture filtrates assayed directly for activity or dialysed overnight in running tap water. They may then be cooled to 0°C and the enzymes precipitated by addition of two volumes of chilled ethanol (−20°C), followed by standing for 4 hr and refrigerated centrifugation at 3000G. After redissolving in distilled water, the enzymes should be freeze-dried.

5.5. Activity Against Polysaccharides

5.5.1. *Release of Reducing Sugars*

Reaction mixtures consisting of 1·0 ml of culture filtrate or an aqueous solution of freeze-dried material, 5·0 ml of a 0·2% solution of substrate and 5·0 ml 0·1 M citrate buffer (pH 5·4) are incubated at 37°C for 3 hr, along with controls containing enzymes inactivated in a boiling water bath. The reaction is stopped by immersion in boiling water (15 min), following which duplicate 1·0 ml samples are withdrawn and reducing sugars are estimated by the Somogyi–Nelson colorimetric method (Whistler and Wolfrom, 1962), standard calibration graphs being prepared using glucose (where starch, carboxymethyl-cellulose or β-glucan is the substrate) or xylose (where arabinoxylan is the substrate) in 0·05 M citrate buffer (pH 5·4). The substrates are the same as those used for the growth of fungi, except that sodium carboxymethyl-cellulose (d.s. 0·5–0·8) is used instead of crystalline cellulose.

5.5.2. *Chromatographic Examination of Enzyme Digests*

The residue of the reaction mixture is de-ionised by the addition of granules of mixed-bed resin, filtered and concentrated by rotary evaporation. It is then spotted, with standards, on paper chromatograms which are subsequently developed by descending chromatography for 18 hr in ethyl acetate:pyridine:water (10:4:3 v/v). Reducing sugars are then detected by the silver nitrate dip method of Trevelyan *et al.* (1950). Although the method is qualitative, some indication of the relative proportion of oligosaccharides to monosaccharides is gained, and in reaction mixtures containing arabinoxylan the release of only arabinose indicates the possible presence of arabinofuranosidases, whilst the release of xylose and/or xylo-oligosaccharides indicates xylanases.

5.6. Proteolytic Activity

Reaction mixtures containing 1·0 ml of culture filtrate or enzyme solution and 4·0 ml of a 0·3% solution of casein (Hammarsten) in 0·1 M acetate buffer (pH 5·4) are incubated for 1 hr at 37°C. To stop the reaction and precipitate unreacted protein, 5·0 ml of 5·0% trichloroacetic acid (TCA) is added. After 15 min the mixture is centrifuged and duplicate 1·0 ml aliquots of the supernatant assayed for TCA-soluble material by the modified Lowry method (Miller, 1959). Reaction mixtures containing inactivated enzymes serve as controls. Results are calculated from a standard curve prepared using bovine serum albumin (Fraction V, 0–100 μg/ml).

5.7. Lipolytic Activity

Following the method of Somkuti *et al.* (1969), the substrate is prepared by homogenising a mixture containing 10·0 g corn oil, 4·0 g sodium taurocholate, 10 ml 0·1 M calcium chloride and 50 ml 0·1 M acetate buffer (pH 5·4) for 15 min. Reaction mixtures containing 5 ml of this mixture and 1 ml of culture filtrate or enzyme solution are incubated in 50 ml conical flasks for 2 hr at 37°C in a reciprocal shaking water bath (120 strokes/min). The reaction is stopped by the addition of 1 ml of 1 N sulphuric acid and the whole is transferred to a 100 ml separating funnel. The residue in the reaction flask is washed into the funnel with 25 ml of petroleum ether which is used to extract the liberated fatty acids. This is repeated with fresh petroleum ether, the upper (solvent) phases being combined and titrated against 0·01 N ethanolic potassium hydroxide, using phenolphthalein as indicator. Reaction mixtures containing inactivated culture filtrate or enzyme solution are used as controls. The amount of fatty acids liberated (in micromoles) is $10V$, where V is the volume of alkali.

C. *IDENTIFICATION*

Since the range of fungi encountered in grain is relatively wide, it is not possible in this text to include a complete list of publications which would be necessary for identification of all species which can be found. However, some publications which are of use to both the non-specialist and the specialist are listed below. Those marked with an asterisk are particularly helpful in that they include numerous photographs, rather than drawings, of diagnostic features:

Ainsworth, G. C., Sparrow, F. K. and Sussman, A. S. (eds.) (1973). *The Fungi: An Advanced Treatise*, vol. IV A: *Ascomycetes and Fungi Imperfecti;* vol. IV B: *Basidiomycetes and Lower Fungi*. Academic Press, London.

Barnett, H. L. and Hunter, B. B. (1972). *Illustrated Genera of Imperfect Fungi*. Burgess, Minneapolis.

*Barron, G. L. (1968). *The Genera of Hyphomycetes from Soil*. Williams & Wilkins, Baltimore.

Cooney, D. G. and Emerson, R. (1964). *Thermophilic Fungi*. Freeman, San Francisco.

Ellis, M. B. (1971). *Dematiaceous Hyphomycetes*. Commonwealth Mycological Institute, Kew.

*Malone, J. P. and Muskett, A. E. (1964). *Seed-borne Fungi: Description of 77 Fungus Species*. Handbook on Seed Health Testing, Series 4(1). International Seed Testing Association, Wageningen, Netherlands. This is a reprint of *Proc. Int. Seed Testing Assoc.*, **29** (1964), 179–384.

*Smith, G. (1969). *An Introduction to Industrial Mycology*, 6th ed. E. J. Arnold, London.

Most of the above publications list monographs and papers on particular genera and other lists are to be found elsewhere, *e.g.* in the *Guide to the Literature for the Identification of British Fungi*, 3rd ed., prepared by Miss M. Holden and reprinted from the *Bulletin of the British Mycological Society*, **9** (2) (1975), 67–106. It should not be construed that the list is restricted to publications dealing exclusively with species found in the UK, since American and European publications are included.

REFERENCES

Aspinall, G. O. and Ferrier, R. J. (1957). The constitution of barley husk hemicellulose. *J. Chem. Soc.*, 4188–94.

Aspinall, G. O. and Mahomed, R. S. (1954). The constitution of wheat straw xylan. *J. Chem., Soc.*, 1731–4.

Clarke, J. H., Hill, S. T., Niles, E. V., Williams, R. O. and Howard, M. A. R. (1968). Ecology of the microflora of moist barley. *Pest Infestation Res.*, 1967, 15–16.

Clarke, J. H. Niles, E. V. and Hill, S. T. (1967). Ecology of the microflora of moist barley. *Pest Infestation Res.*, 1966, 14–16.
Cook, A. H. (1962). *Barley and Malt*. Academic Press, New York and London, 740 pp.
Flannigan, B. (1973). An evaluation of dilution plate methods of enumerating fungi. *Lab. Pract.*, **22,** 530–1.
Flannigan, B. (1974). The use of pour plates for yeast counts. *J. Stored Prod. Res.*, **10,** 61–4.
James, N. and Sutherland, M. L. (1939). The accuracy of the plating method for estimating the numbers of soil bacteria, actinomycetes, and fungi in the dilution plated. *Can. J. Res.*, Part C, **17,** 72–86.
Jensen, V. (1962). Studies on the microflora of Danish beech forest soils. I. The dilution plate count technique for the enumeration of bacteria and fungi in soil. *Zentr. Bakteriol. Parasitenk.*, Abt. II, **116,** 13–32.
Malone, J. P. (1962). The application of heat to seed oats as an aid in the detection of *Pyrenophora avenae* by the Ulster method. *Proc. Int. Seed Testing Assoc.*, **27,** 856–61.
Miller, G. L. (1959). Protein determination for large numbers of samples. *Anal. Chem.*, **31,** 964.
Pomeranz, Y. (1971). *Wheat, Chemistry and Technology*, 2nd ed. American Association of Cereal Chemists Inc., St Paul, Minn., 821 pp.
Preece, I. A. and MacKenzie, K. G. (1952). Non-starchy polysaccharides of cereal gums. I. Fractionation of the barley gums. *J. Inst. Brewing*, **58,** 353–62.
Somkuti, G. A., Babel, F. J. and Somkuti, A. C. (1969). Lipase of *Mucor pusillus*. *Appl. Microbiol.*, **17,** 606–10.
Trevelyan, W. E., Procter, D. P. and Harrison, J. S. (1950). Detection of sugars on paper chromatograms. *Nature* (*London*), **166,** 444–5.
Whistler, R. L. and Wolfrom, M. L. (1962). *Methods in Carbohydrate Chemistry*, vol. I. Academic Press, New York and London, 589 pp.

12

Techniques Available for Microbiological Investigation of Grassland Vegetation and Produce

A. E. APINIS

38 *Groesffordd, Bryn Crug, Tywyn, Gwynedd, Wales, UK*

1. INTRODUCTION

Results of microbiological investigation of grassland vegetation and its crops constitute an important aspect in our understanding of numerous natural phenomena in the field. Furthermore, harvested feeds and foods are also affected by microflora in storage, food production and distribution under various environmental conditions. Such activities of Actinomycetes, bacteria and fungi in the field, in stores and in production are invariably accompanied by certain animal populations. However, only those considerations relevant to microorganisms will be given below.

In this short chapter it is not possible to describe in detail all the various field and laboratory techniques employed so far, including those which may be useful in future investigations; it is intended only to outline a few general procedures and to quote selectively relevant references.

2. PRELIMINARY OBSERVATIONS

Interpretation of laboratory results frequently depends to a large extent upon observations and tests made *in situ*. In certain cases this involves description of the site and its vegetation where the produce is grown [25, 48, 59, 69, 80, 86, 93, 129, 143, 162]. In applied and fundamental microbiology there is at present a distinct trend towards approaching investigations from the total organism population (biocoenose) aspects and the respective environment (ecosystem) as related to any given problem.

Field tests [21, 27, 31, 32, 37, 60–62, 79, 80, 90, 91, 146, 161] and the preliminary microscopic examination [55, 71, 123, 125, 139, 146] of sampled material may be helpful in setting up specific laboratory experiments, thus saving time as well as laboratory materials needed for such examinations.

3. FIELD TECHNIQUES FOR MICROBIOLOGICAL EXAMINATION

Useful techniques for field examination were developed at an early stage in soil microbiology by Cholodny [32], Kubiena [124], Rossi [27] and others. Chesters [31] and Thornton [154] added new techniques which enabled fungi and other microorganisms actively growing from soil to be isolated via immersion tubes, and also by placing screened plates in soil. Waid and Woodman [161] introduced the nylon gauze technique, and Lowrey [100] reported the nylon bag technique for estimation of forage quality. So far, few or no attempts have been made to adapt such field techniques to microbiological examination of agricultural produce such as hay, silage, straw and other feeds, apart from the techniques developed for trapping air spora [7, 60, 105] and a number of techniques employed by plant pathologists [57a, 69, 80, 82, 85, 94, 122a, 123, 162].

4. GENERAL QUANTITATIVE TECHNIQUES

For quantitative estimates of taxonomic and ecological groups of microorganisms present in the material, various techniques may be employed (*see* Table I):

(a) Total suspension of disintegrated material examined by microscopy.
(b) Suspensions plated out on water agar or any other suboptimal medium, such as soil extract agar or hay extract agar, and counts carried out after incubation at a certain temperature. The latter incubated material may also be used for isolations.
(c) Washings obtained from material shaken with sterile water may be similarly examined by direct microscopy and plating out of dilutions on suboptimal media. The washed substrate (solids of the suspension) may be further homogenised or plated on suboptimal

media or placed in a moist chamber. The moist chamber incubation of untreated or washed material for a shorter or longer period is very useful, and could be extended by adopting differential temperature and humidity treatments which may easily be combined with quantitative procedures.

TABLE I

A SCHEME FOR MICROBIOLOGICAL EXAMINATION OF PLANT MATERIAL

Preparation for investigation

↓

Description of material and site

↓

Field tests and sampling

↓

Transport and storage of samples

↓

Botanical and microscopic examination of material

↓

Determination of chemical and physical parameters

↓

Selection of three equal lots for microbiological laboratory examination

↓	↓	↓
(a) Grinding or homogenisation: Microscopic examination of suspension	(b) Washing: Microscopic examination of washings	(c) Surface sterilisation: Incubation of surface-sterilised and washed material in moist chambers at defined humidity levels and at certain temperatures
↓	↓	
Dilutions plated with general and selective agar media	Dilutions plated with agar media	

↓

Incubation at 3–5°, 10–12°, 18–20°, 30°, 38°, 45° and 55°C

↓

Counting, isolation and identification

↓

Slides, specimen and subcultures

↓

Evaluation of results

In the preparation of dilutions from homogenised suspensions the dipper technique may be very useful [107]. Direct microscopic counts may be carried out as described by Conn [37], or according to Jones and Mollison [81], Strugger [146], or by other techniques [5, 66, 80, 124, 137]. Addition of dispersing compounds (sodium pyrophosphate), shaking for a specified period, or applying sonic vibration [144] is often essential [124]. For counts from dilution plates in agar media the use of optical counting lenses or a binocular microscope with magnification ×20 or even ×50 may be helpful in making counts of very small colonies densely crowded in water agar or soil extract agar. At the same time it is possible to evaluate whether colonies have originated from mycelium fragments or spores.

In addition to homogenates and suspensions obtained from shaken and washed material for quantitative estimates of microorganisms, a suitably modified percolation technique may also be used.

5. MICROBIOLOGICAL MEDIA

In microbiological investigation of plant material, the use of enriched media for quantitative and isolation purposes is to be avoided as far as possible. Water agar, silica gel, soil extract agar, dilute extracts from respective plant material with agar (or without) and similar media of low nutritional value are preferable. This helps to establish selective microbial preference for the substrates provided while discouraging the growth of other microorganisms present. This principle is also adopted for material incubated in moist chambers or in water. Overcrowding of substrate surfaces is thus reduced and a greater variety of species can find space for development which enables better direct examination, identification and isolation, especially of Actinomycetes and fungi [8–10, 13, 15, 73, 74, 84, 102, 123, 156, 165, 166, 169].

Use of enriched natural or synthetic media should be confined to enrichment culture and for differential diagnostic media to isolate certain groups of microorganisms, such as certain Actinomycetes [38, 80, 88, 89, 97, 103, 124, 137, 140, 167], certain bacteria [4–6, 47, 56, 66, 75, 76, 80, 88, 92, 97, 103, 124, 125, 137, 140, 152] and fungi [24, 43, 62, 80, 97, 99, 104, 112, 123, 125, 131, 137]. Additions of antibiotics (actidione 50 μg/ml, chloramphenical 0·2 mg/ml, nystatin 50 μg/ml, penicillin 5–50 μg/ml, streptomycin 20–30 μg/ml, tetracycline hydrochloride 2–3 μg/ml), organic or inorganic acids (hydrochloric acid, lactic acid, Rose Bengal, etc.), salts (bile salt, potassium tellurite, sodium azide, etc.), stains [66, 103, 124], as well as fungicides, pesticides and other biocides are widely used as selective inhibitors incorporated at certain concentrations in liquid or semi-liquid media.

6. IDENTIFICATION

Correct taxonomic identity of any microorganism is of primary importance in testing work. The large number of genera and species of Actinomycetes, bacteria and fungi [1, 3, 20, 68, 88, 123, 140] makes this task difficult and specialist assistance is frequently required. Identification down to species level may be adequate, but often precise identification of a strain may be necessary.

In routine work, however, general or provisional taxonomic identification up to taxonomic group or generic level may be adequate for microorganisms not directly involved in the particular biodeteriorative process under investigation. In such cases, the use of accepted selective media is advisable to exclude the less important microorganisms which may be present in large numbers and make isolation and identification of any particular species difficult. Knowledge of the general composition of microflora on any particular substrate may be helpful in identification of organisms isolated in culture. For this purpose, the references in Table II may be of assistance.

TABLE II

REFERENCES TO CERTAIN PLANT MATERIALS INVESTIGATED

A. Living grasses and herbs:
   (a) Leaf surface [11, 14, 18, 19, 36, 41–44, 68, 82, 85, 94, 95, 106, 118, 123, 125, 133, 139, 145, 152, 159, 160, 168].
   (b) Colonisers of plant tissue [8–11, 13, 14, 19, 36, 40, 47, 55, 57a, 68, 82, 85, 123, 139, 142, 145, 152, 153, 155, 163–166, 168].
   (c) Rhizosphere and root colonisers [11, 15a, 19, 26, 36, 40, 65, 80–82, 85, 104, 107, 110, 119, 123, 124, 143, 157, 162].
B. Dead plants [8–11, 13–15, 19, 36, 40, 54, 63, 64, 69, 79, 98, 108, 109, 123, 127, 128, 139, 141, 142, 155, 163, 165, 166, 169].
C. Hay and straw [10, 11, 13, 15, 19, 38, 49, 52, 54, 61–63, 67, 74, 76, 79, 96, 100, 101, 108, 110, 123, 135, 141, 152, 161, 165, 166].
D. Composts and litter [9–11, 13, 15, 17, 19, 30, 36, 54, 59, 73, 74, 80, 81, 108, 110, 124, 155, 158, 163–166, 169].
E. Grain and seeds [4, 12, 29, 52, 53, 55–57, 67, 72, 102, 112, 116, 117, 126, 132, 135–137, 151].
F. Flour [4–6, 21, 29, 33, 56, 58, 66, 72, 102, 112, 116, 117, 126, 132, 135–137, 151].
G. Bread and other dough foods [5, 29, 56, 58, 66, 67, 75, 103, 126, 137, 138, 149, 151].
H. Various feeds [12, 17, 29, 31, 57, 61, 67, 98, 100, 103, 108, 112, 114, 126, 141, 152, 153, 161].

TABLE III

SOURCES OF PROCEDURES AND TECHNIQUES ADOPTED FOR MICROBIOLOGICAL EXAMINATION

A. Field techniques [7, 25, 27, 31, 48, 60–62, 80, 82, 86, 90, 93, 100, 101, 105, 122a, 123, 129, 143, 150, 154, 161].

B. Microscopic tests:

(a) Light microscopy [4, 5, 8, 9, 12–15, 18, 29, 37, 55, 56, 59, 61, 62, 66, 68, 73, 80, 82, 103, 123–125, 137, 139, 150, 162].
(b) Fluorescent microscopy [68, 80, 103, 123, 124, 146, 157].
(c) Electron microscopy [68, 78, 98, 119].

C. Quantitative estimates:

(a) Numbers from dilutions or suspensions [4, 5, 26, 29, 55, 66, 80, 81, 103, 105, 107, 109, 124, 143, 146].
(b) Plate count [4–6, 8, 26, 29, 33, 36, 58, 66, 77, 80, 82, 85, 103, 104, 110, 123–125, 137, 145, 149, 168].
(c) Most probable number technique [66, 103, 148].
(d) Air spora [7, 38, 60–62, 90, 94, 95, 105, 123, 125].
(e) Other quantitative techniques [31, 32, 37, 43, 44, 65, 66, 80–82, 85, 94–96, 100, 103, 116, 123–125, 137, 154].

D. Chemical and biochemical examination [29, 45, 49, 63, 67, 80, 82, 85, 120, 122, 124, 135, 158].

E. Microorganisms grown on incubated natural substrates on semi-liquid (agar) or in liquid media [4, 5, 23, 29, 80, 82, 85, 103, 112, 116, 123, 124, 127, 128, 137].

F. Microorganisms grown on natural substrates in moist chambers [8, 9, 12–15, 73, 80, 82, 84, 85, 102, 116, 123, 165, 166, 169].

G. Isolation in pure culture [4, 5, 23, 29, 31, 66, 69, 80, 82, 85, 92, 103, 123, 137, 167].

H. Identification [1, 3, 5, 16, 20, 22, 23, 26, 28, 37, 46, 50, 51, 68, 69, 83, 87–89, 99, 103, 113, 115, 123, 130, 131, 140, 147, 162, 167, 170].

I. Ecological resp. physiological groups:

(a) Animal pathogens [2, 8, 10, 12, 20, 38, 62, 67, 90, 91, 103, 138, 140].
(b) Plant pathogens [1, 11, 12, 24, 46, 57a, 69, 80, 82, 85, 90, 102, 117, 122a, 123, 139, 142, 162].
(c) Toxin producers [12, 34, 35, 49, 67, 79, 91, 123, 134–136].
(d) Thermophiles [1, 8, 10, 12, 13, 30, 38, 53, 54, 108, 110, 132, 164].
(e) Mesophiles and psychrophiles [4, 5, 10, 12, 29, 47, 82, 103, 121, 123, 137, 138].
(f) Moisture requirements, water stress [4, 5, 10, 12, 15, 56, 66, 70, 74, 103, 137, 141, 143, 149].
(g) Oxygen requirements [4, 5, 29, 66, 103, 137, 140].
(h) Specific substrates or nutritional requirements [4–6, 29, 66, 68, 75, 85, 103, 124, 126, 137, 140, 149, 150].

7. ECOLOGICAL OR PHYSIOLOGICAL GROUPS

Lists of species colonising various plant materials when grown on organic substrates do not show the characteristic dynamic features of microorganisms when active in a particular environment; therefore any species complex must be further analysed according to environment relationships and biochemical or physiological abilities. Often, as in various groups of bacteria, their biochemical and environmental potentials coincide with certain species characters, as for example in relation to water and oxygen. Other ecological features may not be immediately obvious, or such features may show certain narrower or wider ecological amplitude.

Because of this, an attempt has been made to indicate the value of such ecological groups (*see* Table III) in the interpretation of microbiological data. Two major classes of such groups may be recognised:

(a) The host–substrate relationship as in animal and plant pathogens, toxin-producing microorganisms, and the various nutritional substrate relationships due to the biochemical potential of particular organisms.

(b) Environmental relationships as characterised by temperature and water relations and relations to oxygen. The relevant references given in Table III are indicative, but further inquiries may be needed in some cases.

8. EXAMINATION OF STANDING CROP

Microbiological examination of standing crop or herb and grass vegetation in fields, meadows and pastures is complex. It involves considerations of the whole biocoenose of a particular ecosystem [10, 11, 69]. Because of this, particular attention must be paid to accurate description of the type of vegetation in which specific investigation is carried out [25, 48, 69, 80, 86, 93, 129, 143, 162]. Several fields of investigation have been developed using specific techniques in regard to microorganisms present as colonisers of host plants on or within certain organs (roots, culms or stems, leaves, inflorescences) at various stages of development, senescence and decay, such as rhizosphere and root colonisers [3, 11, 15a, 26, 36, 119, 124, 157], phyllosphere and leaf colonisers [14, 18, 41–44, 57, 95, 106, 109, 118, 127, 133, 145, 159, 168], colonisers of culms and stems [9, 12, 13, 40, 54, 73, 74, 128, 132, 152, 153, 155, 159, 165, 166, 169], pathogens [2, 46, 57a, 81, 85,

122a, 123, 142, 162], and decomposers and toxin producers [11, 13–15, 19, 30, 34, 49, 54, 57, 59, 63, 76, 79, 127, 134, 135, 155, 158].

In the preparation of plant material the total sample is usually divided in the field into certain larger portions (inflorescence, leaves, leaf sheath of upper, lower and medium part of the culm or stem, nodes and internodes, rhizomes, roots) and each portion is placed into a sterile polythene bag. During washing with sterile water, surface sterilisation or homogenisation, such portions are treated separately because they represent different substrates with different populations of microorganisms; some parts may be further examined directly under the microscope or plated out on agar media, in moist chambers or in sterile water. Usually culms, stems, leaves, rhizomes and roots are cut into 3–5 cm or 2–3 mm pieces for ease of handling and recording of organisms growing upon them. So far a considerable number of field and laboratory techniques have been used to reveal the presence of such organisms or to demonstrate their probable activities upon or within certain grass or herb species.

The most widely used laboratory technique for quantitative, qualitative and isolation purposes has been the *dilution plate* technique, of which two basic modifications are in use:

(a) That of *washing* after shaking a certain amount of plant material in a certain volume of sterile water from which dilutions are prepared for direct observation and plating out on non-selective or selective (diagnostic) media for quantitative and qualitative observations [4, 5, 26, 29, 33, 44, 56, 65, 66, 94, 95, 103, 109, 111, 123–125, 133, 137, 145, 152, 157].
(b) Plating out of dilutions from *homogenised* material obtained by the grinding of plant material with sterile water or of finely milled material [8, 33, 55, 58, 77, 80, 81, 107, 145, 150, 157].

In the first case the washed plant material may be further incubated in moist chambers, while in the second case, direct microscopic observations and plating on non-selective or diagnostic media can be carried out from the same diluted suspension.

Another generally used laboratory technique, found to be useful in handling plant material either directly as collected, or washed, or surface-sterilised, is by placing it in filter-paper-lined *moist chambers* [8, 9, 13, 15, 73, 74, 165, 166] or filter-paper-lined, sterile petri dishes [8, 9, 12, 13, 57, 84], in water [15], on water agar, or upon selective agar media [33, 40, 102, 112, 116, 118, 127, 128].

In the study of plant surface microflora a number of interesting techniques have been developed, such as surface impression technique [43, 44, 125], leaf imprint technique [18, 41, 42, 125], spore fall technique [18, 41–43], leaf disc technique [158], as well as the abovementioned surface washing and dilution technique [18, 41, 42, 125], which are applicable also to other surface microfloras in factories, shops and stores. In the study of air microbiology [59, 61, 62, 82, 85, 90, 105, 123, 156] certain techniques have been developed which are adopted for the study of plant and animal bio-surfaces [52, 62, 91].

9. STUDY OF HARVESTED PRODUCE, COMPOSTS AND LITTER

In studies of decomposition of senescent dead standing vegetation, plant litter, plant residues left in the field after harvest, hay, silage and straw, or deteriorating grain and other feeds, numerous techniques have been adopted in the past. For grain and seed testing, well-developed standard techniques are in use in testing establishments [12, 29, 53, 102, 112, 116, 123]. Some special comments might be made regarding decaying standing vegetation, plant residues, as well as hay, straw and similar feeds [8–10, 13–15, 17, 19, 30, 36, 38, 40, 55–57, 59, 63, 66, 67, 71, 73, 74, 79, 80, 82, 100, 101, 108, 114, 117, 118, 123, 127, 128, 141, 142, 163–166, 169]. In general, a similar approach should be adopted as in growing standing vegetation, except that special investigations of the surface microflora may be cut to a minimum and the tissue-colonising microorganisms considered as the main target.

The choice of procedures and techniques depends, as before, upon the problem to be resolved (*see* Tables I and III). Shaking of the material with sterile water, plating out dilutions, and incubation of the washed solid substrates in moist chambers should be obvious for quantitative and qualitative investigation. Homogenisation by grinding, milling or other means may be the best direct way of obtaining total suspension of material from which dilutions are obtained for general or specific treatments. Surface sterilisation, direct microscopic observation and other procedures have been employed with much success. Self-heating of hay in Dewar flasks [63] and activities of thermophilic microflora in composts [163, 164] have been studied.

Estimation of respiration rates by manometric techniques [45, 124, 143, 158] in studies of decomposition of plant litter from certain types of vegetation has produced interesting results. The use of percolation techniques [96] may also be advantageous in studies of certain biochemical changes in decomposing material.

10. FLOUR AND BREAD

Fine flour is suitable for investigation by the dilution plate technique, whereas medium fine and coarse flour is different and here the suspension plate technique would be more appropriate. Use of dispersing agents, including saline, is advantageous to obtain even distribution of particles. However, correct sampling procedures [80, 150] should be scrupulously observed, including sterility precautions and handling of samples before laboratory procedure is started. Moisture content determination is essential before microbiological tests are carried out, as is careful direct microscopic examination of stained preparations from known dilutions to obtain a direct microscopic count of all the major groups of microbes (Actinomycetes, bacteria, moulds and yeasts) present. For this a number of simple or elaborate techniques are now available.

Impurities in flour, such as dirt and insect remains, may be separated by sieving followed by microscopic examination [4, 5, 21, 29, 33, 56, 58, 66, 72, 103, 126, 136–138, 150, 151, 165]. For microbiological examination of dough products, certain hygienic conditions on the premises, in the storage, manufacture, distribution and sale of these foods, are important. Field tests, such as air spora counts [7, 60] and other observations, may be useful in making a proper assessment of laboratory results, especially in sorting out sources of any secondary or surface infection, since the primary infection of flour may be derived from standing crop [12] and grain [136].

11. MICROBIOLOGICAL STANDARDS

Attempts have been made to introduce microbiological standards for foods [75, 138, 149, 151]. In general, such attempts are based upon (a) microbial content, (b) presence of potential pathogens, (c) presence of toxin producers, (d) microbes indicating contamination during manufacture, handling or storage, and (e) unfavourable or objectionable changes in foods and feeds. At present, however, no agreed standards exist for the various agricultural grassland produce, including cereals and the various foods and feeds manufactured. However, numerous published investigations on potential pathogens and toxin-producing microorganisms on plants under natural conditions and in a wide variety of agricultural produce [2, 8, 12, 17, 33–35, 49, 52, 57, 63, 67, 79, 91, 114, 126, 132, 134, 135, 138, 151] indicate a need for a general consideration of this problem by the relevant governmental and international bodies.

The presence of plant pathogens in agricultural crops and produce is best cared for by institutions concerned with plant protection [36, 41–44, 46, 57a, 82, 85, 102, 116, 117, 122a, 123, 125, 142, 162]. To achieve acceptable microbiological standards, two basic aspects must be considered:

(a) The presence of pathogens, toxin producers and other undesirable microbes, as well as microbes indicating contamination in handling.
(b) Agreed microbiological procedures and techniques in field and laboratory examination of agricultural produce, feeds and foods.

Furthermore, any undesirable biochemical changes and characteristics in the materials tested should also be considered in such standard examinations.

REFERENCES

1. Ainsworth, G. C. and Bisby, G. R. (1971). *Dictionary of the Fungi*. Commonwealth Mycological Institute, Kew, Surrey.
2. Ainsworth, G. C. and Austwick, P. K. C. (1973). *Fungal Diseases of Animals*. Commonwealth Agricultural Bureaux, Farnham Royal, Bucks.
3. Ainsworth, G. C., Sparrow, F. K. and Sussman, A. S. (eds). (1973). *The Fungi*, vols. IV A and IV B. Academic Press, London.
4. American Association of Cereal Chemists (1962). *Cereal Laboratory Methods*. AACC, St Paul, Minn.; Academic Press, London.
5. American Public Health Association (1958). *Recommended Methods for the Microbiological Examination of Foods*. APHA, New York.
6. Amos, A. J. and Kent-Jones, D. W. (1931). The 'rope' spore content of flour and its significance. *Analyst*, **56,** 572–86.
7. Andersen, A. A. (1958). New sampler for the collection, sizing and enumeration of viable airborne particles. *J. Bacteriol.*, **76,** 471–84.
8. Apinis, A. E. (1963). Thermophilous fungi of coastal grasslands. In: *Soil Organisms* (ed. J. Doeksen and J. van der Drift), pp. 427–38. North-Holland Publ. Co., Amsterdam.
9. Apinis, A. E. (1964). On fungi isolated from soils and *Ammophila* debris. *Kew Bull.*, **19,** 127–31.
10. Apinis, A. E. (1970). Das Verhalten der Pilze in bestimmten Grassland-Gesellschaften. In: *Gesellschaftsmorphologie* (ed. R. Tüxen), pp. 172–86. Vlg Dr W. Junk NV, Den Haag.

11. Apinis, A. E. (1972). Facts and problems. *Mycopathol. Mycol. Appl.*, **43,** 93–109.
12. Apinis, A. E. (1972). Mycological aspects of stored grain. In: *Biodeterioration of Materials*, vol. 2 (ed. A. H. Walters and E. H. Hueck-van der Plas), pp. 493–8. Applied Science Publishers, London.
13. Apinis, A. E. and Chesters, C. G. C. (1964). Ascomycetes of some salt marshes and sand dunes. *Trans. Brit. Mycol. Soc.*, **47,** 419–35.
14. Apinis, A. E., Chesters, C. G. C. and Taligoola, H. K. (1972). Colonization of *Phragmites communis* leaves by fungi. *Nova Hedwigia*, **23,** 113–24.
15. Apinis, A. E. and Taligoola, H. K. (1975). Biodegradation of *Phragmites communis* Trin. by fungi. In: *Biodégradation et Humification* (ed. G. Kilbertus, O. Reisinger, A. Mourey and J. A. Cancela de Fonseca), pp. 24–32. Pierron, Sarreguemines, France.
15a. Baker, K. F. and Snyder, W. C. (eds.) (1965). *Ecology of Soil-borne Plant Pathogens*. John Murray, London.
16. Barnett, H. L. (1960). *Illustrated Genera of the Imperfect Fungi*. Burgess Publ. Co., Minneapolis.
17. Beck, Th. (1969). Der gegenwärtige Stand der Mikrobiologie der Einsäuerung und Trocknung von Futterpflanzen. *Proc. 3rd Gen. Meeting of European Grassland Federation*, pp. 207–19. Braunschweig, West Germany.
18. Beech, F. W. and Davenport, R. R. (1971). A survey of methods for the quantitative examination of the yeast flora of apple and grape leaves. In: *Ecology of Leaf Surface Microorganisms* (ed. T. F. Preece and C. H. Dickinson). Academic Press, London.
19. Bell, M. K. (1974). Decomposition of herbaceous litter. In: *Biology of Plant Litter Decomposition* (ed. C. H. Dickinson and G. J. F. Pugh), pp. 37–67. Academic Press, London.
20. *Bergey's Manual of Determinative Bacteriology* (1975) (ed. R. E. T. Buchanan and N. E. Gibbons). Williams & Wilkins, Baltimore.
21. Bevan, D. and Bond, J. (1971). Microorganisms in field and mill: a preliminary survey. *Proc. Queensland Soc. Sugar Cane Technologists*, pp. 137–43.
22. Bisby, G. R. and Mason, E. W. (1940). List of British Pyrenomycetes. *Trans. Brit. Mycol. Soc.*, **24,** 127–243.
23. Booth, C. (1971). *Methods in Microbiology*, vol. 4. Academic Press, London.
24. Booth, C. (1971). *The Genus Fusarium*. Commonwealth Mycological Institute, Kew, Surrey.
25. Braun-Blanquet, J. (1964). *Pflanzensoziologie*. Wien.
26. Brisbane, P. G. and Rovira, A. D. (1961). A comparison of methods for classifying rhizosphere bacteria. *J. Gen. Microbiol.*, **26,** 379–92.
27. Brown, J. C. (1958). Fungal mycelium in dune soils estimated by a modified impression slide technique. *Trans. Brit. Mycol. Soc.*, **41,** 81–8.
28. Buchanan, R. E., Bolt, J. G. and Lessell, E. F. (1966). *Index Bergeyana*. E. & S. Livingstone, Edinburgh and London.
29. *Cereal Laboratory Methods* (1947). Am. Assoc. Cereal Chemists, St Paul, Minn.
30. Chang, T. and Hudson, H. J. (1967). The fungi of wheat straw compost. I. Ecological studies. *Trans. Brit. Mycol. Soc.*, **50,** 649–66.

31. Chesters, C. G. C. (1940). A method for isolating soil fungi. *Trans. Brit. Mycol. Soc.*, **24,** 352–5.
32. Cholodny, N. (1930). Über eine neue Methode zur Untersuchung der Bodenmikroflora. *Arch. Mikrobiol.*, **1,** 620–52.
33. Christensen, C. M. (1946). Quantitative determination of moulds in flour. *Cereal Chem.*, **23,** 322–9.
34. Ciegler, A. and Lillehoy, E. B. (1968). Mycotoxins. *Adv. Appl. Microbiol.*, **10,** 155–219.
35. Ciegler, A. and Pitt, J. I. (1970). Survey of the genus *Penicillium* for tremorgenic toxin production. *Mycopathol. Mycol. Appl.*, **42,** 119–24.
36. Clark, F. E. and Paul, E. A. (1969). The microflora of grassland. *Adv. Agron.*, **22,** 375–435.
37. Conn, H. J. (1918). The microscopic study of bacteria and fungi in the soil. *N.Y. Agr. Exp. Sta. Tech. Bull.* No. 64.
38. Cross, T., MacIver, A. M. and Lacey, J. (1968). The thermophilic actinomycetes of mouldy hay: *Micropolyspora faeni* sp. nov. *J. Gen. Microbiol.*, **50,** 351–9.
39. Dennis, R. W. G. (1968). *British Ascomycetes.* J. Cramer, Weinheim.
40. Dickinson, C. H. (1965). The mycoflora associated with *Halimione portulacoides.* III. Fungi on green and moribund leaves. *Trans. Brit. Mycol. Soc.*, **48,** 603–10.
41. Dickinson, C. H. (1967). Fungal colonization of *Pisum* leaves. *Can. J. Bot.*, **45,** 915–27.
42. Dickinson, C. H. (1973). Effect of ethirinol and zineb on phylloplane microflora of barley. *Trans. Brit. Mycol. Soc.*, **60,** 423–31.
43. Dickinson, C. H., Watson, J. and Wallace, B. (1974). An impression method for examining epiphytic microorganisms and its application to phylloplane studies. *Trans. Brit. Mycol. Soc.*, **63,** 616–19.
44. Diem, H. G. (1974). Microorganisms of the leaf surface: estimation of the mycoflora of the barley phyllosphere. *J. Gen. Microbiol.*, **80,** 77–83.
45. Dixon, M. (1952). *Manometric Methods as Applied to the Measurement of Cell Respiration and Other Processes.* Cambridge Univ. Press.
46. Dowson, W. J. (1957). *Plant Diseases Due to Bacteria.* Cambridge Univ. Press.
47. Druce, R. G. and Thomas, S. B. (1970). An ecological study of the psychrotrophic bacteria of soil, water, grass and hay. *J. Appl. Bacteriol.*, **33,** 420–35.
48. Ellenberg, H. (1963). *Vegetation Mitteleuropas mit den Alpen in kausaler, dynamischer und historischer Sicht.* Stuttgart.
49. Ellis, J. J. and Yates, S. G. (1971). Mycotoxins of fungi from fescue. *Econ. Bot.*, **25,** 1–5.
50. Ellis, M. B. (1971). *Dematiaceous Hyphomycetes.* Commonwealth Mycological Institute, Kew, Surrey.
51. Fennell, D. I. (1973). Plectomycetes: Eurotiales. In: *The Fungi*, vol. IV A (ed. G. C. Ainsworth, F. K. Sparrow and A. S. Sussman), pp. 45–68. Academic Press, London.
52. Festenstein, G. N., Lacey, J., Skinner, F. A., Jenkins, P. A. and Pepys, J. (1965). Self-heating of hay and grain in Dewar flasks and the development of farmer's lung antigens. *J. Gen. Microbiol.*, **41,** 389–407.

53. Flannigan, B. (1970). Comparison of seed-borne mycofloras of barley, oats and wheat. *Trans. Brit. Mycol. Soc.*, **55,** 267–76.
54. Forbes, R. S. (1974). Decomposition of agricultural crop debris. In: *Biology of Plant Litter Decomposition*, vol. 2 (ed. C. H. Dickinson and G. J. F. Pugh), pp. 723–42. Academic Press, London.
55. Frankland, J. C. (1974). Importance of phase-contrast microscopy for estimation of total fungal biomass by the agar-film technique. *Soil Biol. Biochem.*, **6,** 409–10.
56. Frazier, W. C. (1967). *Food Microbiology*. McGraw-Hill, New York.
57. Fritzsch, W. and Abadjieff, W. (1967). Mykologische Untersuchungen an Malzkeimproben. *Arch. Tierernährung*, **17,** 463–77.
57a. Gäumann, E. (1945). *Pflanzliche Infektionslehre*. Birkhäuser, Basel.
58. Graves, R. R. and Hesseltine, C. W. (1965). Fungi in flour and refrigerated dough products. *Mycopathol. Mycol. Appl.*, **29,** 277–90.
59. Gray, K. R. and Biddlestone, A. J. (1974). Decomposition of urban waste. In: *Biology of Plant Litter Decomposition*, vol. 2 (ed. C. H. Dickinson and G. J. F. Pugh), pp. 743–55. Academic Press, London.
60. Gregory, P. H. (1961). *Microbiology of the Atmosphere*. Leonard Hill, London.
61. Gregory, P. H. and Lacey, M. E. (1963). Liberation of spores from mouldy hay. *Trans. Brit. Mycol. Soc.*, **46,** 73–80.
62. Gregory, P. H. and Lacey, M. E. (1963). Mycological examination of dust from mouldy hay associated with farmer's lung disease. *J. Gen. Microbiol.*, **30,** 75–88.
63. Gregory, P. H., Lacey, M. E., Festenstein, N. G. and Skinner, F. A. (1963). Microbial and biochemical changes during the moulding of hay. *J. Gen. Microbiol.*, **33,** 147–74.
64. Hansen, H. N. and Snyder, W. C. (1947). Gaseous sterilization of biological materials for use as culture media. *Phytopathol.*, **37,** 369–71.
65. Harley, J. L. and Waid, J. S. (1955). A method of studying active mycelia on living roots and other surfaces in the soil. *Trans. Brit. Mycol. Soc.*, **38,** 104–18.
66. Harrigan, W. F. and McCance, M. E. (1966). *Laboratory Methods in Microbiology*. Academic Press, London.
67. Harrison, J. (1971). Food moulds and their toxicity. *Trop. Sci.*, **13,** 57–63.
68. Hawker, L. E. and Linton, A. H. (1971). *Microorganisms*. E. Arnold, London.
69. Hawksworth, D. L. (1974). *Mycologist's Handbook*. Commonwealth Mycological Institute, Kew, Surrey.
70. Heintzeler, I. (1939). Das Wachtum der Schimmelpilze in Abhängigkeit von den Hydraturverhältnissen unter verschiedenen Aussenbedingungen. *Arch. Mikrobiol.*, **10,** 92–132.
71. Hering, T. F. and Nicholson, P. B. (1964). A clearing technique for the examination of fungi in plant tissue. *Nature* (*London*), **201,** 942–3.
72. Hinton, H. E. and Corbet, A. S. (1965). *Common Insect Pests on Stored Food Products*. Trustees of British Museum, London.
73. Hudson, H. J. and Webster, J. (1958). Succession of fungi on decaying stems of *Agropyron repens*. *Trans. Brit. Mycol. Soc.*, **41,** 165–77.

74. Hudson, H. J. (1968). The ecology of fungi on plant remains above the soil. *New Phytologist*, **67**, 837–74.
75. Ingram, M. (1961). Microbiological standards for foods. *Food Technol.*, (Feb.), 4–16.
76. Iswaran, V. and Harris, J. R. (1968). The bacteria of decomposing cereal straw, and their effects on plant growth. *9th Int. Congr. Soil Sci. Trans.*, **3**, 501–9.
77. James, N. and Smith, H. N. (1948). Studies on the microbial flora of flour. *Can. J. Res.*, Part C, **26**, 479–85.
78. Jeanson, C. (1975). Décomposition expérimentale de la paille de blé dans un sol de limon. Etude à la microsonde et au microscope à balayage. In: *Biodégradation et Humification* (ed. G. Kilbertus, O. Reisinger, A. Mourey and J. A. Cancela de Fonseca), pp. 184–94. Pierron, Sarreguemines, France.
79. Joffe, A. Z. (1961). The mycoflora of overwintered cereals and its toxicity. *Bull. Res. Council of Israel*, Sect. D: *Botany*, V. GD, 101–26.
80. Johnson, L. F., Curl, E. A., Bond, J. H. and Fribourg, H. A. (1960). *Methods for Studying Soil Microflora–Plant Disease Relationships*. Burgess Publ. Co., Minneapolis.
81. Jones, P. C. T. and Mollison, J. E. (1948). A technique for the quantitative estimation of soil microorganisms. *J. Gen. Microbiol.*, **2**, 54–69.
82. Kelman, A. (editor and chairman) (1967). *Sourcebook of Laboratory Exercises in Plant Pathology*. W. H. Freeman & Co., San Francisco and London.
83. Kendrick, W. B. and Carmichael, J. W. (1973). Hyphomycetes. In: *The Fungi*, vol. IV A (ed. G. C. Ainsworth, F. K. Sparrow and A. S. Sussman), pp. 323–509. Academic Press, London.
84. Keyworth, W. G. (1951). A Petri-dish moist chamber. *Trans. Brit. Mycol. Soc.*, **34**, 291–2.
85. Király, Z., Klement, Z., Solymosi, F. and Vörös, J. (1974). *Methods in Plant Pathology*. Elsevier, Amsterdam.
86. Klapp, E. (1971). *Wiesen und Weiden: Eine Grünlandlehre*. Parey, Berlin and Hamburg.
87. Korf, R. P. (1973). Discomycetes and Tuberales. In: *The Fungi*, vol. IV A (ed. G. C. Ainsworth, F. K. Sparrow and A. S. Sussman), pp. 249–329. Academic Press, London.
88. Krassilnikov, N. A. (1949). *Guide to the Identification of Bacteria and Actinomycetes* (ed. J. B. Routien). Chas. Pfizer & Co. Inc., New York.
89. Küster, E. and Williams, S. T. (1964). Selection of media for isolation of *Streptomycetes*. *Nature* (*London*), **202**, 928–9.
90. Lacey, J. (1975). Airborne spores in pastures. *Trans. Brit. Mycol. Soc.*, **64**, 265–81.
91. Lacey, J. (1975). Potential hazards to animals and man from microorganisms in fodders and grain. *Trans. Brit. Mycol. Soc.*, **65**, 171–84.
92. Lapage, S. P., Shelton, J. E. and Mitchell, T. G. (1970). Media for the maintenance and preservation of bacteria. In: *Methods in Microbiology*, vol. 3A (ed. J. R. Norris and D. W. Ribbons), pp. 1–133. Academic Press, London and New York.

93. Large, E. C. (1954). Growth stages in cereals. *Plant pathol.*, **3,** 128–9.
94. Last, F. T. (1955). Seasonal incidence of *Sporobolomyces* on cereal leaves. *Trans. Brit. Mycol. Soc.*, **38,** 221–39.
95. Last, F. T. and Deighton, C. F. (1965). The nonparasitic microflora on the surfaces of living leaves. *Trans. Brit. Mycol. Soc.*, **48,** 83–9.
96. Lees, H. (1949). The soil percolation technique. *Plant and Soil*, **1,** 221–39.
97. Levine, M. and Schoenlein, H. W. (1930). *A Compilation of Culture Media for the Cultivation of Microorganisms.* Williams & Wilkins, Baltimore.
98. Locci, R. (1972). Direct examination of biodeteriorated material microflora by scanning electron microscopy. In: *Biodeterioration of Materials* (ed. A. H. Walters and E. H. Hueck-van der Plas), pp. 416–21. Applied Science Publishers, London.
99. Lodder, J. (ed.) (1970). *The Yeasts.* North-Holland Publ. Co., Amsterdam.
100. Lowrey, R. S. (1969). The nylon bag technique for the estimation of forage quality. *Proc. National Conference on Forage Quality Evaluation and Utilization*, 3–4 Sept.
101. McLauchlin, N. B. (1974). An inexpensive manual coring device for baled hay. *Can. J. Plant Sci.*, **54,** 839–40.
102. Malone, J. P. and Muskett, A. E. (1964). Seed-borne fungi. *Proc. Int. Seed Testing Assoc.*, **29** (2).
103. *Manual of Microbiological Methods* (1957). McGraw-Hill, New York, for Society of American Bacteriologists.
104. Martin, J. P. (1950). Use of acid, Rose Bengal and streptomycin in the plate method for estimating soil fungi. *Soil Sci.*, **69,** 215–32.
105. May, K. R. (1945). The cascade impactor: an instrument for sampling coarse aerosols. *J. Sci. Instr.*, **22,** 187–95.
106. Menna, M. E. di and Parle, J. N. (1970). Moulds on leaves of perennial ryegrass and white clover. *N.Z. Agric. Res.*, **13,** 51–68.
107. Menzies, J. D. (1957). A dipper technique for serial dilutions of soil for microbial analysis. *Soil. Sci. Soc. Am. Proc.*, **21,** 660.
108. Miehe, H. (1907). *Die Selbsterhitzung des Heues. Eine biologische Studie.* G. Fischer, Jena.
109. Minderman, G. and Daniels, L. (1967). Colonization of newly fallen leaves by microorganisms. In: *Progress in Soil Biology* (ed. O. Graff and J. E. Satchell), pp. 3–9. North-Holland Publ. Co., Amsterdam.
110. Mishustin, E. N. (1950). *Thermophilic Microorganisms in Nature and Practice.* Institute of Microbiology, Academy of Sciences, Moscow and Leningrad. (In Russian.)
111. Montégut, J. (1960). Value of the dilution method. In: *Ecology of Soil Fungi* (ed. D. Parkinson and J. S. Waid). Liverpool Univ. Press.
112. Mulinge, S. K. and Chesters, C. G. C. (1970). Methods of isolating the microflora of moulding, high-moisture barley in partially sealed silos. *Ann. Appl. Biol.*, **65,** 285–92.
113. Müller, E. and von Arx, J. A. (1973). Pyrenomycetes: Meliolales, Coronophorales, Sphaeriales. In: *The Fungi*, vol. IV A (ed. G. C. Ainsworth, F. K. Sparrow and A. S. Sussman), pp. 87–132. Academic Press, London.

114. Müller, G. (1961). Mikrobiologische Untersuchungen über die 'Futterverpilzung durch Selbsterhitzung'. 2. Mitteilung: Untersuchungen verschiedener verpilzter Futtermittel landwirtschaftlicher Betriebe. *Zbl. Bakteriol.*, Abt. II, **114,** 388–425.
115. Munk, A. (1957). Danish Pyrenomycetes. *Dansk Bot. Arkiv*, **17,** 1–491.
116. Muskett, A. E. (1938). Technique for the examination of seeds for the presence of seed-borne fungi. *Trans. Brit. Mycol. Soc.*, **30,** 74–83.
117. Noble, M. and Richardson, M. J. (1968). *An Annotated List of Seed-borne Diseases*. Commonwealth Mycological Institute, Kew, Surrey.
118. Norse, D. (1972). Fungi isolated from surface-sterilized tobacco leaves. *Trans. Brit. Mycol. Soc.*, **58,** 515–78.
119. Old, K. M. and Nicolson, T. H. (1975). Electron microscopy of the microbial colonization of roots of sand dune grasses. In: *Biodégradation et Humification* (ed. G. Kilbertus, O. Reisinger, A. Mourey and J. A. Cancela da Fonseca). Pierron, Sarreguemines, France.
120. Paech, K. and Tracey, M. V. (eds.) (1956). *Modern Methods of Plant Analysis*, vols. 1–4. Springer, Berlin.
121. Panasenko, V. T. (1934). Ekologiya plesnevych grybov. *Mikrobiologiya*, **13,** 158–70. (In Russian.)
122. Piper, C. S. (1947). *Soil and Plant Analysis*. Interscience Publ. Inc., New York.
122a. Plank, J. E. van der (1963). *Plant Diseases*. Academic Press, New York and London.
123. *Plant Pathologist's Pocketbook* (1968). Commonwealth Mycological Institute, Kew, Surrey.
124. Pochon, J., Tardieux, P. and d'Aguilar, J. (1969). Methodological problems in soil biology. In: *Soil Biology Reviews of Research*, pp. 13–63. UNESCO, Paris.
125. Preece, T. F. and Dickinson, C. H. (eds.) (1971). *Ecology of Leaf Surface Microorganisms*. Academic Press, London.
126. Prickett, P. S. (1964). Role of microbiological methods in the sanitary control of cereals and cereal products. *Cereal Sci.*, **9,** 192–7.
127. Pugh, G. J. F. (1958). Leaf litter fungi found on *Carex paniculata* L. *Trans. Brit. Mycol. Soc.*, **41,** 185–95.
128. Pugh, G. J. F. and Mulder, J. L. (1971). Mycoflora associated with *Typha latifolia*. *Trans. Brit. Mycol. Soc.*, **57,** 273–82.
129. Rabotnov, T. A. (1974). *Lugovedenie* (*Grassland Science*, in Russian). Univ. of Moscow Press.
130. Raper, K. B. and Fennell, D. I. (1965). *The Genus Aspergillus*. Williams & Wilkins, Baltimore.
131. Raper, K. B. and Thom, C. (1968). *Manual of Penicillia*. Williams & Wilkins, Baltimore.
132. Rautenstein, J. I. (1939). Microbiological processes in wheat grain collected with combines during its post-ripening, drying and storage. *Microbiologiya*, **8,** 211–24.
133. Ruinen, J. (1961). The phyllosphere. I. An ecologically neglected milieu. *Plant and Soil*, **15,** 81–109.
134. Scott, De B. (1965). Toxigenic fungi isolated from cereal and legume products. *Mycopathol. Mycol. Appl.*, **25,** 213–22.

135. Scott, P. M., van Walbeck, W., Kennedy, B. and Anyeti, A. (1972). Mycotoxins (ochratoxin A, citrinin and sterigmatocystin) and toxigenic fungi in grains and other agricultural products. *J. Agr. Food Chem.*, **20,** 1103–9.
136. Semeniuk, G. (1954). Microflora. In: *Storage of Cereal Grains and Their Products* (ed. J. A. Anderson and A. W. Alcock), pp. 77–151. Am. Assoc. Cereal Chemists Monogr. series V(2), St Paul, Minn.
137. Sharf, J. M. (ed.) (1966). *Recommended Methods for the Microbiological Examination of Foods*. Am. Publ. Health Assoc. Inc., New York.
138. Shiffman, M. A. and Kronick, D. (1963). The development of microbiological standards for foods. *J. Milk Food Technol.*, **26,** 110–14.
139. Shipton, W. A. and Brown, J. F. (1962). A whole-leaf clearing and staining technique to demonstrate host–pathogen relationships of wheat stem rust. *Phytopathol.*, **52,** 1313.
140. Skerman, V. B. D. (1967). *The Genera of Bacteria.* Williams & Wilkins, Baltimore.
141. Snow, D., Chrichton, M. H. G. and Wright, N. C. (1944). Mould deterioration of feeding-stuffs in relation to humidity in storage. I. The growth of moulds at low humidities. *Ann. Appl. Biol.*, **31,** 102–10.
142. Sprague, R. (1950). *Diseases of Cereals and Grasses in North America.* Ronald Press, New York.
143. Steubing, L. (1965). *Pflanzenökologisches Praktikum.* Parey, Berlin and Hamburg.
144. Stevenson, I. L. (1958). The effect of sonic vibration on the bacterial plate count of soil. *Plant and Soil*, **10,** 1–8.
145. Stout, J. D. (1960). Bacteria of soil and pasture leaves at Claudelands Showgrounds. *N.Z. J. Agr. Res.*, **3,** 413–30.
146. Strugger, S. (1949). *Fluoreszenzmikroskopie und Mikrobiologie.* Schaper, Hannover.
147. Sutton, B. C. (1973). Coelomycetes. In: *The Fungi*, vol. IV A (ed. G. C. Ainsworth, F. K. Sparrow and A. S. Sussman), pp. 513–82. Academic Press, London.
148. Swaroop, S. (1951). The range of variation of the most probable number of organisms estimated by the dilution method. *Indian J. Med. Res.*, **39,** 107.
149. Tanner, F. W. (1944). *Microbiology of Foods.* Garrard Press, Champaign, Ill.
150. Thatcher, F. S., Coutu, C. and Stevens, F. (1953). The sanitation of Canadian flour mills and its relationship to the microbial content of flour. *Cereal Chem.*, **30,** 71–102.
151. Thatcher, F. S. (1955). Microbiological standards for foods: their function and limitations. *J. Appl. Bacteriol.*, **18,** 449–61.
152. Thomas, S. B. and McQuillin, J. (1952). *Coli-aerogenes* bacteria isolated from grass. *Proc. Soc. Appl. Bacteriol.*, **15,** 41–52.
153. Thomas, S. B. and Hobson, P. M. (1955). *Coli-aerogenes* bacteria isolated from ears and panicles of cereal crops. *J. Appl. Bacteriol.*, **18,** 1–8.
154. Thornton, R. H. (1952). The screened immersion plate. *Research*, **5,** 190–1.

155. Töth, J. A., Papp, L. B. and Lenkey, B. (1975). Litter decomposition in an oak forest ecosystem (*Quercetum petraeae cerris*) of northern Hungary studied in the framework of 'Sikfokut' projects. In: *Biodégradation et Humification* (ed. G. Kilbertus, O. Reisinger, A. Mourey and J. A. Cancela de Fonseca), pp. 41–58. Pierron, Sarreguemines, France.
156. Tribe, H. T. and Mabadeje, S. A. (1972). Growth of moulds on media prepared without organic nutrients. *Trans. Brit. Mycol. Soc.*, **58,** 127–37.
157. Trolldenier, G. (1967). Vergleich zwischen fluoreszenzmikroskopischer Direktzählung, Plattengussverfahren und Membranfiltermethode bei Rhizosphärenuntersuchungen. In: *Progress in Soil Biology* (ed. O. Graff and J. E. Satchell), pp. 59–71. North-Holland Publ. Co., Amsterdam.
158. Visser, S. and Parkinson, D. (1975). Litter respiration and fungal growth under low temperature conditions. In: *Biodégradation et Humification* (ed. G. Kilbertus, O. Reisinger, A. Mourey and J. A. Cancela de Fonseca), pp. 88–97. Pierron, Sarreguemines, France.
159. Voznyakovskaya, Y. M. and Khudyakov, Y. P. (1960). Species composition of the epiphytic microflora of living plants. *Microbiologiya*, **29,** 97–103.
160. Voznyakovskaya, Y. M. (1962). Epiphytic yeasts. *Microbiologiya*, **31,** 616–22.
161. Waid, J. S. and Woodman, M. F. (1957). A method of estimating hyphal activity in soil. Symposium sur les Méthodes d'Etudes Microbiologiques du Sol, June. Louvain, Belgium.
162. Walker, J. C. (1969). *Plant Pathology*. Tata/McGraw-Hill, Bombay and New Dehli.
163. Webley, D. M. (1947). The microbiology of composting. I. The behaviour of the aerobic mesophilic bacterial flora of composts and its relation to other changes taking place during composting. *Proc. Soc. Appl. Bacteriol.*, **2,** 83–9.
164. Webley, D. M. (1947). Activity of thermophilic bacteria in composts of fresh green material. *Nature* (*London*), **159,** 35.
165. Webster, J. (1956). Succession of fungi on decaying cocksfoot culms. I. *J. Ecol.*, **44,** 517–44.
166. Webster, J. (1957). Succession of fungi on decaying cocksfoot culms. II. *J. Ecol.*, **45,** 1–30.
167. Williams, S. T. and Cross, T. (1971). Actinomycetes. In: *Methods in Mycology*, vol. 4 (ed. C. Booth), pp. 295–334. Academic Press, London.
168. Wöller, H. (1929). Über die epiphytische Bakterienflora gesunder, grüner Pflanzen. *Zbl. Bakteriol.*, Abt. II, **79,** 173–7.
169. Yadaw, A. S. and Madelin, M. F. (1968). The ecology of microfungi on decaying stems of *Urtica dioica*. *Trans. Brit. Mycol. Soc.*, **51,** 249–59.
170. Zycha, H., Siepmann, R. and Linneman, G. (1969). *Mucorales*. J. Cramer, Weinheim.

13

Techniques for Selective Isolation of Fungi from Aircraft Fuel

T. RUBIDGE

Materials Quality Assurance Directorate, Ministry of Defence Procurement Executive, Royal Arsenal East, London SE18 6TD, UK

1. AIM

Rubidge (1974) described an agar medium containing aviation kerosene and Bushnell–Haas mineral salts solution (Bushnell and Haas, 1941) which was adapted from a method of incorporating liquid hydrocarbons in aqueous media developed by Barvah *et al.* (1967). The Avtur–Bushnell–Haas agar (ABHA) has been shown to be selective for kerosene fungi when streaked with naturally and artificially contaminated fuels, but in all cases contamination of the fuels was of a high order. With the growing recognition of aircraft fuel biodeterioration, 'good housekeeping' and biocide application have been promoted and the degree of contamination of aircraft fuel systems greatly reduced (Hitzman, 1964; Elphick and Hunter, 1968; Scott, 1971). Consequently the concentration of deteriogens in most samples now tends to be low and, so that the presence of such organisms can be detected and monitored at such levels, a selective medium for their isolation assumes greater importance. The following method is aimed at such isolation, and with particular reference to *Cladosporium resinae*, the predominant fuel spoilage fungus. Thermotolerant fuel fungi, which are currently arousing much interest, can also be applied to this medium.

2. MATERIALS AND APPARATUS

Colloidal silica (BDH Chemicals Ltd).
Diethyl ether.

Ethanol.
Aircraft kerosene.
Bushnell–Haas broth (Difco Laboratories Inc.).
Agar.
Polystyrene or glass petri dishes.
Sterile membrane filters, pore size 0·45 μ, 47 mm diameter.
Filter holder.
Water tap vacuum pump.
Forceps.

The apparatus is illustrated in Fig. 1.

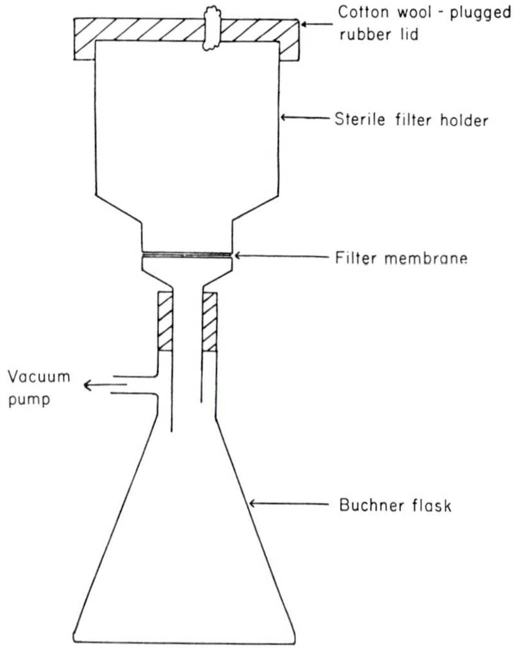

FIG. 1. Membrane filtration apparatus.

3. METHOD

1. Weigh 15 g of colloidal silica into a beaker. (*Caution:* Due to the extremely fine nature of the silica powder the weighing should be carried out in a fume cupboard and rubber gloves should be worn.)

2. To 4·5 ml of diethyl ether add 30 ml of aircraft kerosene and then add the mixture to the colloidal silica contained in the beaker.

3. Stir the contents of the beaker until the liquid is evenly dispersed throughout the silica and a fluffy powder of reduced volume is formed, then set aside to allow the ether to evaporate. (Evaporation may be assisted by placing the vessel on a steam-heated water bath placed inside a spark-proof fume cupboard.)

4. When all the ether has been driven off, transfer the silica/kerosene mixture to a beaker containing 1 litre of Bushnell–Haas broth, add 20 g of agar and set aside for 15 min.

5. Stir well to disperse the kerosene/silica mixture throughout the liquid, then dispense into suitably sized bottles and sterilise in an autoclave at 70×10^3 N/m^2 (10 lb/in^2) for 10 min. (Rubber liners should not be used in the bottle tops because of the reaction between kerosene and rubber.)

6. Allow the medium to cool and then pour into glass or polystyrene petri dishes. (In the case of polystyrene petri dishes the medium should be poured as cool as possible to prevent clouding of the lids. Should this occur, however, fresh sterile lids can be substituted when setting is complete.) (*Caution:* Due to the flammability of the kerosene the mouths of the bottles should not be flamed.)

7. Assemble the filter holder and with all apertures plugged with cotton wool and capped with tin foil, sterilise by autoclaving at 105×10^3 N/m^2 for 15 min.

8. Install a sterile filter membrane of pore size 0·45 μ in the sterilised holder and place the holder in a Buchner flask connected to a water pump.

9. Draw a suitable volume of fuel sample through the membrane, leaving the water pump on for about 15 sec after the fuel has passed through. Remove the membrane with forceps dipped in ethanol and flamed and invert on to the medium in the petri dish.

10. Incubate at 25°C and observe daily for growth appearing on the upper side of the membrane.

4. RESULTS

Cladosporium resinae (Lindau) de Vries is the fungus most often responsible for aircraft fuel spoilage. After about 5 days it is visible to the naked eye as a brown patch on the upper side of the inverted membrane. When examined at $\times 50$ magnification the brown patch can be seen as consisting of conidiophores standing upright on the membrane. Vegetative hyphae are not visible. When the membrane is lifted off the agar surface, the underside will be seen to be essentially free of growth except for a small number of isolated conidiophores, while on the surface of the agar directly beneath each patch of growth on the membrane a densely compact sporulating mycelium will be visible. In all cases where colonies develop on top of the membrane, this compacted growth can be seen below. However, during trials with this medium using artificially contaminated fuels, no growth was visible above the membrane but was present on the medium in approximately 10% of the cases. Hence absence of viable *Cladosporium resinae* should only be concluded after an examination of the medium surface.

5. DISCUSSION

Membrane filtration is well established as a technique for concentrating microbes from infected liquids. When hydrophobic liquids are examined in this way, it is generally necessary to wash the membranes with a detergent, followed by a saline rinse to remove detergent traces (Mulvany, 1969). In addition to this being a somewhat tedious process, the surfactant may adversely affect spore viability. Additionally, the author has found that a water tap vacuum pump is unable to draw many detergents through the membrane. Elimination of a detergent wash thus has obvious attractions; the procedure becomes simpler and the microbes are less severely handled.

The method described obviates the need for a detergent wash provided the excess kerosene is drawn off during the filtration and the membrane is inverted. The thin spreading nature of fungal growth on this medium does not allow for accurate counting, but it is theoretically possible to detect the presence of a single viable spore.

Table I shows the selective nature of the medium when inoculated with actual samples of aircraft fuel. In none of the cases was *C. resinae* not detected by ABHA when it was with malt extract agar and potato dextrose agar. However, fungi present in the samples merely as non-growing contaminants did not grow on ABHA, whereas they did on the more complex media which do not reflect the fuel tank environment. Additionally, in one case (sample 2), *C. resinae* was not detected with the complex media but was with ABHA, perhaps because of overgrowth of developing *C. resinae* by the yeasts, bacteria and *Penicillium* sp. Since these microbes could not grow on kerosene, *C. resinae* was able to grow unhindered on ABHA.

TABLE I

GROWTH ON MALT EXTRACT AGAR (MEA), POTATO DEXTROSE AGAR (PDA) AND ABHA INOCULATED WITH AIRCRAFT FUEL
(Rubidge, 1974)

Fuel sample number	*Growth analysed* 10 *days at* 25°*C on*		
	MEA	*PDA*	*ABHA*
1	Yeasts	Yeasts	Actinomycete
	Bacteria		
	Unidentified fungus	Bacteria	
2	Yeasts	Yeasts	*C. resinae*
	Bacteria	Bacteria	
	Penicillium sp.	*Penicillium* sp.	
3	Yeasts	Yeasts	*C. resinae*
	Bacteria	Bacteria	
	Penicillium sp.	*Penicillium* sp.	
	C. resinae	*C. resinae*	
4	*C. resinae*	*C. resinae*	*C. resinae*
5	Yeasts	Yeasts	
	Bacteria	Bacteria	*C. resinae*
	Penicillium spp.	*Penicillium* spp.	
	C. resinae	*C. resinae*	
6	*C. resinae*	*C. resinae*	*C. resinae*

The medium, being selective for kerosene fungi, does not allow non-fuel biodeteriogenic fungi to flourish, and although the technique is somewhat time-consuming there is no necessity to inoculate isolates into kerosene/water liquid media to confirm potential biodeterioration capability.

REFERENCES

Barvah, J. N., Alroy, Y. and Mateles, R. I. (1967). Incorporation of liquid hydrocarbons into aqueous media. *Appl. Microbiol.*, **15** (4), 961.

Bushnell, L. D. and Haas, F. H. (1941). The utilization of certain hydrocarbons by microorganisms. *J. Bacteriol.*, **41** (5), 653–73.

Elphick, J. J. and Hunter, S. K. P. (1968). Evaluating biocidal fuel additives for intermittent use in aircraft fuel systems. In: *Biodeterioration of Materials*, vol. 1 (ed. A. H. Walters and J. J. Elphick), pp. 364–70. Applied Science Publishers, London.

Hitzman, D. O. (1964). The control of bacterial and fungal growth in jet fuels by use of a fuel additive. *Devel. Ind. Microbiol.*, **6,** 105–15.

Mulvany, J. G. (1969). Membrane filter techniques in microbiology. In: *Methods in Microbiology*, vol. 1 (ed. J. R. Norris and D. W. Ribbons), pp. 205–53. Academic Press, London.

Rubidge, T. (1974). A new selective medium for the screening of aircraft fuel for biodeteriogenic fungi. *Int. Biodetn. Bull.*, **10** (2), 53–5.

Scott, J. A. (1971). Microbiological contamination of aircraft fuel tanks: airframe considerations. Paper presented at National Air Transportation Meeting of the Society of Automotive Engineers, Atlanta, Ga.

14

Evaluation Techniques for Biodeterioration of Water-Miscible Metalworking Fluids

H. W. ROSSMOORE

Department of Biology, Wayne State University, Detroit, Michigan 48202, *USA*

1. INTRODUCTION

In this chapter, methods for demonstrating relative bioresistance of water-miscible metalworking fluids will be described.

Such methods should be a reliable prediction of biodeterioration and also be of utility in evaluating the efficacy of antimicrobial agents incorporated in the fluids. One of the more frequently used tests in the past (Yanis and Wolfe, 1960), which has had widespread use in industry in the USA, has fallen into disfavour recently. Although it is a fairly simple test to perform and evaluate, it lacks both an acceptable degree of repeatability and reliability to actual use. Its principal advantage lay in the reliance on organolepsis. However, more knowledgeable and sophisticated industrial practitioners have recognised the need for applying microbiological techniques in order to establish biodeterioration criteria. The potential for extrapolation from the laboratory to the field has already been demonstrated (Rossmoore and Williams, 1971; Rogers *et al.*, 1975) if care is taken to simulate, where possible, all the variables involved in the actual metalworking operations.

2. MATERIALS AND METHODS

It is possible easily to standardise all aspects of this test. The procedures to be described in this chapter are suitable not only for water-miscible metalworking fluids but also for similar fluids used in rolling mills and for spin finish fibre lubrication.

Water-soluble fluids are generally classified into three general types, all of which are alkaline with a pH of as high as 9·5 at use concentrations:

(i) *Soluble oils.* Essentially mineral oils, with or without fatty oils, with an emulsifier such as petroleum sulphonate. These form oil-in-water emulsions and are diluted in use to give final oil concentrations that vary from 1 to 20%. They have an opaque or milky appearance.

(ii) *Semi-chemical fluids or preformed emulsions.* These fluids are in reality emulsions that are diluted with water by their manufacturers during compounding in order to incorporate certain water-soluble additives into the formulation. This could include wetting agents, corrosion inhibitors and biocides as well as emulsifying agents. Preformed emulsions are usually translucent in appearance owing to the lower oil concentrations.

(iii) *Synthetic.* These compounds contain no oil, are transparent, and may contain a variety of wetting agents and lubrication additives. Many of the latter may be dispersed colloidally and not truly in solution.

A more detailed specific classification can be found in ASTM D-2881 (1973).

3. TEST PROCEDURES

3.1. The Microorganisms

There are three major types of microorganisms involved in fluid deterioration, each primarily responsible for a unique aspect of spoilage:

(i) *Aerobic bacteria.* This group is dominated by *Pseudomonas* species and less frequently by members of the Enterobacteriaceae, *e.g. Escherichia coli*, *Enterobacter aerogenes* and *Klebsiella pneumoniae*. They are highly oxidative and adapt extremely well to the wide variety of organic molecules found in cutting fluids. Their growth leads to emulsion separation, loss of lubricity qualities and corrosion.

(ii) *Anaerobic bacteria.* Essentially these are organisms that reduce sulphate to sulphide, causing fluid blackening, foul odours and anaerobic corrosion of tools and work pieces. They are mostly members of the genus *Desulfovibrio*.

(iii) *Fungi.* Filamentous fungi grow well in higher concentration fluid where surface tension and osmotic conditions are less favourable for bacteria. The fungi are slime-producers and physically plug lines, thereby reducing fluid flow.

Although all three groups contribute to biodeterioration, in the long run there is a population succession in which aerobic bacterial growth precedes both anaerobic bacterial growth (Guynes and Bennett, 1959) and fungi (Rossmoore and Holtzman, 1974). Therefore, the initial inoculum need only contain sufficient numbers of active aerobic bacteria with an appropriate protocol simulating field conditions. The other groups will appear in susceptible fluids.

FIG. 1. Physical set-up with individual aeration control.

3.2. Method

The test is carried out in a 1 litre container with a maximum height to surface area ratio (Fig. 1). In each container, the following are added in order: 10 g of metal chips (the type appropriate to the operation), 100 ml of inoculum and sufficient diluted test fluid to reach the neck. The contents are carefully mixed and set aside for 64 hr. This approximates a weekend shutdown period and is the first of eight such quiescent periods during the test. During quiescence, anaerobic and facultative bacteria are encouraged to grow and the fungi present initiate sites of attachment. After the shutdown period, aliquots are removed for bacterial and fungal viable counts and pH measurements made. Then each system is exposed to a regimen (Fig. 1) of compressed air at a rate of *c*. 25 litres/hr via suitable glass capillary tubing. No attempts should be made to filter the air since adventitious biological and chemical contaminants will add to the bio-deteriogen burden, further parallelling field conditions. Aeration even at low levels may produce excessive foaming in some fluids. If this occurs, add a suitable silicone-based antifoam agent (in sufficient quantity to depress the foaming). Aeration is continued for about $5\frac{1}{2}$ days followed by a 64 hr shutdown.

The procedure following the second and subsequent shutdowns differs from the first as follows: fluid evaporation is made up with distilled or deionised water to the original level prior to removing aliquots for microbial evaluation. Then 100–120 ml of fluid is removed, just sufficient to fill a suitable screw-cap bottle to the top. This removal simulates average drag-out during a week of operation and is replaced with fresh fluid. The used fluid, sequestered in the screw-cap bottle (Fig. 2), is set aside and will gradually become anaerobic. If viable sulphate-reducers are present they will become evident in time by slowly blackening the fluid, beginning at the bottom of the bottle.

FIG. 2. Two emulsion samples showing differences in sulphate reduction.

4. THE INOCULUM

The selection of the inoculum is most important since it is obvious that no test can realistically measure absolute biodeterioration resistance. Performance is a function of the particular environment. Each milieu has its own adapted flora and therefore, to be meaningful in a practical sense, the inoculum used in a particular environment should be taken from that environment. Recently collected spoiled cutting fluid from an operation or site concerned with the test fluid is used as the specific biodeteriogen. This fluid should contain no less than 10^7 viable aerobic bacteria/ml, so that when diluted in the test the initial count will not be less than 10^6/ml. If fungi and sulphate-reducing bacteria have been a problem in this area, then the inoculum will reflect this fact and during the 8 weeks of aeration/shutdown both these groups will appear if the fluid is susceptible.

In addition to serving as a source of microbial species, the spoiled fluid also adds some metabolites that will shorten the lag period in the new environment. Frequently fungi and sulphate-reducer species may be seen during the course of the test, even when not originally detected in the inoculum. Undoubtedly these come from the contaminated compressed-air lines and reflect the importance of the aeration process in biodeterioration induction.

During the last 15 years two genera of imperfect fungi, *Cephalosporium* and *Fusarium*, have been consistently isolated, and in situations where one group has been purposefully added as a part of the inoculum, the other has been isolated also before the test period ended. The experience has been that if these fungal groups are to become serious biodeteriogens in a fluid, they will appear adventitiously. However, if an antimicrobial agent is a putative fungicide or fungistat, then it would be necessary to add known fungal inocula at the beginning of the test procedure along with spoiled fluid in order to establish efficacy. Since the primary purpose of this chapter is not biocide evaluation, this will not be discussed here.

5. EVALUATION OF GROWTH

The selection of retrieval media for estimating viable numbers of organisms can be very critical. Most of the aerobic species growing in metalworking fluids are not discriminating nutritionally, but for maximal colony development in the 48 hr incubation period, soybean–casein digest agar is most successful. This medium is made by a number of manufacturers in the USA and the UK.

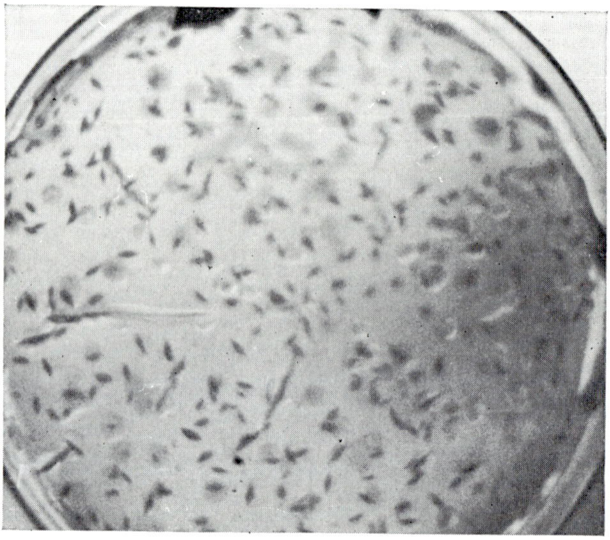

(a)

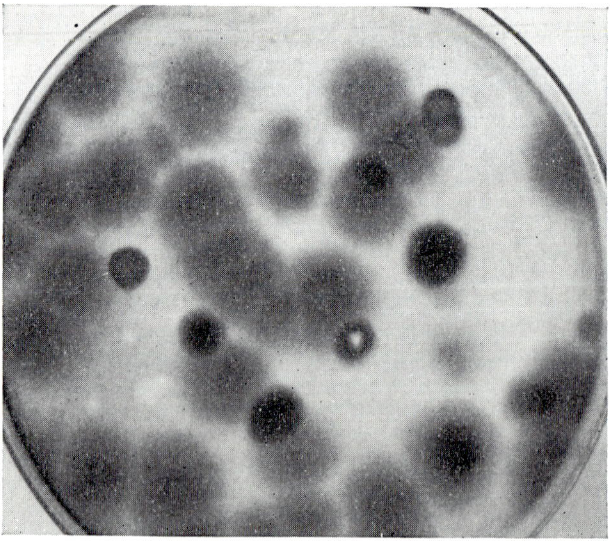

(b)

FIG. 3. (a) Potato dextrose agar plate from cutting fluid sample showing bacteria-like colonies. (b) Potato dextrose agar plate with 5 μg gentamicin/ml showing fungal colonies. Same cutting fluid sample as (a).

Soybean–casein digest agar USP.

(Trypticase soy agar, tryptic soy agar, tryptone soya agar, casein soy peptone agar, etc.)

Pancreatic digest of casein USP	15 g
Papaic digest of soy meal USP	5 g
Sodium chloride	5 g
Agar	15 g
Distilled water	1 litre

Final pH 7·3

Heat with agitation until the medium boils. Dispense and autoclave at 118–121°C for 15 min.

Determination of fungal levels requires the use of a more selective medium, since fungi are usually present in much smaller numbers than bacteria and tend to be overgrown by them. Traditional media for fungi rely on low pH (*c.* 5·6) for selectivity. However, it has been found that certain bacterial species, *e.g. Klebsiella* sp., will grow at pH 5·6 and crowd out the fungi or make them difficult to detect. In Figs. 3(a) and 3(b) the effects of gentamicin on detection of fungal populations is readily seen. These two plates both contain potato dextrose agar at pH 5·6 and have been mixed with an equal amount of the same metalworking fluid. Figure 3(b) has had 5 μg of gentamicin/ml added. This made it possible to detect the fungi. The bacterial colonies growing in the absence of gentamicin in this case were determined to be *Klebsiella pneumoniae*. Fungal agar with gentamicin gives maximal fungal yields even with bacterial levels in excess of 10^8/ml (Rossmoore and Treusch, 1975). In a comparative study, Czapek medium proved most successful and is available commercially from a number of manufacturers.

Czapek–Dox solution agar + Gentamicin:

Gentamicin reagent solution 1 %*	10 ml
Sucrose	30 g
Sodium nitrate	2 g
Dipotassium phosphate	1 g
Magnesium sulphate	0·5 g
Potassium chloride	0·5 g
Ferrous sulphate	0·01 g

* Gentamicin Reagent Solution 50 mg/ml—Schering Corporation.

Agar	15 g
Demineralised water	1 litre

Final pH 7·3

Mix and heat to boiling to dissolve completely.
Dispense and autoclave at 121°C for 15 min.

For qualitative detection of sulphate-reducing bacteria the sequestered screw-cap bottles mentioned earlier will suffice. For semi-quantitative estimation, several liquid and solid media are commercially available (Sulfate API Broth—Difco Corp., and Iron Sulphite Agar—Oxoid Ltd). The sulphate-reducer medium is placed in 15 ml quantities in 20 ml screw-cap tubes and sterilised according to directions. The iron sulphite agar is maintained between 45° and 50°C after sterilisation and used as soon as possible. For each sample to be tested, five tubes are used and inoculated according to the following protocol:

Inoculate one set with 0·1 ml used coolant.
Inoculate one set with 0·5 ml used coolant.
Inoculate one set with 1·0 ml used coolant.
Inoculate one set with 2·5 ml used coolant.
Inoculate one set with 5·0 ml used coolant.

In both the liquid and solid media add sufficient sterile medium to reach the top of the tube. Close tightly, mix thoroughly and incubate for 1–5 days.

5.1. Standard Microbiological Procedure

In the counting of bacterial and fungal populations, standard plate count techniques are used and the minimal equipment needed is as follows:

1. Sterile bacteriological pipettes. Pipettes made to deliver contents rapidly are preferred. One such pipette calibrated to deliver 0·1 ml and 1 ml in duplicate is highly desirable (2·2 ml bacteriological pipette).
2. 100 × 15 sterile petri dishes, glass or plastic.
3. One illuminated colony counter.
4. One calibrated water bath to maintain sterile agar between 45° and 50°C.
5. 99 ml sterile water blanks in closed containers suitable for adequately mixing dilutions of fluid to be counted.
6. One incubator that can be set at 28° ± 1°C.

For counting bacterial populations, and particularly in trying to assess the rate of biodeterioration, it is most valuable to come as close as possible to the actual viable count. Therefore, it is recommended that pour plates for total aerobic counts be done in duplicate on the undiluted fluid and on decimal dilutions up to $1{:}10^6$. For fungal counts it is not necessary to go beyond a dilution of $1{:}10^3$ since levels of fungal colony-forming units rarely exceed 10^5/ml.

Although all three groups of organisms will grow at ambient room temperature (*c.* 25°C), it is preferable to incubate them at a slightly higher temperature, 28–30°C. Aerobic bacterial counts should be made after 48 hr, fungi at 72 hr and sulphate-reducers from 1 to 5 days, and evaluated according to the key in Table I.

TABLE I

KEY FOR SULPHATE-REDUCER LEVELS IN METALWORKING FLUIDS

Sample size		*Incubation time*			
	Positive in:	*24 h*	*48 h*	*72 h*	*120 h*
0·1		4+	3+	2+	+
0·5		3+	2+	2+	+
1·0		3+	2+	1+	+
2·5		2+	2+	1+	+
5·0		2+	1+	1+	+

6. EVALUATION OF RESULTS

There are no fixed microbial levels at which it can be said that the fluid has become deteriorated. A number of factors must be considered in assessing biodeterioration. These include pH, colour, odour and appearance, and it is possible to correlate many of these changes with increases in microbial levels. As a rule, when the aerobic bacterial population reaches 10^7/ml, most probably the fluid is becoming biodeteriorated. However, many fluids in practical use conditions continue to function at this level, the organisms no doubt feeding on extraneous organic material added to the sumps containing the metalworking fluid.

Figure 4 depicts a typical aerobically spoiled fluid. The emulsion has split and the milky-white fluid has turned brown. The bacterial count in this fluid when the photograph was taken was 2×10^8/ml. Figure 5 shows the results of bacterial growth microscopically. In the background there are large numbers of rod-shaped cells and throughout the field various-sized oil droplets, a sign of emulsion breakdown since as droplet size increases, emulsion stability decreases.

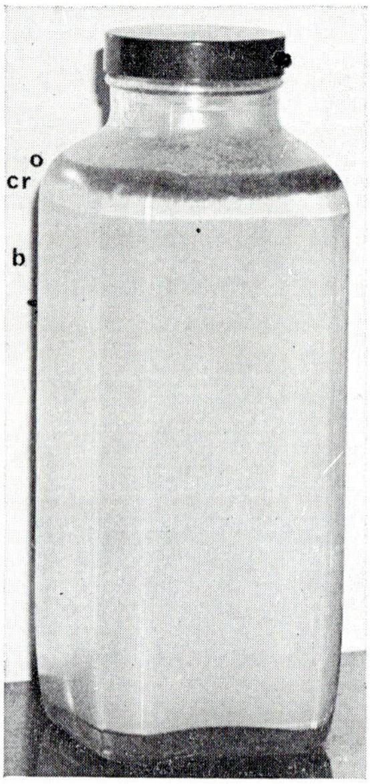

FIG. 4. An emulsion cutting fluid deteriorated from aerobic bacterial growth. o = oil line, cr = cream line, b = body of fluid which is turned tan-brown.

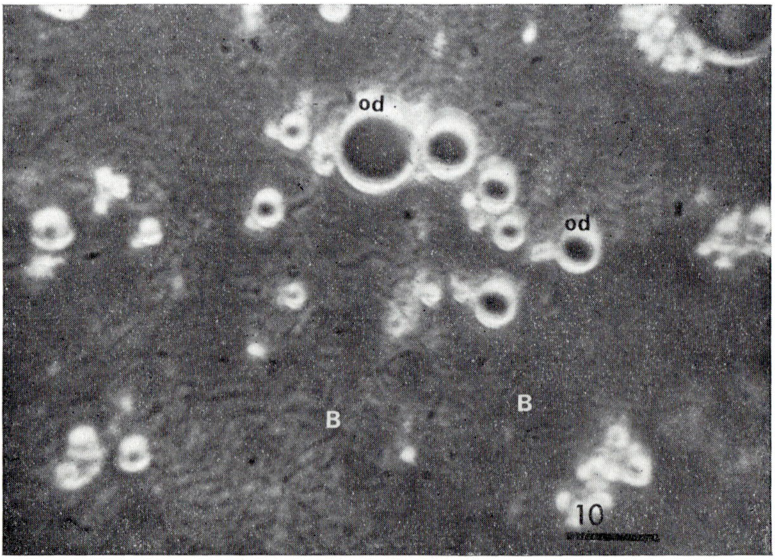

FIG. 5. Phase contrast (×667) of spoiled emulsion in Fig. 4. od = oil droplets, B = bacteria.

As the bacterial levels increase, the pH usually falls. However, this is not always the case, and although a drop in pH is indicative of bacterial activity resulting from oxidation of hydrocarbon to organic acid, there may in fact be incomplete oxidation with no consequential drop in pH. At the same time, in those fluids in which sodium nitrite is used as a corrosion inhibitor, there may in fact be a loss in corrosion protection due to the conversion of nitrite to assimilable ammonia. In these cases the ammonia would tend to raise the pH. If the mixed population contains *Proteus* sp. or *Coliform* sp., other unpleasant odours commonly referred to as 'outhouse (privy) smells' may be detected. These too are signs of biodeterioration.

The role of fungi in biodeterioration is not so easily evaluated with plate counts since this group can yield colonies from many kinds of biological particles (propagules), including micro- and macroconidia, hyphal strands or mycelial clumps, as well as sclerotia, which may yield no more colonies than a conidium with a great deal less biomass. Colony-forming units can be viewed as merely indicative as far as fungi are concerned, since quite often counts are low or non-existent even when fungal masses are grossly visible (Figs. 6 and 7) and when examined microscopically (Fig. 8) usually exhibit structural characteristics associated with *Fusarium* sp.

FIG. 6. Fungal mass growing attached to tip of aeration tube in an emulsion cutting fluid.

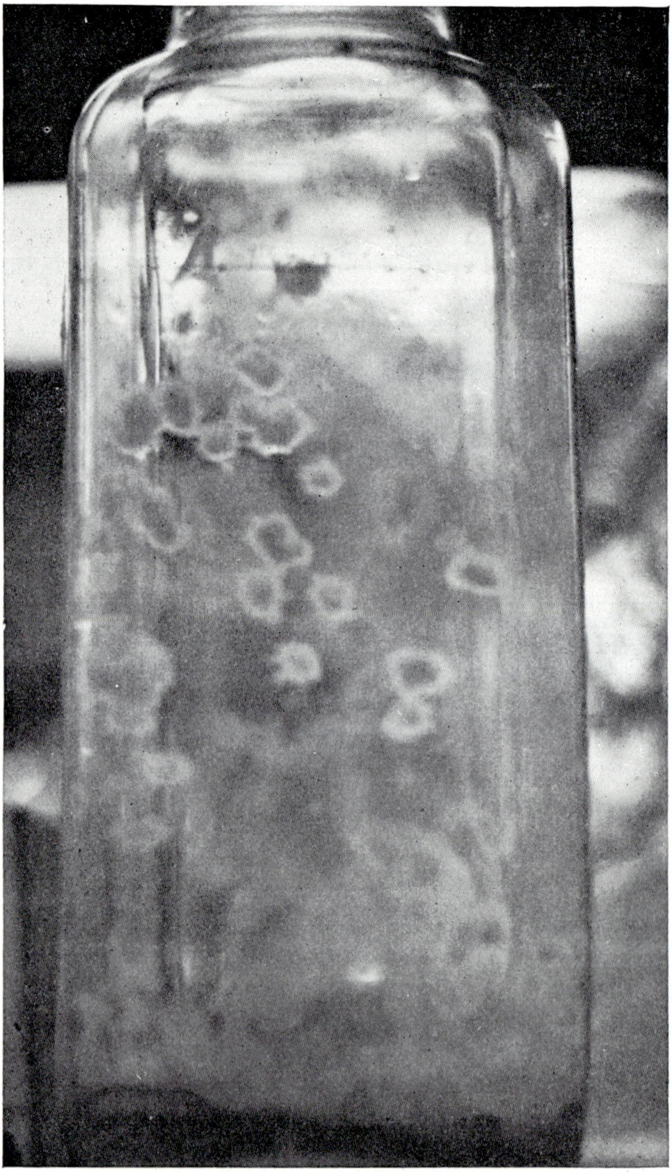

FIG. 7. Fungal colonies growing in synthetic cutting fluid and attached to inside of glass container.

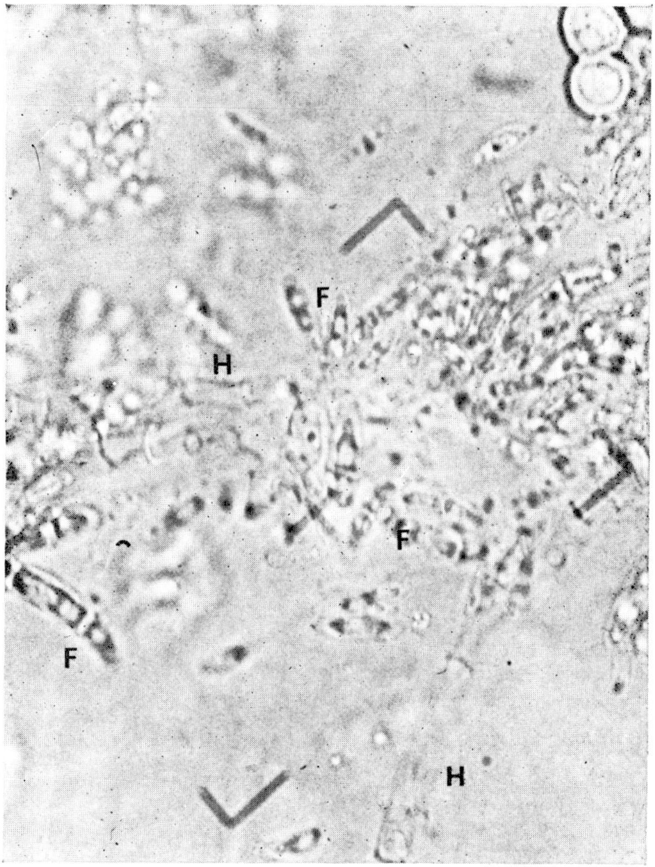

FIG. 8. Phase contrast (×360) of fungal mass from Fig. 6. F = *Fusarium* macroconidia, H = hyphae.

Notice the fungal mass at the tip of the aerated tube and the fungal colonies surrounding the inside of the glass container which had been used for one test system. In both cases the fungal levels were less than 30/ml when first detected. In subsequent weeks, the flaking-off of filamentous particles increased the suspended amounts of fungi to give more representative counts. In some studies the entire contents of the 1 litre system have been subjected to 15 sec in a commercial blender prior to taking counts in order to arrive at a more exact level of fungal biomass. With fungal involvement the large amount of slime and the musty odour are perhaps the primary concerns in biodeterioration, since the mustiness seems to be more objectionable to some workers even than 'privy' smells, and the slime can cause overt mechanical problems. Suspended fungal levels in excess of 10^3/ml should be considered serious, with the system progressing quickly towards deterioration.

The anaerobic sulphate-reducers cause problems, even when present in small amounts, because hydrogen sulphide odours are quite objectionable. These bacteria tend to be insidious, forming at the bottom of large central circulating systems where the oxidation reduction potential is low. A good example of this can be seen in Fig. 9 in which several systems are showing

FIG. 9. Eight-week samples with varying levels of sulphate reduction.

various degrees of sulphate reduction after a weekend shutdown towards the end of an 8 week test. During the test period the sequestered screw-cap bottles will give qualitative sulphate-reducer results; obviously, the sooner after sequestering and the earlier in the test period reduction is noticed, the greater is the sulphate-reducer problem. For a more quantitative evaluation the sulphate reducer media can be interpreted from Table I. Systems with readings above 2+ should be considered biodeteriorated from anaerobic sulphate-reducer activity and would reflect innate susceptibility of the particular fluid to those microorganisms.

TABLE II

MICROBIAL LEVELS USING SYNTHETIC METALWORKING FLUID

Week	*Aerobic bacteria/ml*	*Fungi/ml*	*Estimated sulphate-reducers*
1	40×10^6	86×10^3	+
2	7×10^6	31×10^3	++++
3	35×10^6	27×10^3	+++
4	44×10^6	95×10^3	++
5	28×10^6	43×10^3	++
6	77×10^6	49×10^3	+++
7	50×10^6	70×10^3	+++
8	90×10^6	4×10^3	++
9	160×10^6	3×10^3	+++
10	400×10^6	12×10^3	+++

Thus all three types of organisms can cause specific deterioration problems. In Tables II and III, data from actual tests are shown. In Table II, where the aerobic counts are higher than in Table III, the fungal and sulphate-reducers are also higher. Notice weeks 8–10 in Table II; these have the highest aerobic counts and the lowest fungal counts. This is not atypical and is to be expected with bacterial/fungal interaction. Despite the apparent interdependence of fungi and anerobic sulphate-reducers on aerobic bacterial levels, each group independently can produce a level of biodeterioration resulting in failure and rejection of a metalworking fluid.

TABLE III

MICROBIAL LEVELS IN A SOLUBLE OIL METALWORKING FLUID

Week	*Aerobic bacteria/ml*	*Fungi/ml*	*Estimated sulphate-reducers*
1	15×10^6	100	+
2	19×10^6	4800	++
3	23×10^6	200	++
4	10×10^6	900	++
5	$5{\cdot}2 \times 10^6$	100	+
6	$15{\cdot}5 \times 10^6$	10	++
7	12×10^6	40	++
8	1×10^6	10	+
9	$0{\cdot}2 \times 10^6$	10	+
10	15×10^6	300	++

REFERENCES

American Society for Testing and Materials (1973). *Standard Classification for Metalworking and Related Materials*, **28,** 1060–5.

Guynes, G. C. and Bennett, E. O. (1959). Bacterial deterioration of emulsion oils. I. Relationship between aerobes and sulfate-reducing bacteria. *Appl. Microbiol.*, **7,** 121–5.

Rogers, M. R., Kaplan, A. M. and Beaumont, E. (1975). A laboratory in-plant analysis of a test procedure for biocides in metalworking fluids. *Lubr. Eng.*, **31,** 301–10.

Rossmoore, H. W. and Holtzman, G. H. (1974). Growth of fungi in cutting fluids. *Devel. Ind. Microbiol.*, **15,** 273–80.

Rossmoore, H. W. and Treusch, P. J. (1975). Comparison of several media for enumeration of cutting fluid fungi. *Devel. Ind. Microbiol.*, **16,** 475–82.

Rossmoore, H. W. and Williams, B. W. (1971). An evaluation of a laboratory and plant procedure for preservation of cutting fluids. *Int. Biodetn. Bull.*, **7,** 55–60.

Yanis, R. J. and Wolfe, G. F. (1960). Test procedures for the evaluation of cutting fluids. *Lubr. Eng.*, **16,** 164–70.

15

Methods for Determining Microbial Biodeterioration in Cellulose

FRED A. ROSENBERG and LINDA VALENTINE OBERKOTTER

Department of Biology, Northeastern University, Boston, Massachusetts 02115, *USA*

1. INTRODUCTION

Microorganisms have little difficulty in digesting cellulose, accomplishing it readily and effectively via their cellulase enzyme systems. The measurement of cellulase activity on various substrates has been performed by means of a number of techniques. Major emphasis has been placed on degradation of cellulose substrates through fungal action; extracellular enzymes secreted by these organisms hydrolyse cellulose, yielding utilisable sugars. Numerous bacteria have likewise been shown to have cellulolytic capabilities but, because of lesser yields of readily available enzyme, as well as the fact that substrate–organism contact is often a requirement for hydrolytic activity, the extant literature generally deals with fungal rather than bacterial systems.

It is neither the objective of this presentation specifically to discuss the theory or biochemistry of cellulase activity, nor is it meant to serve as a review of the literature. Rather, it is intended as a guide for the laboratory worker interested in measuring cellulose degradation. To this end, we have attempted to confine ourselves to the description of a number of methods applicable to various situations in which cellulase activity is measured. In order to facilita.e actual laboratory analyses, we have considered it pertinent to describe not merely the methodology involved, but to include formulations of constituents, as well as relevant precautions to be considered while performing the test procedures.

2. SCREENING TECHNIQUES

2.1. Degradation of Filter Paper

The qualitative determination of the presence of cellulolytic micro-organisms may be accomplished by a variety of methods. Rosenberg and Breiter (1969) showed the presence of cellulolytic bacteria in the intestinal tract of marine invertebrates using the following technique.

The stomach, intestine and caecum of representative organisms are removed and rinsed gently in 95% ethyl alcohol. The individual organs are then placed into test tubes containing a sterile mineral salts medium of the following composition:

KNO_3	1·0 g
K_2HPO_4	1·0 g
$CaCl_2$	0·1 g
$FeCl_2$	trace

Lyman and Fleming (1940) artificial seawater to 1000 ml*

Rectangular pieces of sterilised Whatman No. 1 filter paper are placed into each tube to serve as the sole carbon source (Fig. 1). The tissue of the

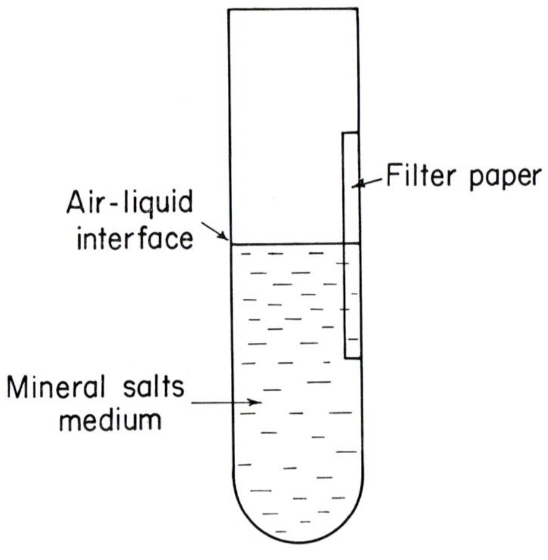

FIG. 1. Test tube method for isolation of cellulolytic bacteria.

* Lyman and Fleming (1940) artificial seawater: NaCl, 23·476 g; Na_2SO_4, 3·917 g; $NaHCO_3$, 0·192 g; KCl, 0·664 g; KBr, 0·096 g; $MgCl_2$, 4·981 g; $CaCl_2$, 1·102 g; $SrCl_2$, 0·024 g; H_3BO_3, 0·026 g; distilled H_2O to 1000 ml. Final pH 7·78. (Modified by Greenfield, 1964.)

organs is gently macerated with a sterile pipette, and the tubes incubated at the temperature of the water from which the organisms under test were isolated. Browning and eventual shredding of the filter paper at the air–water interface indicates degradation of the cellulose fibres (Fig. 2).

For isolation of the cellulolytic organisms, loopfuls of the medium are gently scraped from the surface of the decomposing filter paper and transferred to plates of Bacto-Marine Agar 2216E (Difco) and incubated. Finally, the organisms growing on the marine agar are streaked on to cellulose agar plates with the following composition:

$NaNO_3$	0·5 g
K_2HPO_4	1·0 g
$MgSO_4 \cdot 7H_2O$	0·5 g
$FeSO_4 \cdot 7H_2O$	0·01 g
Powdered cellulose	6·0 g
Bacto-Agar	8·0 g

Lyman and Fleming artificial seawater to 1000 ml

These plates are periodically examined for cellulolytic activity, as evidenced by liquefaction, cratering or weathering of the medium. Duplicate plates are inoculated without cellulose to determine whether agarolytic bacteria are responsible for liquefaction.

The method has been adapted to determine cellulolytic activity of bacteria isolated from beetles. Tap water is substituted for artificial seawater and marine agar is replaced by an appropriate nutrient agar (F. A. Rosenberg and E. Klausner, unpublished).

Berg *et al.* (1972) describe a comparable medium:

$NaNO_3$	2·0 g
K_2HPO_4	0·5 g
$MgSO_4$	0·2 g
$CaCl_2 \cdot 2H_2O$	0·02 g
$MnSO_4 \cdot H_2O$	0·02 g
$FeSO_4 \cdot 7H_2O$	0·02 g

Distilled H_2O to 1000 ml
pH after autoclaving = 7·5

By adding various insoluble cellulose sources (*e.g.* microcrystalline cellulose, newspaper pulp, filter paper) in concentrations of 0·4%, which may be autoclaved with the media, it can be enriched for cellulolytic

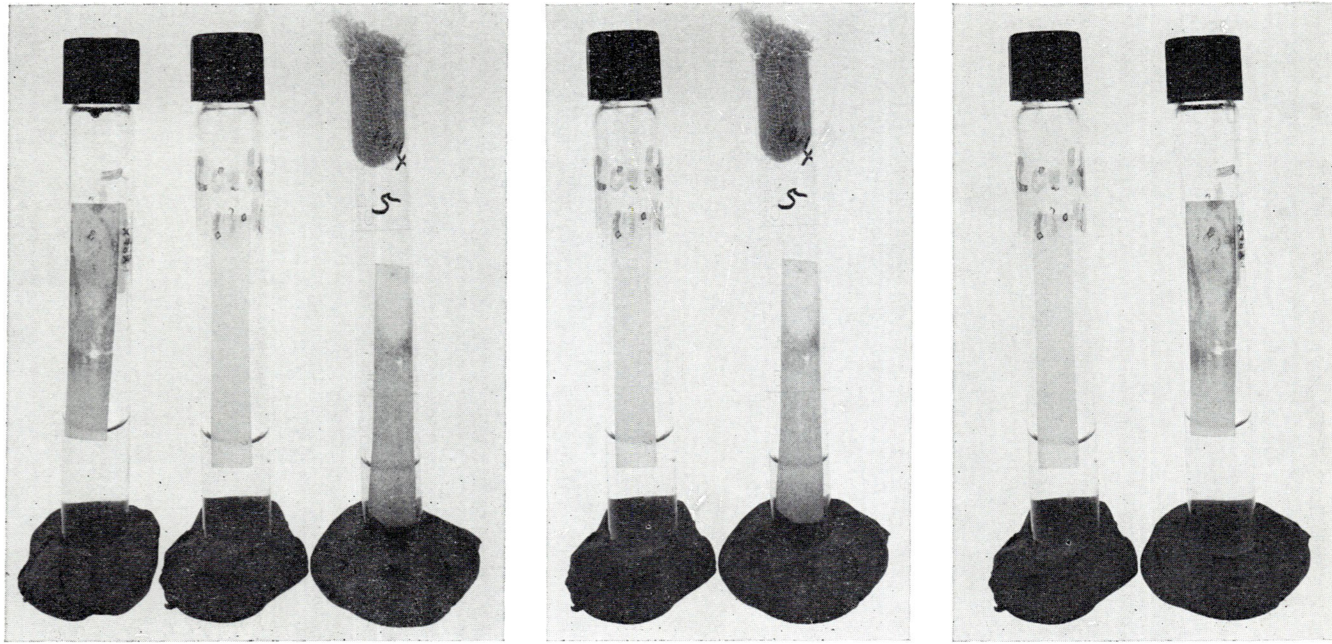

(a) (b) (c)

FIG. 2. Shredding of filter paper through bacterial action. In (a) the central tube is the control, in (b) and (c) the left-hand tube.

bacteria from environmental samples. However, water samples should first be concentrated via Millipore filtration.

A variety of other modifications to these media may be found throughout the literature.

2.2. Agar Diffusion Method

Rautela and Cowling (1966), in a method later modified by Tansey (1971) in studies on thermophilic fungi, used an agar column containing a partially crystalline cellulose preparation to determine cellulolytic activity of fungi. The method is based on the principle that dissolution of cellulose is an essential part of its degradation. Fungi, growing on the surface of the column, secrete cellulases which diffuse into the agar, dissolving the particles and creating a zone of clearing.

Cellulose prepared according to the method of Walseth (1952) gave optimally distinct clearing, as opposed to either Whatman powdered cellulose or Avicel.

Walseth cellulose is prepared using Whatman powdered cellulose, which is swollen in 85% orthophosphoric acid for 2 hr at 4°C, regenerated and washed in the cold by repeated suspending and decanting with distilled water, then 1% (w/v) Na_2CO_3, followed by distilled water again until neutral. Five grammes of the resulting suspension (dry weight) of cellulose particles is combined with 2·0 g of $NH_4H_2PO_4$, 0·6 g KH_2PO_4, 0·4 g K_2HPO_4, 0·89 g $MgSO_4 \cdot 7H_2O$, 100 μg thiamine HCl, 0·5 g yeast extract, 4·0 mg adenine, 8·0 mg adenosine, 17 g of agar and distilled water to 1 litre. Sterile, uniformly opaque, vertical columns (3–6 cm in height) of this medium are prepared in 18 mm test tubes and inoculated with discs of fungal mycelium and agar from petri-plate cultures of the organisms to be tested. Depth of clearing in the tubes is measured at 7 day intervals against a black background, using a millimetre ruler.

Larger values of degree of clearing (DC) were obtained with concentrations of 0·125 and 0·25% than with 0·50 or 1·0% (w/v) Walseth cellulose. Batches of Walseth cellulose prepared at different times showed a variation in DC values with the same fungus and incubation time. It is suggested that comparisons of relative cellulolytic activity be made on the same batch of substrate.

3. ASSAYS FOR CELLULASE

Ideal assays are simple and closely approximate the conditions for which the enzyme is being studied. As with any other enzyme assay, it is essential that substrate concentration be high enough to maintain saturation throughout the incubation period.

Fungal extracellular cellulases may be easily obtained by centrifuging samples to remove solids; the filtrate is then assayed. Bacterial cellulases may be more complex to assay, since they are found as in the above-mentioned extracellular manner, as well as intracellularly and bound within the cell envelope (Yoshikawa *et al.*, 1974). In the latter case, a variety of cell-rupturing techniques may be employed to facilitate examination.

In all cases, blanks of enzyme without substrate, and substrate without enzyme, should be included with all assays in order to correct samples by the blank values.

Methods most commonly used may be grouped under five major categories:

Method	*Enzyme activity measured*
1. Reduction of viscosity in solution of a cellulose derivative.	1. B-glucanase (C_x) activity independent of C_1 action.
2. Production of reducing sugar from soluble cellulose derivative.	2. Same as above; considerably more activity is required to show rise in reducing sugar than change in viscosity.
3. Production of reducing sugar from insoluble cellulose.	3. Role of C_1 becomes more important as resistance of the substrate increases.
4. Weight loss of cellulose.	4. Same as above.
5. Reduction in optical density of sol or suspension of cellulose.	5. Usually measures C_x; with hydrocellulose, can be used to measure C_1.

3.1. Method 1: Reduction of Viscosity

In one method used in our laboratory, sodium carboxymethylcellulose (CMC) broth is used as a culture medium for the organisms under study. The composition of the broth is as follows:

KNO_3	1·0 g
$MgSO_4 \cdot 7H_2O$	1·0 g
K_2HPO_4	1·0 g
$FeCl_3$	trace
CMC	4·0 g
Tap H_2O to 1000 ml	

For a solid medium, 3% purified agar is added.

Inoculated broth is kept on a shaker for 48 hr (80 rev/min, 25°C), after which time the cultures are centrifuged at 10 000 G (IEC International Refrigerated Centrifuge, Model B-20) for 10 min. This treatment deposits a satisfactorily large pellet without the cells being overly clumped and thus difficult to resuspend. Cells are washed twice using 25 ml portions of pH 8·4 alcoholic phosphate buffer and resuspended.

Alcoholic phosphate buffer:

K_2HPO_4	3·560 g
KH_2PO_4	0·580 g
Ethyl alcohol (30 %)	500 ml

Cell suspensions are diluted where necessary with the abovementioned buffer to a density of approximately 10^8 cells/ml (McFarland standard No. 4).

Five millilitres of the 10^8 buffered cell suspension are added to flasks containing approximately 80 ml 0·4 % CMC in broth. The cultures are kept on a shaker table for 72 hr at 25°C and 15 ml samples are withdrawn every 24 hr during this period and filtered through Millipore (Bedford, Mass., USA) membrane filters (pore size 0·45 μ HAWP 04700). The resulting filtrate is used as the crude enzyme sample. If the culture becomes extremely turbid, centrifugation may be substituted for filtering, the supernatant then being used as the enzyme sample.

Reaction mixture:

(1) 2 ml cellulase solution
(2) 5 ml 1 % CMC in buffer
(3) 2 ml McIlvaine citrate PO_4 buffer, pH 7·0
(or pH being tested)
Solution 1: 0·2 M Na_2HPO_4
Solution 2: 0·1 M citric acid

Mix appropriate amounts of solution 1 and solution 2; dilute to 100 ml with distilled water (*see* McIlvaine (1921) for appropriate proportions).

The reaction mixture is placed in an Ostwald viscometer (set level in a water bath, free from draughts) at a temperature of 30°C or suitable test temperature.

Viscosity is determined at time $t = 0$; measurements are repeated at intervals of 5, 10 and 15 min, timing with a stopwatch. All tests are run in duplicate. Controls follow identical procedure except that heat-inactivated samples (100°C, 20 min) are used. Relative viscosity is determined by the formula:

$$\text{Relative viscosity} = \frac{\text{Flow time of reaction solution}}{\text{Flow time of water}}$$

This value may be multiplied by 100 to obtain a percentage expression.

Whereas the abovementioned value is a satisfactory expression for comparative purposes, either specific viscosity or specific fluidity (its reciprocal) is used as a measure for cellulase activity in most cases where a more definitive value is desired.

The values obtained may be plotted against time to follow initial rate of enzymatic activity. Further mathematical treatments of the data may be found in Thomas (1956), Yoshikawa *et al.* (1974) and Eriksson (1969). Viscosity of CMC solutions is affected by pH value and ionic strength of solvent (Iwasaki *et al.*, 1964) and must therefore be fixed when comparisons of cellulase activity are carried out with different CMC solutions. In addition, Reese *et al.* (1950) have shown that the degree of substitution (DS) of cellulose derivatives considerably influences the rate of enzymatic hydrolysis—the higher the DS, the lower the rate of hydrolysis. The degree of substitution, if not listed on the product packaging, can generally be obtained from the manufacturer.

The viscometers used require frequent, extensive rinses, preferably with hot water, owing to the viscous, particulate nature of the CMC. Residual CMC should be washed out thoroughly before acetone is added to the apparatus (for the purpose of drying), or a cloudy precipitate will result, further complicating the wash. Meticulously dry and clean viscometers are extremely important to the accuracy of this method. Several viscometers (corrected for differences in flow time for distilled water) should be employed, allowing for adequate drying time between runs. Viscometers not in use should be stored in a desiccator. This methodology is not recommended for a large number of samples, and is most useful when samples are weakly cellulolytic, and are not readily assayed by other methods (Valentine, 1973).

3.2. Method 2: Production of Reducing Sugar from Soluble Cellulose Derivatives

The most widely used assay for measuring reducing sugars resulting from cellulolysis is the DNS method, as adapted by Mandels (1974) from Miller (1959).

Preparation of the DNS reagent:

Mix:		
	Distilled water	1416 ml
	3,5-dinitrosalicylic acid	10·6 g
	NaOH	19·8 g

Dissolve the above constituents and then add:

Rochelle salts (NaK tartrate)	306·0 g
Phenol	7·6 ml
Na metabisulphite	8·3 g

Titrate 3 ml of sample with phenolphthalein and 0·1 N HCl. This procedure should require approximately 5–6 ml HCl. Add NaOH if required (2 g = 1 ml of 0·1 N HCl).

3.2.1. *Fungal Cellulases*

(i) *Presence of reducing sugars in media.* The sample should contain 0·2–1·0 mg of glucose per ml. Place 1 ml of sample into test tube, add 3 ml DNS reagent, and place in boiling-water bath for 5 min. Allow to cool to room temperature. The percentage transmittance is read at 550 nm with a water blank set to 100% transmittance. For a colorimeter with a very short light path the sample may be read directly. If the light path is 1 cm, the sample is diluted with distilled water to 20 ml after boiling. If cellulose or derivatives are the only carbon source included in the medium, this assay will reflect *in vivo* C_x activity.

Standards are plotted on semilog paper (log %T versus concentration). This should give a straight line intersecting the abscissa at 0·04 mg of glucose. The 0·04 figure represents the glucose lost by oxidation. For an accurate determination of low concentrations of glucose, 0·1 mg of glucose is added to each sample.

Colour (gold to golden-brown) will develop only under alkaline conditions so that acidic samples should always be neutralised. The method is non-specific and will measure any reducing compound.

Since 3 ml of DNS reagent will react with 10 mg of glucose, samples which are too dark to read may be diluted up to tenfold with water. Dilutions are made following boiling and results multiplied by the dilution factor. Boiled samples may be kept for a reasonable time before reading, whereas unboiled samples will tend to deteriorate gradually.

(ii) *Carboxymethylcellulose assay* (Mandels, 1974):

Reagents:

1% carboxymethylcellulose (CMC)

Dissolve 10 g CMC in 800 ml of hot (80–90°C) water by adding the dry powder slowly with continuous agitation. Add 100 ml 0·5 M citrate buffer, pH 4·8, and 10·0 ml 1% merthiolate. Dilute to 1 litre and refrigerate (2–4°C). Warm to 50°C with shaking before using.

0·05 M citrate buffer pH 4·8 (final concentration)
Glucose standards in buffer
DNS reducing sugar reagent
Procedure: Combine 0·5 ml enzyme with 0·5 ml of 1% CMC. Mix well and incubate for 30 min at 50°C. Add 3 ml DNS reagent to stop the reaction. Place tubes in boiling water for 5 min and determine reducing sugar as glucose.

If the glucose value exceeds 0·6 mg, repeat the procedure using diluted enzyme. Most active cellulase culture filtrates have been shown to require a dilution of from 1:100 to 1:500 for this assay. For an optimal reaction rate, the degree of substitution (DS) should be less than 1·0.

This assay measures initial rate of enzyme (C_x) activity *in vitro*, using reducing sugars produced within 30 min incubation time as indicative of the CMCase activity present.

3.2.2. *Bacterial Cellulases*

The following modifications of the method of Berg *et al.* (1972) are used in our laboratory for measuring C_x activity of culture filtrates or various cell fractions. C_1 activity may be similarly measured by substituting filter paper or crystalline cellulose for the CMC, and extending the 20 min incubation period given here.

Reagents: 1% CMC in 0·05 M Sorensen's potassium phosphate buffer, pH 7·0. Stir until CMC is totally dissolved (warming increases rate of solubilisation); solution should be homogeneous.
0·01% Na azide is added to the above. Store at 5°C and use within a month.
Procedure: 0·3 ml of sample (enzyme) and 2·0 ml CMC–buffer solution (40°C) are placed in a test tube and incubated at 40°C for 20 min.
The reaction is stopped by transferring tubes to an ice-water bath and then adding 3·0 ml DNS reagent per tube; the samples are next immersed in a boiling-water bath for 15 min to permit colour development.
Samples are allowed to cool under running tap water and are then centrifuged at approximately 3000 rev/min for 5 min in a table-top centrifuge. The supernatant from each tube is drawn off using a long-tipped Pasteur pipette and transferred to a cuvette. Absorbence is read spectrophotometrically at 600 nm.

If samples fall beyond the linear region of the standard curve (formulated with known glucose concentrations in place of enzyme samples), the enzyme should be diluted and the procedure repeated.

It is advisable to run controls to determine the extent of reducing sugar in the filtrate at zero time (*i.e.* a control run eliminating the 20 min incubation period), as some organisms may build up reducing sugars in the medium (and, in fact, their growth may be monitored in this manner).

3.3. Method 3: Production of Reducing Sugar from Insoluble Cellulose

The following methods are described by Mandels (1974).

3.3.1. *Amorphous Cellulose Assay for Cellulase*

Reagents: 1% Walseth cellulose. The method for preparation of the Walseth cellulose is slightly different from that given by Rautela and Cowling (1966) in Section 2.2 above. 20 g of cellulose powder is added from a sieve to 550 ml cold 85% orthophosphoric acid in a large mortar which is kept in a cracked-ice bath. Following intermittent stirring for 30–60 min, the mixture is poured into a large volume of ice water, washed by decantation and filtered on a large Buchner funnel. The cellulose is resuspended, blended and washed until neutral in pH. The suspension is adjusted to 1% in 0·05 M citrate buffer at pH 4·8.
0·05 M citrate buffer
Glucose standards in buffer
DNS reagent for reducing sugar

Procedure: 1 ml of enzyme is mixed with 1 ml of a 1% Walseth cellulose suspension and incubated for 1 hr at 50°C. Undigested solids are centrifuged down and reducing sugar concentration in the supernatant is determined by the DNS reagent method.

Walseth cellulose must be kept in suspension under refrigeration if stored. Recrystallisation occurs if it is dried or frozen.

3.3.2. *Filter Paper Assay for Cellulase*

Reagents: Whatman No. 1 filter paper cut into 1 × 6 cm strips
0·05 M citrate buffer pH 4·8
Glucose standards in buffer
DNS reagent for reducing sugar

Procedure: 0·5 ml enzyme, 1·0 ml buffer and a strip of filter paper are mixed in a Vortex blender, allowing the paper to coil in solution. The mixture is incubated for 1 hr at 50°C, and 3 ml DNS reagent is added to stop the reaction. Tubes are placed in boiling water for 5 min and reducing sugar measured as glucose.

The result in terms of milligrammes of glucose produced in this procedure is referred to as the filter paper (FP) activity. The latter is roughly quantitative up to about 3 mg of glucose and is adequate for monitoring fermenters, screening mutants, etc. FP activity should be expressed relative to concentration as 'at 1 mg/ml the powder has a FP activity of 1·0', or 'at full strength the culture filtrate has a FP activity of 3·5'. For use in quantitative procedures, the assay is repeated using diluted enzyme in those cases where FP activity exceeds 2·0.

In the above 0·5 ml assay, units per ml = mg glucose × 0·185 × Dilution factor

$$\textit{or} \text{ units per ml} = \frac{0{\cdot}37}{\text{Dilution to give 2·0 ml glucose}}$$

$$\text{In general, units per ml} = \frac{\text{mg glucose} \times 0{\cdot}0925}{\text{ml enzyme}}$$

3.3.3. *Cotton Assay for Cellulase*

Reagents: Absorbent cotton

1·0 M citrate buffer pH 4·8

Glucose standards in 0·05 M citrate buffer pH 4·8

DNS reagent for reducing sugar

Procedure: To 100 mg cotton in a test tube, add 0·10 ml buffer and 2·0 ml enzyme. Mix well, being careful to avoid introduction of air bubbles into the cotton. Incubate for 24 hr at 50°C and mix well. Reducing sugar as glucose is determined on an aliquot of the supernatant by the DNS reagent method.

Cotton activity is the number of milligrammes of glucose produced per millilitre of sample. For quantitative studies, the enzyme is diluted to give 2·0 mg of glucose per millilitre.

Cotton activity is a measure of C_1, but the C_x component must be present for sugar to be produced.

3.4. Method 4: Weight Loss of Cellulose

Semi-micro determination of residual cellulose (30–100 mg), adapted from Fahraeus (1947):

Reagents:
12% HCl
5% NH_4OH
99% EtOH
Anaesthesia-grade ether
Gooch crucibles (30 ml, pore size 15–40 μ; for cotton fibres, use a pore size of 40–90 μ)

Procedure: One Gooch crucible, pre-weighed and stored in a desiccator, is used per sample.

Sample cultures are prepared in 125 ml Erlenmeyer flasks with accurately weighed, dry substrate (insoluble cellulose), concentration 0·3–0·4%, in Berg's (1972) salts medium. They are autoclaved, suitably inoculated and incubated for the time-period and temperature under study. If the sample is alkaline at the time of assay, it should be initially acidified with dilute HCl. Flask contents are poured into the Gooch crucible (Fig. 3).

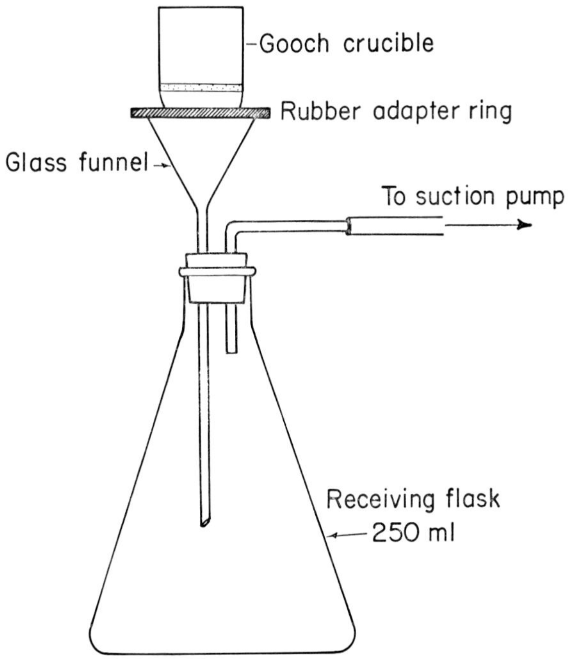

FIG. 3. Gooch crucible assembly for semi-micro determination of residual cellulose.

In order to remove adhering cellulose, successive washes with the following substances are added to the sample flask, swirled and transferred to the crucible:

(a) 100 ml 12% HCl
(b) 100 ml distilled water
(c) 100 ml 5% NH_4OH
(d) 100 ml distilled water
(e) 100 ml 99% ethyl alcohol
(f) 100 ml ether

The crucible is then placed into an oven at 50°C for 1 hr, followed by drying at 105°C for 1 hr. Upon removal, it is cooled in a desiccator and weighed to a constant value. Crucibles should be handled with metal tongs or forceps rather than by hand.

The above procedure has the limitation of being restrictive with regard to the quantity of cellulose which may be assayed. Cellulose in amounts greater than 100 mg tends to clog the crucible filter; using less than 30 mg leads to errors in accuracy due to adsorption of water.

In order to overcome these restrictions, quantities in excess of 100 mg may be assayed using the following modifications (L. V. Oberkotter, unpublished).

The procedure is identical to that outlined above except that the sample is poured into a tared centrifuge bottle and centrifuged at 10 000 G for 15 min. The supernatant is then carefully decanted into the Gooch crucible. Successive washes involve resuspension of the 'pellet', centrifugation, and decanting of the supernatant.

Following the wash with 99% ethyl alcohol, the cellulose pellet is resuspended in 50 ml of ether, and poured directly into the crucible. The centrifuge bottle is rinsed twice with 25 ml portions of ether, and the washings poured into the crucible and filtered off. The crucible is then dried and weighed as mentioned above.

Controls employing known quantities of cellulose may be used to determine percentage loss in transfer, but preliminary experiments indicate that recovery is good. Upper weight limits for insoluble substrate which may be used in this method have not been determined, but would obviously depend on the size of the centrifuge bottle and crucible, and the time required for complete drying of the cellulose.

An alternative method (Halliwell, 1963) involves extraction with 18% caustic soda. The residual cellulose is then washed with water, dissolved in sulphuric acid–dichromate solution and digested for 20 min at 80°C. The amount of cellulose is calculated from the back-titration of excess dichromate with ferrous alum.

3.5. Method 5: Reduction in Optical Density in Sol Suspensions of Cellulose

This C_1 assay by Li *et al.* (1963) is a modification of Norkrans's method (1950); it is carried out with fungal cellulases. Appropriate modification of reagents may be made for bacterial cellulases.

Reagents:
0·1 M sodium citrate, pH 5·5, containing 0·025 M cysteine hydrocellulose
Commercial enzyme preparation for standard curves
Procedure: To 3·0 ml of 0·1 M sodium citrate at pH 5·5 containing 0·025 M cysteine are added 2 mg of a very fine dispersion of highly crystalline hydrocellulose.
The enzyme sample is diluted (if necessary) with water and added to the above tube, to yield a final volume of 6·0 ml; the initial turbidity is then read in a Klett–Summerson colorimeter using the red filter.
After incubation for 1–3 hr the cellulose is resuspended and the turbidity read again. The change in turbidity (ΔT) is interpreted as a measure of C_1 activity using a standard curve.
The standard curve is prepared by setting up a wide range of enzyme concentrations and incubation times. The ΔT values may then be plotted against E × T (ml of enzyme × time of incubation). Li defines one unit of enzyme as the amount of enzyme in 1 ml of solution used to prepare the standard curve (but this is arbitrary).
When dealing with an unknown, the ΔT and time of incubation are recorded; the standard curve is used to determine the E × T value. Dividing this by time of incubation gives units of enzyme in the unknown.

The purposes of this assay, according to Li *et al.* (1963), are '. . . (a) to measure solubilisation of a crystalline substrate, and (b) to be useful when many tubes must be assayed'. It seems that preparing the standard curve would be the most cumbersome part of this method, firstly in terms of the difficulty in determining a suitable substrate-concentration range, and secondly with regard to reproducibility, which is a problem whenever measurement of a suspension is involved. That the tubes be agitated and read in a consistent manner is hence particularly crucial.

3.6. Other Methods

Less clearly defined methods which will not be specifically described here, but for which references are supplied for laboratories requiring a broader spectrum of assays, include:

(a) Breaking strength of thread, yarn or fabric (Selby, 1963).
(b) Swelling of cotton or paper (Chang and Usami, 1967).
(c) Microfragmentation of cellulose micelles (Halliwell, 1963; Liu and King, 1967).

REFERENCES

Berg, B., van Hofsten, B. and Pettersson, G. (1972). Growth and cellulase formation by *Cellvibrio fulvus*. *J. Appl. Bacteriol.*, **35,** 201–14.

Chang, W. S. and Usami, S. (1967). *Hakko Kyokaishi*, **25,** 349.

Eriksson, K.-E. (1969). In: *Cellulases and their Applications* (ed. R. F. Gould), pp. 87–8. Am. Chem. Soc., Publication No. 95, Washington, DC.

Fahraeus, G. (1947). Studies in the cellulose decomposition by *Cytophaga*. *Symbolae Botanicae Uppsaliensis*, **9**.

Greenfield, B. (1964). *Media for Marine Microbiology*. Univ. of Miami Institute of Science, Miami, Fla.

Halliwell, G. (1963). In: *Methods of Enzymatic Analysis* (ed. H. U. Bergmeyer), pp. 64–71. Academic Press, New York.

Iwasaki, T., Kokuyasu, K. and Funatsu, M. (1964). *J. Biochem.* (Tokyo), **55,** 30.

Li, L., Flora, R. M. and King, K. W. (1963). Purification of B-glucosidases from *Aspergillus niger* and initial observations on the C_1 of *Trichoderma koningii*. *J. Ferment. Technol.*, **41,** 98–105.

Liu, T. H. and King, K. W. (1967). *Arch. Biochem. Biophys.*, **120,** 462.

Lyman, J. and Fleming, R. (1940). Composition of sea water. *J. Marine Res.*, **3,** 134–46.

McIlvaine, T. C. (1921). A buffer solution for colorimetric comparison. *J. Biol. Chem.*, **49,** 183.

Mandels, M. (1974). *Production and Applications of Cellulase Laboratory Procedures*. US Army Natick Development Center, Dec.

Miller, G. L. (1959). Use of dinitrosalicylic acid reagent for determination of reducing sugar. *Anal. Chem.*, **31,** 426.

Norkrans, B. (1950). *Symbolae Botanicae Uppsaliensis*, **11,** 5.

Rautela, G. S. and Cowling, E. B. (1966). Simple cultural test for relative cellulolytic activity of fungi. *Appl. Microbiol.*, **14,** 892–8.

Reese, E. T., Siu, R. G. H. and Levinson, H. S. (1950). The biological degradation of soluble cellulose derivatives and its relationship to the mechanism of cellulose hydrolysis. *J. Bacteriol.*, **59,** 485.

Rosenberg, F. A. and Breiter, H. (1969). The role of cellulolytic bacteria in the digestive processes of the shipworm. I. Isolation of some cellulolytic micro-organisms from the digestive system of teredine borers and associated waters. *Material u. Organismen*, **4,** 147–59.

Selby, K. (1963). In: *Advances in Enzymic Hydrolysis of Cellulose and Related Materials* (ed. E. T. Reese), p. 33. Pergamon Press, Oxford.

Tansey, M. R. (1971). Agar-diffusion assay of cellulolytic ability of thermophilic fungi. *Arch. Mikrobiol.*, **77,** 1–11.

Thomas, R. (1956). *Austr. J. Biol. Sci.*, **9,** 159.

Valentine, L. C. (1973). A comparative study of cellulase activity in bacterial isolates from a marine environment, pp. 26–7. M.S. Thesis, Northeastern University, Boston, Mass.

Walseth, C. S. (1952). Occurrence of cellulases in enzyme preparations from microorganisms. *TAPPI*, **35,** 228–33.

Yoshikawa, T., Suzuki, H. and Nisizawa, K. (1974). Biogenesis of multiple cellulase components of *Ps. fluorescens* var. *cellulosa*. I. Effects of culture conditions on the multiplicity of cellulase. *J. Biochem.* (Tokyo), **75,** 531–40.

16

Methodology in Paper and Pulp Microbiology

J. M. SHARPLEY

Sharpley Laboratories Inc., PO Box 846, Belle Plains Road, Fredericksburg, Virginia 22401, USA

1. INTRODUCTION

Microorganisms are important biodeteriogens in the paper and pulp industry. Biodeterioration occurs in the fresh-cut logs at the mill, in the pulp, the papermaking process and in the additives used in various processes.

The methods outlined in this chapter are those used in this laboratory. Many other methods are available and new methods are constantly being developed. However, those outlined have stood the test of time in our hands, particularly when the results from the laboratory have been extrapolated to the operation of the mill.

2. PULPWOOD: MICROBIAL BIODETERIORATION

There is no doubt that significant losses of cellulose occur during the storage of pulpwood. However, the manner in which pulpwood is cut, transported and handled frequently prevents adequate treatment of stored wood. Increased economic pressures are gradually forcing papermakers to examine economic losses in the mill more carefully, including the loss of an indeterminate percentage of the raw material.

The fungal biodeterioration of pulpwood that has been examined most closely is from fungal 'white rots' or 'brown rots'. Wood-staining fungi are also very common but there is doubt regarding the damage they may cause in the final product. The white rots are frequently *Fomes pini* or *F. annosus*. Wood so infested will yield good chemical pulps but poor mechanical pulps (Bjorkman *et al.*, 1949). Far more serious are the brown rots frequently caused by *Polyporus hirsutus*, *P. versicolor* and *P. adustus* on aspen and similar species. On conifers the primary species of brown rots are *Lenzites saepiaria*, *Peniophora gigantea*, *Polyporus abietinus*, *Fomes pini* and *F. roseus*.

The identification of rot in pulpwood is not a difficult laboratory task. The quantification of the degree of biodeterioration and loss of product is sometimes a very difficult task. The following methods for the identification of rots, particularly in conifers, have proved useful: (a) gum guaiac test; (b) lactofuchsin staining; (c) microscopic examination; and (d) alkali solubility.

2.1. Gum Guaiac Test

The white rots produce an active oxidase; the brown rots do not. A 5% alcoholic solution of gum guaiac will produce a blue colour in 3–5 min in the presence of an oxidase. Thus, no colour is produced with brown rots, but white rots will produce a blue colour. This empirical test is reasonably useful in the field or for preliminary laboratory examinations when one is unable to culture and identify the causative fungi.

2.2. Lactofuchsin Staining

Lactofuchsin was developed for use in this laboratory (Sharpley, 1967, pp. 143 ff.) as a modification of the well-known lactophenol-cotton blue stain. The formula is:

Acid fuschsin (C.I. Acid violet 19)	0·3 g
Lactic acid, reagent grade	100·0 g

Dissolve by gentle warming with mechanical stirring. Stain is indefinitely stable.

The primary advantage of lactofuchsin is its differential staining of fungal hyphae in a cellulose fibre matrix. Characteristically, this is a slow process and a wet mount should be allowed to stain for 24 hr or more. At the end of this prolonged staining period the fibre matrix is pink and the hyphae a bright red.

2.3. Microscopic Examinations

To determine whether a fungus observed in pulpwood is cellulolytic, it must be demonstrated that the hyphae have penetrated the wood vessel cell wall. This is sometimes difficult. Free-hand sections stained with lactofuchsin are often sufficient, provided the sections are patiently observed. Failing this, the alternative must be the examination of conventional wood sections prepared with a sliding microtome.

It should be emphasised that the mere observation of fungal hyphae is not sufficient evidence for concluding that the wood sample is rotted. Saprophytic non-cellulolytic fungi are ubiquitous in such environments and will always be seen. However, the hyphae of such saprophytes will not penetrate into the lumen of the wood vessel (Fig. 1).

2.4. Alkali Solubility

The alkali solubility of pulpwood has been used as an index of the degree of rot. A number of bolts of wood must be sampled, generally by boring, to obtain statistically significant data. The sawdust should be ground and screened through a 60 mesh screen.

After screening, the moisture content of the sample must be determined. The method of choice is refluxing with toluene (TAPPI Standard T3m-54), but alternatively may be determined by drying at 105°C for 2 hr. Two grammes of the dry wood are mixed with 100 ml of 1·0% sodium hydroxide. The mixture is heated in a water bath at 100°C for exactly 60 min with periodic stirring. At the end of the heating period the slurry is filtered through a tared glass filter. The residue is washed with hot water followed by 10% acetic acid. The filter and contents are dried to a constant weight at 105°C and weighed. The loss in weight, calculated on a moisture-free basis, is the alkali solubility.

There are no standards for alkali solubility except that the alkali solubility increases with increased decay, and research is needed to produce definitive data concerning this potentially valuable test. When performed on a regular basis at a given mill, the results have been found useful in detecting changes in stored pulpwood.

In summary, it is possible to recognise and partially quantify losses by microbial biodeterioration in pulpwood. Such losses, although they may account for at least 10% of the total value and amount to many millions of dollars, are little investigated on a routine basis.

3. PULP: MICROBIAL BIODETERIORATION

The process of pulping is one of dissolving or decomposing the cementing substance binding the cellulose fibres. The usual processes employ mechanical force with the generation of much heat, as in groundwood pulp, or the action of strongly caustic chemicals. In both instances, the

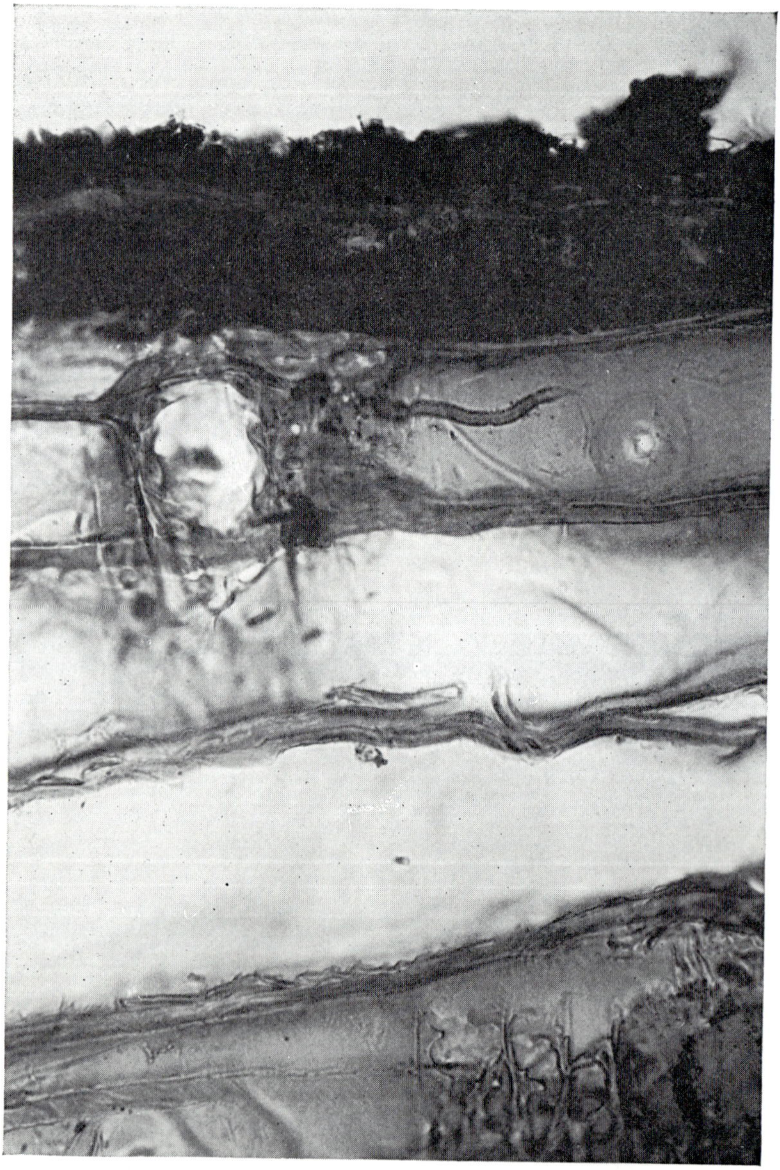

FIG. 1. Fungal hyphae penetrating wood vessels.

action ensures that freshly prepared pulp has a negative or very low microbial population.

The surrounding environment quickly contaminates the pulp when it has cooled. The pulp contains crude sugars from the wood as well as trace elements and some nitrogen. Many common bacteria, perhaps most, are able to grow quite vigorously in pulp slurries, and there is no requirement that the microorganisms must be cellulolytic.

3.1. Lap Pulp

When pulp is manufactured it may be partially dried and shipped in laps to a paper mill for rehydration and conversion to paper. The microbial biodeterioration of lap pulp is primarily a function of moisture and, secondarily, of temperature. These same parameters also govern whether the attack will be bacterial or fungal. With regard to moisture, little or no growth may be expected at less than 12 % moisture. Fungi will start to grow at between about 14 and 20 % moisture with an optimum at about 25–30 %. Above this point, the growth of bacteria will predominate. Since, for reasons of economy, lap pulp is shipped as dry as feasible, the usual biodeteriogen encountered is fungal (Table I).

TABLE I

MICROBIAL DETERIORATION IN THE PAPER AND PULP INDUSTRY

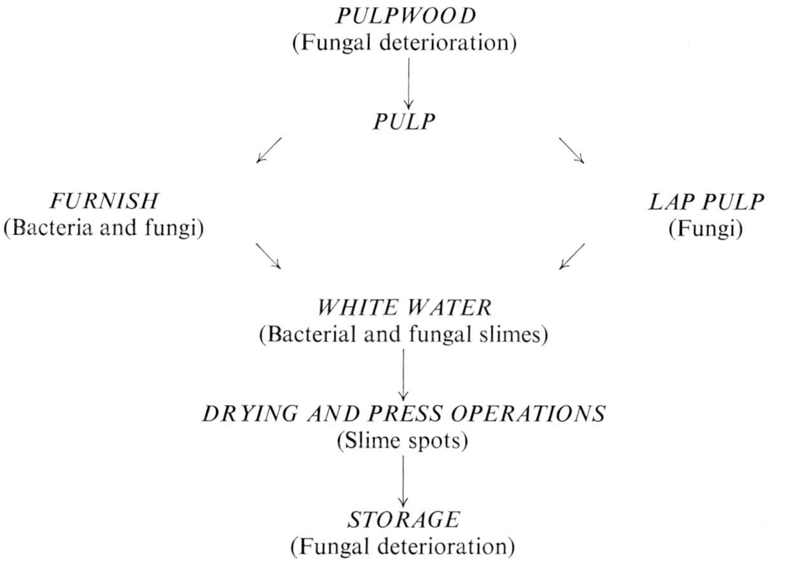

Fungicides are frequently added to lap pulp and tests for resistance to fungal attack are frequently requested. In general, it is considered undesirable to supply any nutrient in these tests other than mineral salts, since the lap pulp will furnish a carbon source.

The strains of fungi that are used for testing are various but they usually consist of a strongly cellulolytic species such as *Chaetomium globosum* or *Myrothecium verrucaria* together with a common saprophyte such as *Aspergillus niger* or *Penicillium citrinum*.

The test specimens are usually 6 cm square or cut to fit a conventional glass petri dish. The pulp samples should be cut with a clean sharp knife under clean conditions; sterile precautions are not required. The test squares are placed in a clean petri dish with provisions for sealing the edge of the dish to retard drying.

In this laboratory the test fungus is grown on commercial rehydrated potato dextrose agar until abundant fruiting is evident. A small amount of mineral salts medium is added to the culture and the fungal spores suspended with the aid of a small glass rod. The suspension of spores is then added to 100 ml of mineral salts solution and used for an inoculum. One of the commonly employed mineral salts solutions is:

Magnesium sulphate ($MgSO_4 \cdot 7H_2O$)	0·2 g
Calcium chloride ($CaCl_2$)	0·02 g
Potassium phosphate, dihydrogen (KH_2PO_4)	0·5 g
Potassium phosphate, monohydrogen (K_2HPO_4)	0·5 g
Ammonium nitrate (NH_4NO_3)	0·5 g
Distilled water, *q.s.*	1000 ml

Note: It is often advantageous to add a drop of a non-toxic wetting agent to the portion of the mineral salts solution used for the original spore suspension.

Sufficient mineral salts–spore suspension mixture is added to the lap pulp to ensure about 50–70% moisture. Free liquid in the petri dish is undesirable. The petri dishes are sealed to retard evaporation and incubated at 28°C for 14 days.

Evaluation of mould resistance is generally made on a 0, 1+, 2+, 3+, 4+ scale where 0 is no growth and 4+ is the growth equivalent to the growth obtained in a control specimen which is known to be free of fungicide. Arbitrary scale grading, in our experience, frequently leads to discussions regarding the amount of growth present, since the evaluation of growth is unavoidably subjective. For this reason such tests are invariably photographed and the photographic record incorporated in the report (Fig. 2).

The test procedure outlined is biased towards cellulolytic fungi since, in theory, no carbon source is available other than the cellulose. In practice it will be found that nearly all lap pulps contain sufficient free wood sugars to support the growth of some non-cellulolytic fungi in the presence of the added nitrogen introduced from the mineral salts. However, a procedure such as outlined, free of added carbon and agar, is well accepted in the industry.

FIG. 2. Mould resistance testing of paperboard.

3.2. Furnish Stock

If the pulp is to be used directly in the manufacture of paper, it is diluted to a satisfactory working consistency of about 1–3 %. Such a slurry is very liable to bacterial deterioration and much effort has been devoted to the control of microbial growth in this area of the paper mill.

In the process of papermaking, the water must be removed from the furnish. This water, commonly termed 'white water', contains much of the soluble nutrient from the original wood and is an excellent substrate for microbial growth. Prior to about 1950, such white water was simply discharged into the nearest stream. The economy and water quality regulations have obviated such a procedure for many years and white

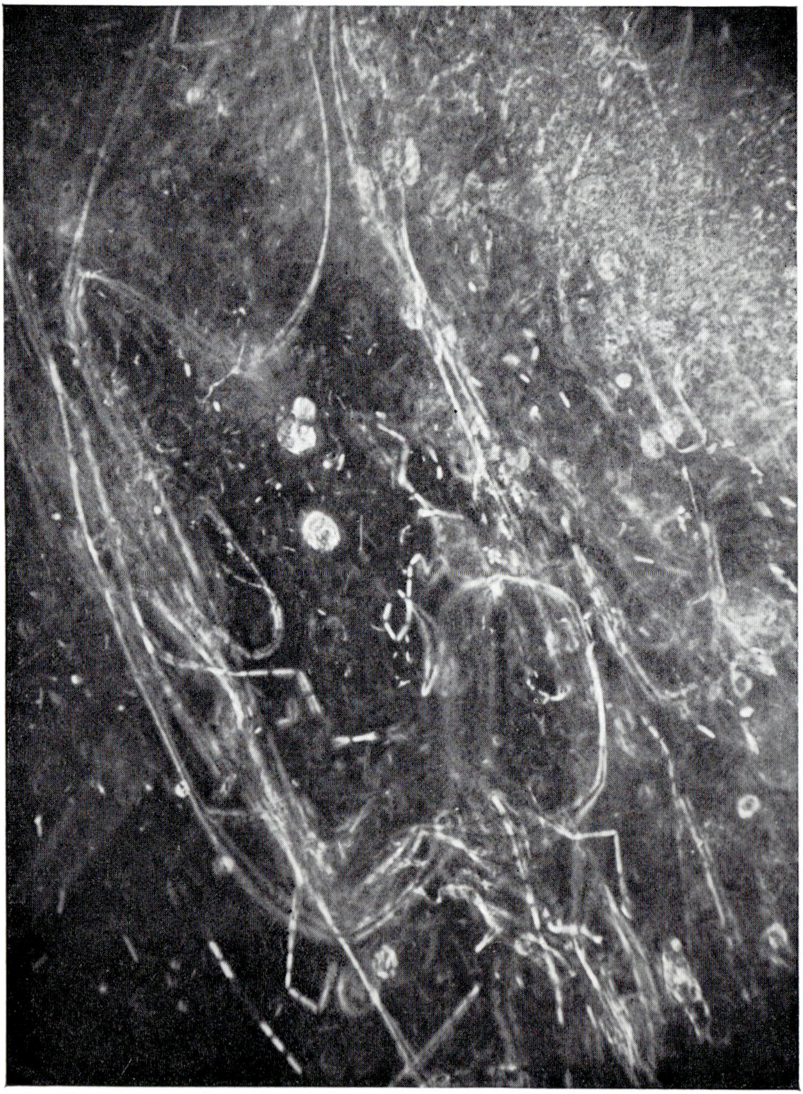

FIG. 3. Mixed slime showing bacteria, *Sphaerotilus* and other microorganisms.

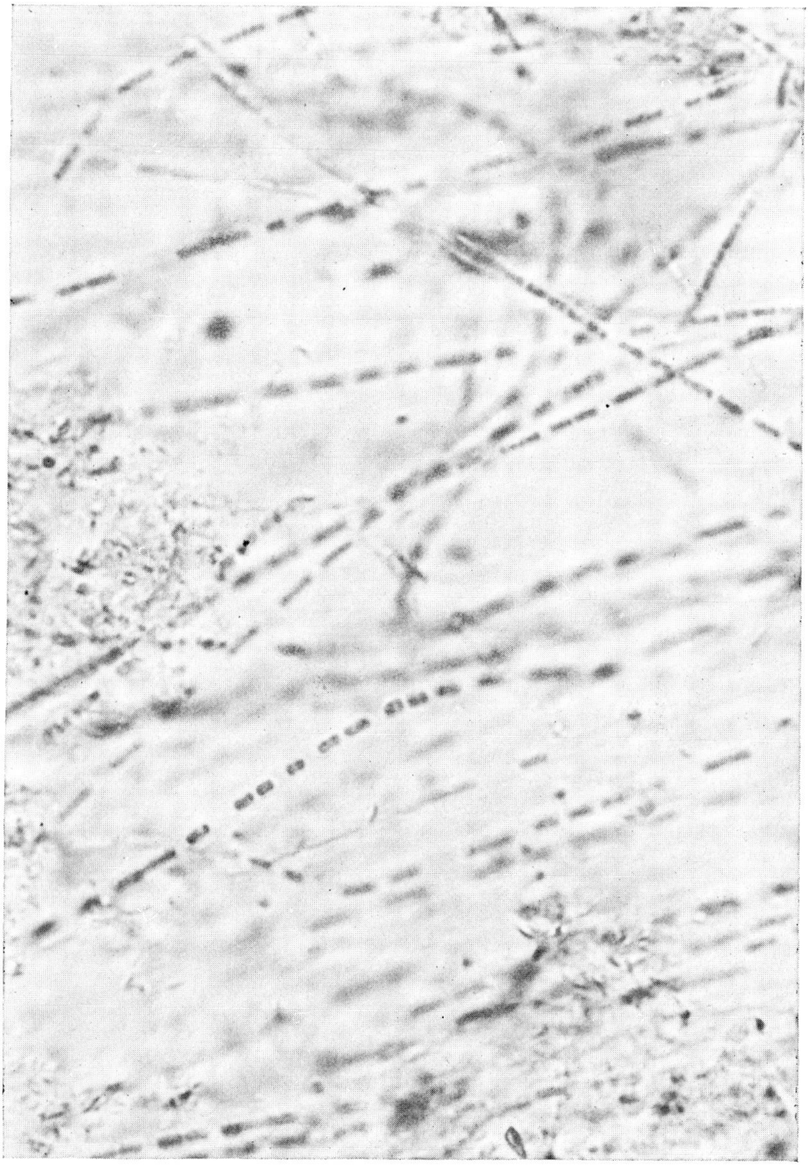

FIG. 4. Paper mill slime composed primarily of *Sphaerotilus*.

water is now recirculated and reused. There is a concomitant increase in the nutrients and a corresponding increase in the microbiological growth.

It should be remembered that the numbers of bacteria in the paper mill system are of no practical importance to the papermaker. As the sheet is dried and calendered, the heat applied will kill the great majority of vegetative bacteria trapped in the fibre. Thus, the microbial population is of pragmatic importance only as it affects the operating characteristics of the machine or damages the end-product. Such damage occurs when slime masses form that either impair the drainage or cause spots and breaks on the paper web.

Much effort has been devoted by microbiologists investigating whether the bacteria observed in paper mills have the capacity to form slime masses. The consensus of opinion seems to be that capsulated or the higher bacteria are responsible for the formation of such slime masses. Thus, the detection of such capsulated organisms is of importance to microbiologists working in the paper industry (Figs. 3 and 4).

The staining of capsules surrounding bacteria is routinely practised in all kinds of bacteriology laboratories and the methods are thoroughly described in many laboratory manuals. Most such techniques suffer from the drying and shrinking of the capsule during the procedure. The following technique was developed by the author many years ago (Sharpley, 1961) and since then has been taught widely. The material to be examined is emulsified with a drop of drawing ink ('india ink') on a microscope slide and a coverslip applied. The specimen is then placed between two pieces of heavy absorbent paper and pressed firmly with the hand. A properly prepared slide will have areas ranging from light grey to quite black. Microscopic examination will show those areas where bacteria are arranged in a single layer. The capsule will be revealed as a clear halo around the capsulated bacteria and will be very obvious. The bacteria are living mucoid colonies (Fig. 5) and the capsule is observed without shrinkage or distortion (Fig. 6).

Several precautions are necessary when using this technique. Obviously it should not be used with pathogenic microorganisms but should be restricted to industrial situations where pathogens do not occur. It will be found that ordinary drawing ink is pH-sensitive and the colloidal suspension is destroyed at about pH 5 or lower. Simple adjustment of the pH of the material to be examined to approximately neutral will solve this problem. Finally, the observations must be made with an adequate microscope and the best results are obtained with dark phase contrast microscopy.

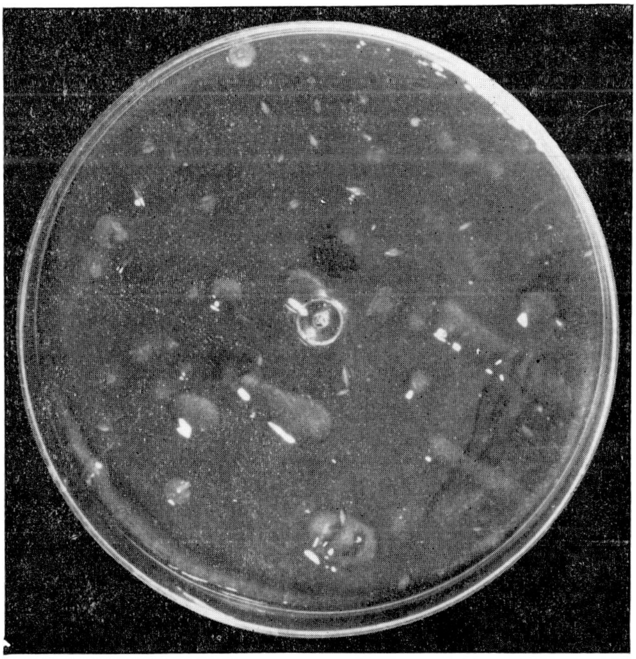

FIG. 5. Mucoid colonies of capsulated bacteria.

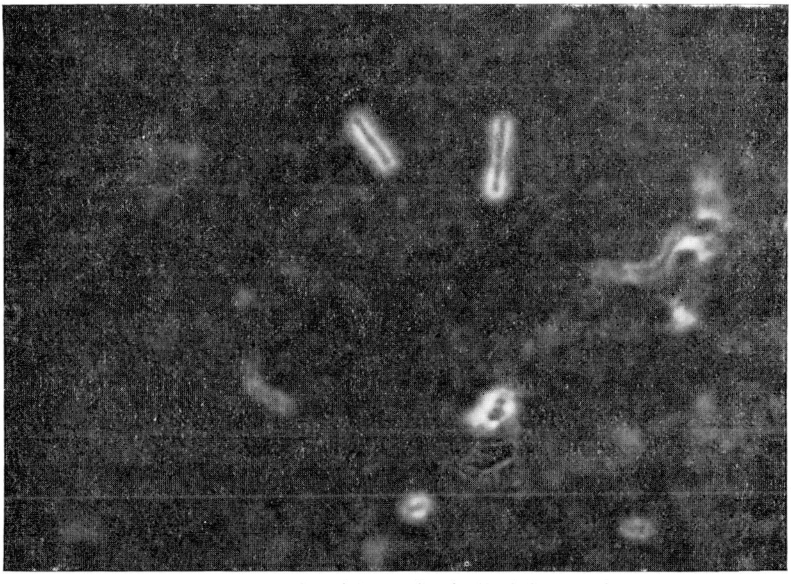

FIG. 6. Capsulated bacteria: india ink technique.

3.3. Examination of Slimes and Deposits

The examination of slimes and deposits from paper mills is certainly the most common microbiological exercise requested and sometimes the most difficult.

The collection and proper preservation of the sample is always of great importance and frequently critical in the examination. If the deposit is microbiological in nature, the appearance and composition will change very quickly if not properly preserved. Shipment of samples in an ordinary small commercial thermos bottle is probably the most convenient method. The containers are available and when samples are packed in ice, they ship very well without significant change. Samples shipped otherwise—as is fairly frequently the case—must be interpreted with great care, keeping in mind the rapid changes that may occur in microbiological samples.

The first question that must be answered is whether the contamination in the specimen is 'microbiological' in origin. Microbiological slimes may be simulated by alum deposits, dirt and foreign materials, hydrated cellulose and many other materials. Microscopic examination is certainly the first method of choice; if fungi are present in a fresh sample or if capsulated bacteria are observed, the assumption is that the contamination is microbiological in nature. An alcoholic solution of ninhydrin is sometimes of benefit in detecting alpha-amino groups indicative of protein, but it must be made very certain that protein-containing additives are not in use in the mill. If microscopic examinations are inconclusive, other chemical tests for identification must be used. For example, the presence of large amounts of aluminium in the deposit might well indicate that the deposit is derived from alum. A semi-quantitative spectrographic analysis of the deposit has proved very useful since this will detect all elements of possible significance.

The biochemist/microbiologist must be very careful when interpreting chemical data obtained from slime deposits to engineers and chemists. Elemental analysis of such samples will often indicate that the biological portion of the deposit is less than 1 % of the total; thus, those not trained in the area are very inclined to consider the microbiological aspect of the deposit unimportant. It has been demonstrated many times that the sticky nature, and probably the surface structure, of industrial bacteria acts as a cementing or binding substance in slime masses (Sharpley, 1967, pp. 14 ff.). This phenomenon is adequately demonstrated each time a paper mill system is treated with a biocide. With no other changes in the operating parameters, the slime masses with their accumulations of inert materials will disappear.

The occurrence of fungal slime masses in paper mill systems operating at an approximately neutral pH should be regarded with suspicion; fungi are the most common shipping artefact in slimes and are not usually found in other than slightly acidic environments. However, the majority of mill systems, in our experience, are operated at a slightly acid pH because of other operational demands and fungal slimes are common. They are among the easiest slime masses to identify.

4. EVALUATION OF BIOCIDES FOR USE IN THE MILL

Commercial independent laboratories, or sometimes captive laboratories within an industrial organisation, are very frequently requested to provide operation management with data reflecting the comparative effectiveness of biocides.

It is reasonably safe to assume that the usual microbiological methods for the evaluation of microbiocides are worse than useless, since very deceptive data may be derived from such tests. To be meaningful in paper operations, such tests must satisfy the following minimal criteria:

1. Tests must be performed in the presence of paper fibre. Many very adequate bactericides and fungicides are adsorbed on the surface of fibre and removed from the system.
2. As far as possible, the additives normally used in the mill must be present in the test system. These usually include rosin, papermaker's size and the adjustment of pH. The latter is very important but far too complex to be discussed in detail. However, consider for a moment the use of alkali metal salts of various toxicants. Their effectiveness is a function of pH, since the parent compound is formed when the pKa is exceeded (Fig. 7).
3. The time of contact in the mill is an important parameter, and also a matter of some dispute in laboratory testing. In this laboratory it is customary to determine the percentage kill of the test organism at 2, 6 and 24 hr. Other laboratories use other contact times.
4. Paper mills employ highly agitated systems, hence test systems that are not agitated are not truly representative of mill conditions. In this laboratory, continuous agitation is used throughout the incubation system.
5. Unquestionably the best substrate for a specific mill is the white water from that mill. Such a substrate may be used without modification

with great success, but it should be remembered that such data are specific for that specific mill, and to extrapolate the results to other systems will involve a good deal of risk.

Appling (1946) published very careful observations on test systems, as described above, and his work has stood the test of time. With suitable modifications, the techniques described can be safely used today.

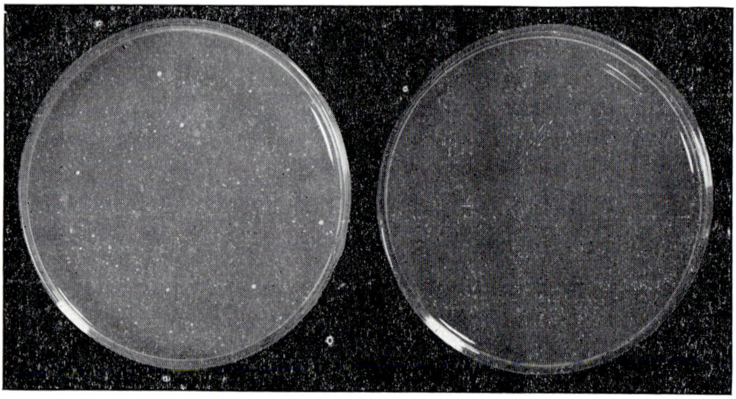

FIG. 7. Contrast of response in same medium and inoculum (*left*) at pH 9·0, (*right*) at pH 6·8.

5. MICROBIOLOGY OF FINISHED PAPER

After the drying, calendering and press operations in a paper mill, the product is essentially free of vegetative bacteria and fungi and no operational problems may be expected.

This does not imply that the paper is sterile, since heat-resistant spores may survive in the paper. These spores may be important, particularly in paper used in contact with food, and the determination of the total bacterial count is a common laboratory operation. Such estimates of total bacterial population are a well-standardised procedure and commonly found in the literature (Appling, 1955). Essentially, the procedure is to emulsify a weighed portion of the paper in buffered sterile distilled water. Aliquots of the emulsion are then used to inoculate a nutrient agar and the total numbers of bacteria calculated on the basis of the original paper.

The storage of paper and the resistance of paper documents to biodeterioration is a problem that has occupied microbiologists for many years. All the nutrients necessary for the growth of fungi are present except moisture. Thus, the easiest and most common method of preservation is simply to keep the paper reasonably dry, *i.e.* less than 12% moisture. If this is not possible or feasible, as in the case of shipping cartons, it has been the practice in past years to add to the pulp a fungicide strongly adsorbent to cellulose. However, in recent years environmental considerations have made this practice difficult, since the most commonly used residual fungicides have been either chlorinated phenols or organic mercury compounds. Some organic compounds of copper are still used for this purpose, but with much care, and the prevention of fungal deterioration of paper products and cellulose film (Fig. 8) in commercial use is rapidly assuming the dimensions of a major problem.

The microbiological aspects of the manufacture of paper are many and varied. They range from the deterioration of wood, the cause of operational problems in the paper mill, to the deterioration of the final product. Many aspects are poorly understood and further investigation into the cause and prevention of biodeterioration of paper is badly needed.

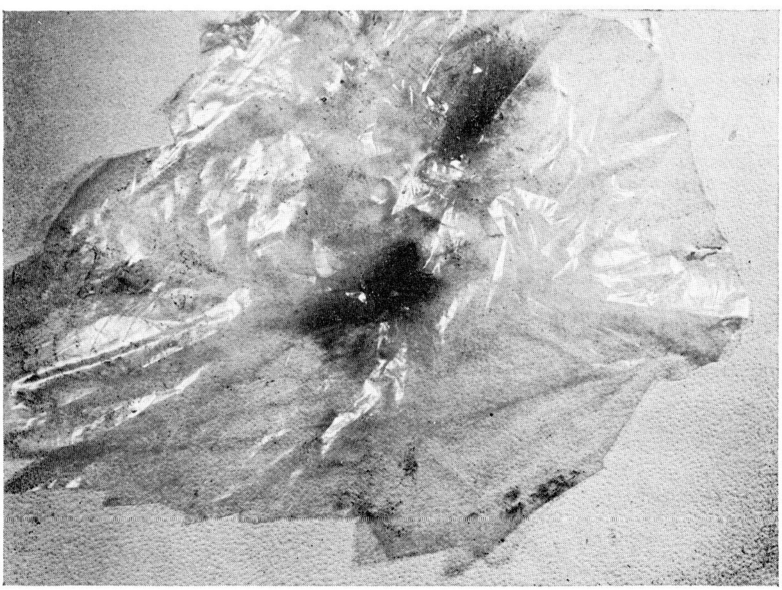

FIG. 8. Fungal growth on cellulose films.

REFERENCES

Appling, J. W. (1946). Relative toxicity of disinfectants recommended for use in the paper industry. *Paper Trade J.*, **119,** 116–18.

Appling, J. W. (1955). *Microbiology of Pulp and Paper*, pp. 218 ff. TAPPI Monograph No. 15 (out of print).

Bjorkman, E. *et al.* (1949). Om rotskador i granskog och deras betydelse vid framstallning av kemish pappersmassa och silkemassa. *Kgl. Skogshogskol Skrifter*, **4,** 1–73.

Sharpley, J. M. (1961). *Applied Petroleum Microbiology*, pp. 136 ff. Gulf Publishing Co., Houston, Texas.

Sharpley, J. M. (1967). *Elementary Petroleum Microbiology*, Gulf Publishing Co., Houston, Texas.

17

Methods for Determining Fungal Biodeterioration in Wood and Wood Products

BJÖRN HENNINGSSON

Department of Forest Products, Royal College of Forestry, Fack, S-104 05 Stockholm 50, Sweden

1. INTRODUCTION

Numerous methods of determining fungal decomposition in wood have been employed in research. An examination of the available literature reveals that almost every author has used his own methods or has modified known methods to a greater or lesser extent. At first glance this wide diversification in methodology seems very irrational and even confusing to the reader; in certain situations it can be confusing indeed, even for a person with long experience within the field. It is, for example, sometimes difficult to compare the results of studies of identical organisms and wood species when the methods differ.

On the other hand, it is often quite logical for different scientists to use different methods. It may even be said that it would have been a mistake to use identical methods in two very similar studies. There are many factors which influence the choice of method and how a method must be modified to fulfil the aims of a study. In one case the absolute optimal rate of wood attack is aimed at, and certain extra nutrients must therefore be introduced into the culture vessel or into the wood. In another case the progressive changes in chemical composition of the wood during decay are to be studied and consequently foreign substances cannot be added to the wood. In still another case, heat sterilisation of the wood is out of the question since the wood has been impregnated with a heat-unstable preservative. In this case, sterilisation by means of an inert gas or non-destructive radiation is to be preferred, provided the preservative or the wood substance itself is not affected, and that there are no secondary effects on the test fungus.

From the above it can be concluded that standardisation of decay test methods is a very delicate task. Standardisation may also be of doubtful value or even disadvantageous for the development of research unless the person who practises the standardised method is fully aware of the limitations of the method and the aim of his study. A correct interpretation of the results further requires a good knowledge of wood chemistry and physics, and of the general biology and biochemistry of microbial wood degradation.

The standardisation of laboratory methods for determining fungal decomposition has mainly been carried out within the field of wood preservation. A number of national and international standard methods have been established, *e.g.* ASTM D 1413, BS 838, DIN 52 176, NWPC 1.4.1.1/70. These standard methods may also be applicable to the testing of wood in general, and not only to treated wood.

In this chapter, a few selected methods will be presented. It must be pointed out, however, that these methods are not necessarily ideal or even suitable for the kind of information it is desired to obtain. They have been selected since they are relatively simple to carry out, they require a minimum of equipment, the results are not too difficult to interpret, and they have been used successfully in our laboratory for several years.

2. THE SOIL JAR METHOD

Soil jar methods have been used to determine fungal wood decomposition for a number of years. The methods employed differ to some extent from each other, depending on the purpose of the test and the organisms involved. For those who wish to examine and compare some variants of soil jar methods, the publications by Duncan (1960), Rennerfelt (1964) and Bravery (1968) can be recommended.

The soil jar method is especially useful when rapid and homogeneous decomposition is desired. It works very well with most of the common wood-destroying Basidiomycetes and can also be used in tests with many common soft rot fungi. It is recommended, however, that different shapes and dimensions of wood samples be used for Basidiomycetes and soft rot fungi.

The method can also be employed to test materials such as plywood, fibreboard and particle board, provided the glue or other additives used in the board do not give rise to volatile compounds of fungicidic or fungistatic nature.

If close chemical analyses of the decayed wood samples are to be made, this method might not be suitable. Substances from the soil can migrate into the samples and significantly influence the lignin analysis, nitrogen compounds can be absorbed from the soil and transported into the wood by the hyphae, etc.

2.1. Test Samples

The samples may have varying designs depending on the purpose of the test, the wood or fibre material to be tested and the organisms used.

When wood is tested against ordinary decay fungi (Basidiomycetes) a 20 × 20 × 20 mm cube is a suitable sample. If a wood species with very narrow sapwood is to be tested, the sample size may be reduced to 20 × 20 × 10 mm. In a test of soft rot fungi the following sample size is recommended: 30 × 15 × 5 mm, with the 15 mm dimension in the radial direction.

If composite materials such as board are to be tested, the size of the sample might be restricted by the thickness of the board. In certain cases, *e.g.* when an oriented attack is desired, it might be necessary to seal certain surfaces of the samples.

The samples must be adequately numbered for identification. The wood samples may be numbered by punching on one of the radial surfaces.

2.2. Exposure Jars

Glass jars should be used with a capacity of about 500 ml and fitted with screw caps with a filter or with cotton plugs.

2.3. The Soil Medium

Fresh garden compost soil with 20–40% content of organic matter and a pH of 5·0–6·5 is a suitable medium. The soil should be sifted through a 5 mm gauge sieve.

For testing of Basidiomycetes the moisture content (MC) of the soil should be approximately 50%. When soft rot fungi are used, a soil moisture content corresponding to the water-holding capacity (WHC) is recommended for softwood samples. For hardwood samples, however, 80% of WHC of the soil is preferable.

2.4. Procedure

The samples are dried to constant weight (usually overnight) at 103°C. They are then transferred to a desiccator and allowed to cool to room temperature before weighing to ascertain the original dry mass. All weighing operations should be carried out to an accuracy of 0·01 g. The weighing of each sample must be done quickly since it starts to absorb moisture immediately after having been taken out of the desiccator.

The jars are filled with lightly packed soil to slightly more than half their capacity. The caps are screwed on and the jars sterilised by autoclaving for 1 hr at approximately 120°C. If cotton-wool plugs are used instead of screw caps, they should be covered with aluminium foil to avoid excessive wetting during autoclaving. The jars are then left until the next day to allow any living heat-resistant spores in the soil to germinate.

The test samples are now introduced into the jars and pressed down into the soil. Samples for testing with Basidiomycetes should be completely covered by soil except for the numbered surface facing the opening of the jar. Samples for soft rot testing are placed in a vertical position and completely covered by soil. The manipulations are carried out in a clean room (not necessary sterile) to avoid further contamination. Normally, four samples are placed in each jar. It is recommended that the samples be put in the soil in a certain order because the identification numbers may be difficult to read on the decayed samples after the test.

The jars are closed and, with the test samples inside, autoclaved for a second 1 hr period at 120°C. This second autoclaving will sterilise the samples and kill any heat-resistant spores remaining in the soil. After cooling, the soil substrate is inoculated with the desired fungus or fungi under sterile conditions. The inoculum used consists of pieces (*c.* 8 × 8 mm) of fresh fungus mycelium growing on an agar plate. It is recommended to use more than one piece per jar, preferably four, each of which is placed on the soil between the test samples.

A mycelium suspension may also be used for inoculation. It is conveniently prepared from the fresh mycelium on agar contained in a 10 mm petri dish. The culture is homogenised together with 75–100 ml of sterile water under sterile conditions. (In our laboratory a MSE AtoMix is used for the preparation of mycelium suspensions. This mixer is equipped with heat-stable homogenisation vessels which can be sterilised in the autoclave.) The fungus culture should not be more than 2 weeks old when used. The inoculation requires 5 ml of mycelium suspension for each jar which is pipetted directly on to the soil and the wood blocks under sterile conditions.

A spore suspension is sometimes to be preferred for inoculation, especially when rich-sporulating species are tested. A fungus culture, 2–4 weeks old and growing on agar in a culture tube or petri dish, is used. Sterile water (approximately 10 ml) is added to the tube or dish and the surface of the fungus culture is carefully scraped and squeezed with a sterile glass rod or spatula. A concentrated spore suspension is thereby obtained. This concentrated spore suspension is transferred under sterile conditions to a flask containing 100–200 ml of sterile water to give the final spore suspension. Each jar is inoculated with 1–2 ml of spore suspension which is pipetted under sterile conditions.

In order to correct for changes in mass of the test samples which are not due to fungus attack, non-inoculated sterile jars must be included in the test. It is recommended that two such sterile control jars (=8 samples) be employed for each combination tested.

After inoculation, the jars are incubated in a room with a suitable temperature and a relative humidity of 75% ± 5%. The optimum temperature may vary from one species to another, but most fungi grow well at temperatures of 23–25°C. Thermophilic fungi, of course, require higher temperatures, *e.g.* 40°C, and low-temperature fungi like *Serpula lacrymans* may grow best at 20°C.

It is important to keep the soil jars under observation during the incubation period. If the temperature is high and/or the air humidity low, or if the incubation is extended over very long periods, the soil may dry out and the addition of sterile water may become necessary. It is therefore advisable to have some extra jars in each test which can be withdrawn at intervals and checked.

After the desired period of incubation, the samples are removed from the jars. Each sample is then weighed immediately to give the final wet mass which is needed for the calculation of the final moisture content of the sample. Before weighing, adherent mycelium and soil must be carefully brushed off. It is recommended that the samples be always removed in a defined order, since it may be difficult to identify the specimen numbers on those which are heavily attacked.

The samples are left to air-dry for a minimum of 24 hr. They are then dried to constant weight at 103°C (normally overnight), transferred to a desiccator, cooled, and weighed again to ascertain the final dry mass.

The loss in dry mass due to fungal activity is equal to the original dry mass minus the final dry mass, minus the correction for change in weight in sterile soil. The correction factor is calculated as an average of the weight changes of the eight sterile samples. The percentage loss of mass is the above-corrected loss of mass as a percentage of the original dry mass.

3. THE VERMICULITE METHOD

As in the soil jar method, several vermiculite methods have been employed and described in the literature. Vermiculite was first used as a substrate for wood decay tests in the 1960s. Among the reports on such methods may be mentioned those by Henningsson (1967*a*), Kaune (1970) and Kerner-Gang and Gersonde (1972). The main methodological differences concern the type of culture vessel, the composition and quantity of the nutrient solution and the size of the test samples. The vermiculite methods are recommended when it is important to have a defined substrate and at the same time a fairly rapid and homogeneous attack is desired. These methods are also useful for studying the effects of different additional nutrients on wood decay, since the composition of the nutrient solution can be varied *ad infinitum*. Soft rot fungi as well as traditional decay fungi (Basidiomycetes) can be used as test organisms. However, the two types of fungi usually require different wood moisture content and different nutrient media.

The methods may also be employed for the testing of products such as fibreboard, particle board and plywood, provided these products do not produce volatile substances of fungicidic or fungistatic nature.

3.1. Test Samples

Different sizes and shapes of test samples are recommended, depending on the purpose of the test and the organisms used. For the testing of wood against Basidiomycetes, a 20 × 20 × 20 mm cube is suitable. If wood with very narrow sapwood is to be tested, a smaller dimension may be required, *e.g.* 20 × 20 × 10 mm.

For testing against soft rot fungi, a thin wood sample 30 × 15 × 5 or 20 × 20 × 5 mm is to be preferred. Unless softwood is of special interest, hardwood is recommended in tests with soft rot fungi. The susceptibility of hardwood to soft rot is usually much higher than that of softwood.

3.2. Exposure Vessels

Jars of the same type as those described in the soil jar method can also be used for the vermiculite methods. However, as a result of the very high water-holding capacity (WHC) of the vermiculite, a smaller volume of the substrate is required and smaller culture vessels are therefore usually needed. For most purposes it has proved quite adequate to use 100–125 ml wide-necked Erlenmeyer flasks.

3.3. Substrate

Vermiculite is a mica material originally prepared for insulation purposes. In addition to its very high water-holding capacity, mentioned above, vermiculite is also an excellent insulator. Several types of vermiculite are marketed. It is recommended that Vermiculite Medium, Exfoliated, which among other things has a convenient particle size, be used.

3.4. Nutrient Solution

When vermiculite is used as a growth medium it is usually necessary to add a suitable nutrient solution in order to increase the decay activity of the fungus being tested. If additional nutrients are undesirable, only distilled water need be added. The water should be buffered to a pH suitable for the test organism (for decay fungi 5·0–6·5). For general tests with Basidiomycetes an aqueous nutrient solution containing 25 mg of malt extract per 1000 ml can be used.

Various solutions containing mineral salts have been suggested by various authors for tests with soft rot fungi. A modified medium after Duncan (1965) has been widely used (cf. Lundström, 1973; Nilsson, 1973). It has the following composition:

NH_4NO_3	6·0 g
K_2HPO_4	4·0 g
KH_2PO_4	5·0 g
$MgSO_4 \cdot 7H_2O$	4·0 g
Glucose	2·5 g
Distilled water to 1000 ml	

Another nutrient solution, which has been used by Kaune (1970) and Kerner-Gang and Gersonde (1972), has the following composition:

KCl	0·25 g
K_2HPO_5	2·56 g
$MgSO_4$	1·02 g
$FeSO_4$	0·001 g
$ZnSO_4$	0·001 g
$MnSO_4$	0·001 g
NaCl	0·005 g
NH_4NO_3	3·00 g
Distilled water to 1000 ml	

This solution is free of carbohydrate.

Results have shown that a reasonable rate of attack is obtained with both these nutrient solutions when most of the common soft rot fungi are used in the tests.

3.5. Procedure

The test blocks are dried to constant weight at 103°C (usually overnight), placed in a desiccator to cool and weighed to an accuracy of 0·01 g, to ascertain the original dry mass. The weighing must be carried out rapidly in order to prevent the blocks from absorbing atmospheric moisture.

The vermiculite (7–10 g) is introduced into each flask. Then 14–20 ml (for Basidiomycetes) or 21–30 ml (for soft rot fungi) of nutrient solution is pipetted into each flask. Two test samples per flask are thereafter buried in the vermiculite with their upper surface level with the vermiculite surface. The flasks are fitted with a cotton-wool plug and autoclaved.

After autoclaving for 20 min at 120°C the flasks are allowed to cool and are then inoculated with the selected fungi. For this purpose either agar mycelium pieces, a mycelium suspension or a spore suspension can be used (*see* p. 280), depending on the type of fungus to be tested.

To correct for changes in dry mass of the samples not due to fungal activity, sterile (non-inoculated) flasks are included in the test. Two such control flasks (=4 samples) per tested combination are recommended.

After inoculation and incubation at a suitable temperature for a desired incubation period (*see* Section 2.4), the samples are removed, weighed in wet condition and again after drying to constant weight at 103°C. The final wet mass and the final dry mass for each sample is thereby obtained. The absolute loss in dry mass as well as the percentage loss in dry mass can then be calculated (*see* p. 281).

4. DECOMPOSITION OF WOOD CHIPS: LABORATORY METHOD

Nowadays much of the raw material for the pulp and paper industry is stored outdoors in the form of wood chips. A special interest in testing the fungal decomposition of industrially produced wood chips has thereby arisen.

The wood chips may vary in size and shape depending on the chipping equipment, the wood species and the pulping process for which it is prepared. A common variation in size is 0·5–2 × 2–4 × 3–5 cm. The chips are stored in large piles varying from a few thousand to several hundred thousand cubic metres. In these piles the temperature normally rises to 50–70°C during storage. This permits the development of a special microbial flora containing numerous thermophilic fungi. Simple laboratory methods have been devised for the testing of wood chip deterioration. A method used by Bergman and Nilsson (1968) will be presented here.

4.1. Test Material

Chips collected from an industrial location are used after light cleansing, during which big knots, bark, needles and other impurities are removed.

The chips are then preferably stored fresh in plastic bags in a refrigerator until required.

4.2. Exposure Jars

500 ml jars (*see* Section 2.2) can be used.

4.3. Procedure

The weighed exposure jars are filled with fresh chips and placed in a heating cabinet at 103°C. After drying to constant weight (usually within 24 hr) and cooling in a desiccator, the jars containing the dry chips are weighed. All weighings are performed to an accuracy of 0·01 g. The original dry mass of the chips in the jars is then obtained by subtraction.

The jars with the dry chips are then filled with water and the chips allowed to soak in the water until a moisture content of around 100% has been reached. The time required for the soaking process varies and depends on the wood species and the size and shape of the chips. The soaking time required must therefore be determined in each individual case. The moisture content of the chips can be checked repeatedly during the soaking period by pouring out the water and weighing the jar.

The jars with the moist chips are then autoclaved for 20 min at 120°C. After cooling at room temperature, the jars are inoculated with a suitable mycelium suspension or spore suspension (*see* p. 280) of the organism(s) to be tested.

The jars are then incubated for a certain period at a temperature suitable for the fungus employed. In these tests, rather high (40–50°C) temperatures will often be used, since thermophilic or thermotolerant fungal species are being tested. The jars can be placed in plastic bags in order to reduce the risk of excessive drying of the chips when tests are conducted at high temperatures for prolonged periods.

After incubation the cotton-wool plugs or screw caps are removed and the jars are weighed and dried to constant weight at 103°C. They are then weighed again, whereupon the final wet mass and final dry mass is obtained. The loss in dry mass can then be calculated in absolute figures or as a percentage of the original dry mass.

5. LOSS IN IMPACT BENDING STRENGTH

Methods employing changes in wood strength properties as a criterion of fungal decomposition have not been widely used in the past. This is probably due to the difficulties of obtaining samples with equal original strength properties and of designing a test where these samples are attacked by the fungi in a homogeneous manner. However, it is possible to design methods where the change in certain strength properties can be adequately determined. The impact bending strength has usually been employed, since this property is very sensitive to enzymatic attack on the wood structure.

Further information regarding strength tests on decayed wood can be found in publications by Scheffer (1936), Liese and von Pechmann (1959), Liese and Ammer (1964), Henningsson (1967*b*) and Hoffmeyer (1975). A modification of the method used by Henningsson will be described in the following.

5.1. Test Samples

The preparation of the samples is (as emphasised above) very important. The following procedure has been found suitable for homogeneously grown, moderately dense (500–700 g/cm^3) hardwood species (*e.g. Betula* and *Fagus*).

Pure sapwood (or pure heartwood) is cut into blocks measuring 80 × 20 × 7 mm. Seven such blocks, adjacent to each other, constitute one test group. Block No. 4 in each group is chosen as the control (*see* Fig. 1). All test groups used in one experiment are best cut from the same tree and taken out as close to each other as possible.

5.2. Exposure Jars

Glass jars with a minimum volume and diameter of 600 ml and 120 mm respectively are required. Wide-mouthed jars are required in order to faciliate the introduction of the samples. The jars are fitted with screw caps.

5.3. Substrate

Soil or vermiculite can be used as growth media for the fungi. The preparation and condition of these substrates is described on pp. 279 and 282.

5.4. Procedure

All samples are dried to constant weight at 103°C and weighed to an accuracy of 0·01 g to ascertain the original dry mass. The jars are filled to just over half their capacity with the growth substrate. If soil is used as a substrate, the jars are autoclaved once *before* inserting the samples (p. 280). The samples are buried horizontally in the substrate. Except for the control, the six blocks belonging to a group are placed in the same jar. The control blocks are buried six together in separate jars.

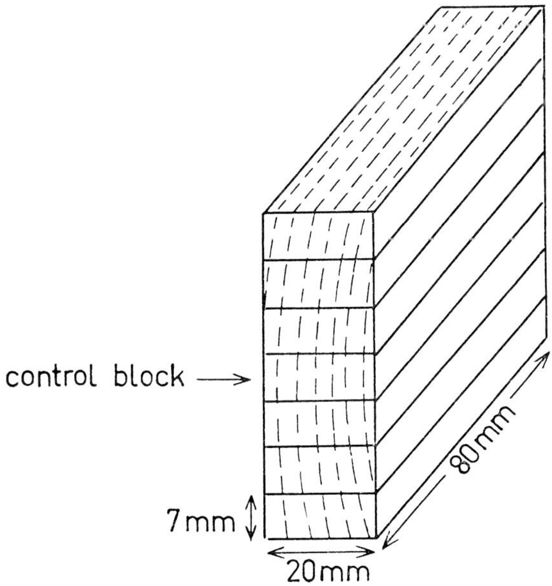

FIG. 1. Size and orientation of the test samples (from Henningsson, 1967*b*).

All the jars are then autoclaved for 1 hr at 120°C. After cooling, the jars containing the test blocks are inoculated with agar-mycelium pieces, mycelium suspension or spore suspension (p. 280). It is recommended that a minimum of three jars be employed for each series tested. Jars with control blocks are left without any inoculation.

After incubation for a certain period at a suitable temperature, the blocks are removed from the jars and immediately weighed to determine the final wet mass. They are then dried to constant weight at 103°C and reweighed to obtain the final dry mass. The final moisture content and losses in dry mass of the samples can then be calculated.

The samples are then conditioned to a constant moisture content of, *e.g.* 15%. Their remaining impact bending strength is determined by the use of a suitable instrument. A simple pendulum hammer can be used to break the test blocks, the impact bending strength being calculated as the energy in Nm absorbed by the sample.

Using the strength obtained for the control sample in the group as a base, and assuming that figure to be 100%, percentage values are calculated for the relative strengths of the decayed samples in the group. Direct comparisons can thereby be made between the various test combinations even if the original strength (strength of the control) varies between groups of samples.

6. MICROSCOPIC DETERMINATION OF SOFT ROT ATTACK

It is possible to study the fungal decomposition of wood cell walls by the use of various microscopic techniques—light microscopy, transmission electron microscopy (TEM) or scanning electron microscopy (SEM). However, in most cases the microscopic techniques do not yield quantitative results. In recent years studies have been made of the relations between soft rot attack measured as a percentage, through cavity formation and disintegrated cell wall area on one hand, and loss of strength properties and loss in dry mass on the other hand. It has thereby been possible to translate microscopic observations into terms of weight and strength changes. It is important to remember, however, that these relations differ between different wood species and that they must be examined closely in each separate case.

6.1. Principle

In a study on the mechanical properties of soft-rot-decayed Scots pine, Hoffmeyer (1975) examined microscopic cross-sections of wood samples, which were also tested as to their bending strength and density. The approximate values extracted from his work are shown in Table I.

TABLE I

Degree of soft rot	*Appearance of cross-sections*	*Area of disintegrated secondary wall (%)*	*Relative bending strength at MC >30%*	*Loss in dry mass (%)*
0	No cavities	0	100	0
1	Only few, scattered cavities	1–10 (5)	77	1
2	Numerous cavities in most cells	10–40 (25)	55	5
3	Cavities abound, often confluent	40–90 (65)	33	17
4	Total or almost total disintegration of secondary wall	90–100 (95)	12	40

With the aid of this table and of Fig. 2 it is possible to translate estimated disintegrated cell wall areas, as examined on microscopic cross-sections of soft-rot-attacked Scots pine wood, into loss in dry mass and loss in relative bending strength. The method is especially suitable for examination of wood in service.

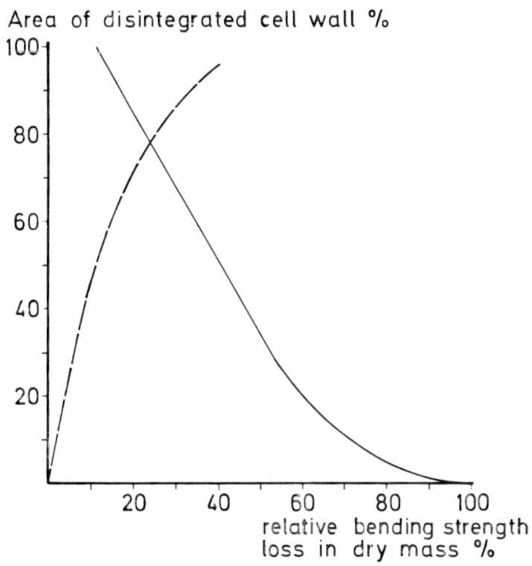

FIG. 2.

6.2. Procedure

With a razor blade or a microtome, cross-sections are cut as thin as possible from a piece of wood. It is desirable to obtain sections with a thickness of 10–25 μm. If the wood samples are very dry and difficult to cut, they can be softened by boiling in a mixture of glycerol and water (1:1) for 10–60 min.

The cross-sections are placed on a glass slide. A drop of Saffranine solution* is added to the section. A cover glass is placed in position, and the section is examined under a microscope in ordinary light. If the wood sections are to be stored for a period up to one year, they can easily be mounted in glycerol gelatine.

During the microscopic examination the cell wall area disintegrated by cavity formation and erosion is estimated as a percentage of the total cell wall area. Training is necessary in order to carry out a good estimation.

* Saffranine solution: 25 ml glycerol, 25 ml 70% ethanol, 50 ml distilled water and 0·1 g Saffranine.

It is recommended that a system of classes of cell wall disintegration be established. As can be seen from Table I, Hoffmeyer employed five classes. These classes were 'standardised' by the use of photographic examples to which the examiner could always refer.

It is necessary to estimate the cell wall disintegration on at least five different parts of the section and to determine an average figure of these different estimates. This procedure is usually absolutely necessary since the degree of soft rot attack may vary considerably within the same section.

When the average percentage disintegration area has been estimated, the curves in Fig. 2 can be used for converting this figure into strength loss or mass loss values. It must be emphasised, however, that Fig. 2 is valid, as far as we now know, *only* for Scots pine (*Pinus silvestris*) attacked by soft rot fungi. It is also important to realise that the figures obtained are of an approximative nature.

7. LOSS IN DRY MASS OF ROUNDWOOD DURING STORAGE

In forestry, wood is often stored in the form of rounds in the forest, at road sides, terminals, etc. It is of interest to determine the fungal degradation of the wood under such storage conditions. Species of wood, cutting period, log length, bark condition, storage conditions, storage period and climatic conditions are all factors which influence the rate of fungal degradation.

A method for determining the changes in basic density during storage will be described in the following. It is, however, important to bear in mind that the registered changes in basic density may influence the yield and quality of the end-product (pulp, paper, fibreboard, etc.) differently, depending on the kind of industrial process used for manufacturing and also on the type of end-product itself. It is therefore often necessary to follow up with various paper technological tests on the stored wood. For example, if the end-product is kraft paper, a reduction in the tear strength will probably not be accepted, while a slight reduction of this type can be tolerated in journal paper. Similarly, a degradation of the hemicelluloses influences the yield much more in a semi-mechanical high-yield pulping process than in a traditional sulphate or sulphite process, where most of the hemicelluloses are still dissolved in the process itself.

7.1. Procedure

Freshly cut trees are divided into logs of a length normally used in practice. Discs about 50 mm thick are cut from both ends of the logs, free from knots, major resin pockets, decay or other defects. These discs are the controls used for calculating the original basic density of the log. The remainder of the log is then stored in the desired way at a given place. The logs are usually stored in a pile.

The control discs are debarked and weighed immediately for later calculation of the original moisture content of the log. The discs are then kept in a fresh condition and transported to the laboratory as soon as possible, where their volume is measured. A convenient method is to weigh the discs above and below a water surface and to use Archimedes' principle (Peterson and Winqvist, 1960). After the volumetric measurement the discs are dried to constant weight at 103°C. The original basic density of the log is then calculated from the basic densities of the two discs.

At intervals, or after a fixed period of storage varying from a few months to several years, depending on the purpose of the experiment and the storage conditions, 10–20 logs are withdrawn. From each of these logs a minimum of five discs is cut, either at fixed, equal distances from the log ends or—if a greater number of logs are available—at random. The discs from the stored logs are treated in the same manner as the control discs above. An average basic density can now be calculated for each log from the five discs. In order to avoid any risk of shrinkage due to drying, the discs can be soaked in water for 12–18 hr before the volumetric determination.

When the original basic density and the basic density after storage for each log are known, the average loss in basic density can be calculated for each series of 10–20 withdrawn logs. If the effect of storage time is to be determined, a number of such series will give a useful curve of the course of decomposition.

The reduction in basic density follows very closely the loss in dry mass of the wood. Only after prolonged storage under moist conditions, when the decay is far advanced, are discrepancies likely to arise between loss in basic density and loss in dry mass during the volumetric determination. This means that in moderately storage-decayed wood the method gives good information about losses in dry mass. However, when relative dry mass losses are determined on individual logs, errors of one to several percentage units can be expected due to limitations in the methods used.

When an experiment is designed, it is important to arrange that the logs included in the experiment have the same storage conditions and are representative of the stored wood. If the logs are stored in a pile, it is therefore recommended that the experimental logs in a pile be included with other logs to avoid drying and temperature influences from the side of the pile. It is also recommended that logs representing different levels in the pile, different diameter classes and different log types be included in each series of logs.

8. LOSS IN DRY MASS FOR WOOD IN CHIP PILES

Before beginning to store wood in the form of chips at a pulp factory, it might be of interest to study the storage defects. Methods have been worked out for determining the dry mass loss in piled wood chips. This work has been carried out mainly in the USA, Canada and Scandinavia, where many of the big pulp mills of the world are also located. The investigations normally published also involve chemical, paper technological and economic analyses, which will not be treated here.

An industrial chip pile normally comprises a minimum of 10 000 m^3 of chips and usually 5–25 times as many. During storage the temperature rises in the pile. The central parts soon reach 60–70°C while the other layers have temperatures close to those of the environment. Thus, temperatures vary widely in the pile, resulting in different microflora activity and different decomposition in various parts. Samples for determining the fungal degradation must therefore be taken from many positions to ensure a total representative of the situation in the pile.

8.1. Procedure

Chips are sampled from the normal production during the build-up of the pile. The chips are mixed thoroughly by intensive shovelling. The sampled chips are then transferred to nylon net bags (mesh size 2 × 2 mm). It is recommended that 3–5 kg of chips be placed in each bag. Weighings are made to an accuracy of 1 g.

At the same time, chips are transferred to plastic bags for immediate transport to the laboratory. There, samples (about 700 g) are taken out in order to determine the original moisture content by weighing in a fresh condition and after drying to constant weight at 103°C. Using the moisture content, the original dry mass of the bag samples can be calculated.

The prepared nylon net bags are then placed at various positions in the pile during the build-up period. It is recommended that the position of each bag be carefully marked so that it can be found when the pile is reclaimed. A total of 50–100 bags is recommended for each pile to be studied.

At the end of the experimental period the pile is carefully searched in those sections where the bag samples are located. The bags are weighed immediately to give the final wet weight of the chips. Two samples of about 700 g are taken out of each bag. These samples are weighed wet and after drying to constant weight at 103°C to give the final moisture content of the chips after storage.

By using the final moisture content, the final dry mass of the bag samples can be calculated. The difference between the original dry mass and the final dry mass is regarded as the loss in dry mass.

Further information regarding chip storage studies can be found in papers by Bergman (1973), Hatton (1970) and Shields and Unligil (1968).

REFERENCES

American Society for Testing and Materials (1956). *Tentative Method of Testing Wood Preservatives by Soil-Block Cultures.* ASTM D 1413-56 T.

Bergman, Ö. (1973). Wood substance losses in some chip storage experiments (in IUFRO Symposium Protection of Wood in Storage). Research Note R 83, Dep. of Forest Products, Royal College of Forestry, Stockholm.

Bergman, Ö. and Nilsson, T. (1968). On outside storage of birch chips at Mörrum's sulphate mill. Research Note R 60, Dep. of Forest Products, Royal College of Forestry, Stockholm.

Bravery, A. F. (1968). Determining the tolerance of soft-rot fungi to wood preservatives: a comparison of test methods. *Material u. Organismen*, **3**, 213–27.

British Standards Institution (1961). BS 838: *Methods of Test for Toxicity of Wood Preservatives to Fungi*. BSI, London.

Deuscher Normenausschuss (1939). *Prüfung von Holzschutzmitteln—Mykologische Kurzprüfung, Klötzchen-Verfahren.* DIN 52 176.

Duncan, C. G. (1960). Wood-attacking capacities and physiology of soft-rot fungi. For. Prod. Lab., US Dep. Agr., Rep. No. 2173.

Duncan, C. G. (1965). Determining resistance to soft-rot fungi. US For. Serv. Res. Paper, FPL 48 (Dec.).

Hatton, J. V. (1970). Precise studies on the effect of outside chip storage on fiber yield: White Spruce and Lodgepole Pine. *Tappi*, **53,** 4.

Henningsson, B. (1967*a*). Physiology of fungi attacking birch and aspen pulpwood. *Stud. For. Suec.* No. 52.

Henningsson, B. (1967*b*). Changes in impact bending strength, weight and alkali solubility following fungal attack on birch wood. *Stud. For. Suec.* No. 41.

Henningsson, B. (1967*c*). Microbial decomposition of unpeeled birch and aspen pulpwood during storage. *Stud. For. Suec.* No. 54.

Hoffmeyer, P. (1975). Mechanical properties of soft rot decayed Scots Pine with special reference to wooden poles. Swedish Wood Preservation Institute, Rep. No. 117.

Kaune, P. (1970). Bedingungen für das Prüfen mit Moderfäulepilzen in Vermiculit-Eingrabe-Verfahren. *Material u. Organismen*, **5**, 95–112.

Kerner-Gang, W. and Gersonde, M. (1972). Untersuchungen zur Prüfung mit Moderfäulcpilzen in Vermiculit-Eingrabe-Verfahren. *Material u. Organismen*, **7**, 241–56.

Liese, W. and Ammer, U. (1964). Über den Einfluss von Moderfäulepilzen auf die Schlagbiegefestigkeit von Buchenholz. *Holz als Roh-u. Werkstoff*, **22**, 97–102.

Liese, W. and von Pechmann, H. (1959). Untersuchungen über den Einfluss von Moderfäulepilzen auf die Holzfestigkeit. *Forstwiss. Zbl.*, **78**, 271–9.

Lundström, H. (1973). Studies of the wood-decaying capacity of the soft rot fungi *Allescheria terrestris*, *Phialophora* (*Margarinomyces*) *luteo-viridis* and *Phialophora richardsiae*. Research Note R 87, Dep. of Forest Products, Royal College of Forestry, Stockholm.

Nilsson, T. (1973). Studies on wood degradation and cellulolytic activity of microfungi. *Stud. For. Suec.* No. 104.

Nordic Wood Preservation Council (1970). NWPC Standard for testing of wood preservatives. Mycological test. Jordburk method—a soil block test with wood-rotting Basidiomycetes. NWPC Standard No. 1.4.1.1/70.

Peterson, O. and Winqvist, T. (1960). Seasonal changes in weight and moisture content of birch. Research Note R 28, Dep. of Forest Products, Royal College of Forestry, Stockholm.

Rennerfelt, E. (1964). Testing of wood preservatives against fungi. Swedish Wood Preservation Committee, Rep. No. 75.

Scheffer, T. C. (1936). Progressive effects of *Polyporus versicolor* on the physical and chemical properties of Red Gum Sapwood, *US Dep. Agr. Tech. Bull.* 527.

Shields, K. and Unligil, H. W. (1968). Deterioration of softwood chips owing to outside storage in New Brunswick. *Pulp Paper Mag. Can.*, **69**, 21.

18

Conservation Techniques for Old Waterlogged Wood from Shipwrecks Found in the Netherlands

J. de Jong

IJsselmeerpolders Development Authority, Smedinghuis, Zuiderwagenplein 2, Lelystad, The Netherlands

1. INTRODUCTION

In 1918 the Dutch Government decided to have the Zuiderzee separated from the North Sea by a barrier dam, and to construct five polders in the resulting freshwater lake, Lake IJssel. The first polder, the Wieringermeerpolder, was drained in 1930 and the barrier dam completed in 1932. The first polder was therefore constructed in a marine environment. The other polders have been reclaimed from the freshwater lake. The second polder, the Noordoostpolder (North-east Polder), was drained in 1942, Oostelijk Flevoland (Eastern Flevoland) in 1957, and Zuidelijk Flevoland (Southern Flevoland) in 1968. Markerwaard is still under development (Fig. 1).

During reclamation, many finds of archaeological and historical value have been made. Remains of neolithic settlements, dated between 3650 and 3350 B.C., have been excavated. Finds of a more recent date are mainly shipwrecks or parts of ships dating from the Middle Ages up to recent times, and crashed aeroplanes from World War II. Besides being an important fishing area, Lake IJssel has been, in previous centuries, a centre of national and international shipping. About 350 shipwrecks have been found there.

A number of these have been excavated to study the history of the shipbuilding trade and of life on board. In several cases the entire wreck, large parts of it, or the ship's inventories appeared to be of such interest and value that it was decided to add them to the collection of the Museum for Ship Archaeology at Ketelhaven. To preserve the excavated remains of the wrecked ships it is necessary to apply certain techniques of wood conservation to prevent extreme shrinkage, cracking and distortion, all common phenomena in old waterlogged wood exposed to drying.

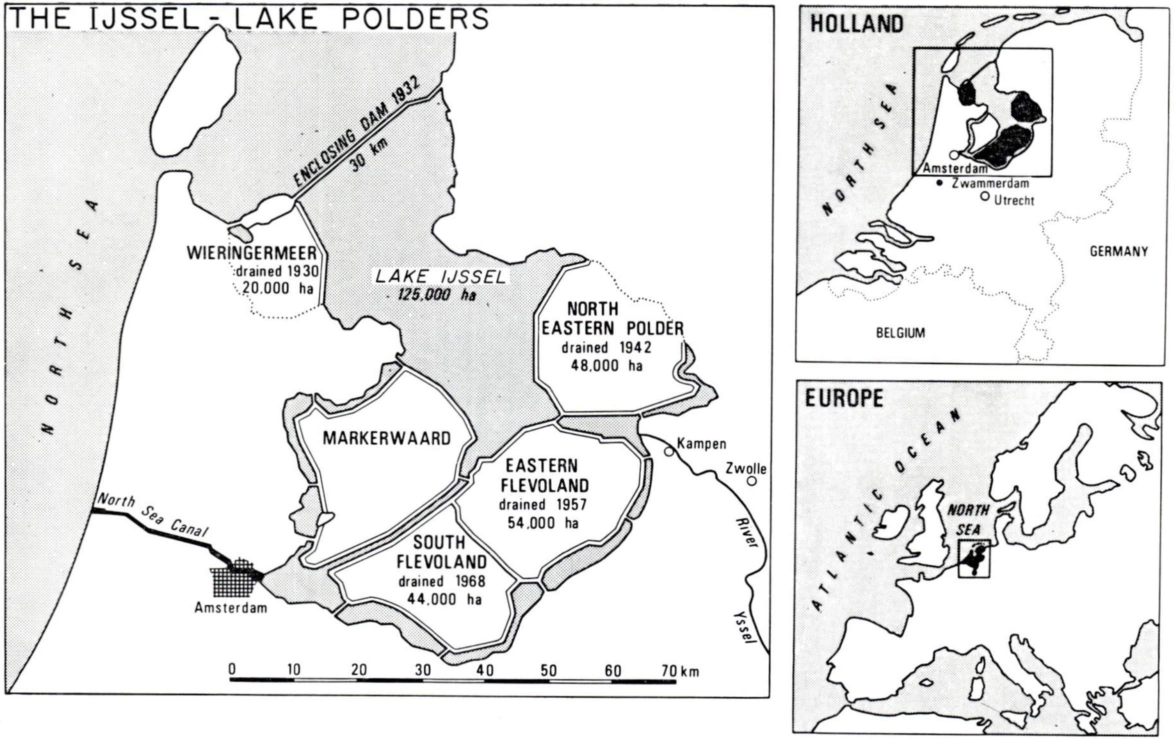

FIG. 1. The Lake IJssel land reclamation and development project.

2. THE PROPERTIES OF OLD WATERLOGGED WOOD

Wood kept in a very moist or wet environment will absorb water, and eventually become waterlogged. In time, changes will occur in the physical structure and chemical properties of the wood. These changes adversely affect the suitability of the wood for construction purposes, and are mainly caused by microbiological attack. Physico-chemical processes also contribute to the breakdown of the wood.

How far wood will disintegrate depends on the wood species, its growth history, the kind of environment in which the wood has been used and the conditions and the length of the period after its use. Important factors in this respect are mechanical damage, type of medium, humidity, temperature, pH and redox potential [1, 2]. Decay may occur under both aerobic and anaerobic conditions. The amount of oxygen in water-saturated soil, in which most of the shipwrecks are found in The Netherlands, is small, hence the disintegration will be mainly anaerobic. Microbiological decay causes more damage than physico-chemical processes.

In many cases, deterioration of the wood may start *before* it becomes waterlogged. Attack by rot-causing fungi and teredos may be taken as examples. Finally, the period of time between the moment the wood became waterlogged and was exposed to decay-causing phenomena, and the moment of excavation, will play a role.

The process of deterioration of wood in wet, anaerobic conditions is as follows. At the outset, the readily water-soluble compounds in the wood will diffuse into the surrounding medium, followed by readily hydrolysed compounds such as the pectins and pentosans. Microbiological degradation of the more stable hemicellulose and cellulose will then follow. Finally, lignin remains, which also can be decomposed slowly by micro-organisms in an anaerobic environment. Both fungi and bacteria cause microbiological breakdown of wood, and both use extracellular enzymes to hydrolyse hemicellulose, cellulose and lignin. In fungal breakdown, degradation spreads throughout the capillary system of the wood, at least as far as the extracellular enzymes can penetrate [3, 4].

The result of the decay of the cell material is a considerable loss of original strength. When drying, old waterlogged wood shows very strong shrinkage, warping, cracking and sometimes complete disintegration. Cell collapse can often be observed when old waterlogged wood is drying. The contracting capillary forces of the evaporating water exceed the strength of the capillary system of the deteriorated wood. Thus, with the decay of cell material the size of the capillaries increases, causing a decrease in the specific surface of the wood. Cellulose breakdown also results in a decrease of adsorptive sites, causing decreased swelling capacity of dried old waterlogged wood as compared with new wood. In practice, dried and shrunken old waterlogged wood swells only to a very limited extent when again in contact with water. Cell collapse is mostly irreversible. Noack [6, 49] recorded 9–10% radial and 20% tangential shrinkage in the slightly

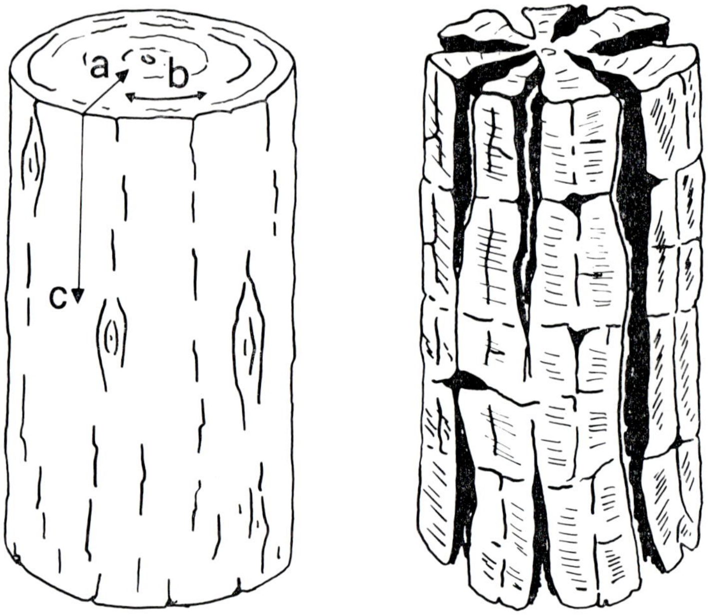

FIG. 2. The three main directions in wood and the damage due to shrinkage after drying a piece of old waterlogged wood. (*left*) Old waterlogged wood before drying: a = radial, b = tangential, c = longitudinal. (*right*) Old waterlogged wood after drying, showing large shrinkages in all directions. The shrinkages are larger in the highly deteriorated areas than in the relatively sound core of the wood.

decayed wood of the *Bremer Kogge*. The longitudinal shrinkage was almost zero. These data are based on drying from waterlogged to oven-dry (approximately 105°C) (Fig. 2).

The Zwammerdam and Utrecht ships showed that even higher degrees of shrinkage may occur (Table I). The more deteriorated the wood, the higher the water content for any given species. The breakdown of the wood also causes considerable changes in chemical composition (Table II).

Table I shows that in many cases old decayed waterlogged wood after drying shows degrees of shrinkage considerably higher than those of dried new waterlogged wood. Such old wood can be allowed to dry only in a few special cases when the wood is very slightly degraded.

TABLE I

DEGREE OF SHRINKAGE (%) IN NEW AND OLD WATERLOGGED WOOD WHEN DRIED AT 105°C

Wood	*Origin*	% *Water content*	% *Shrinkage when dried from waterlogged to oven-dry* (105°*C*)		
			Radial	*Tangential*	*Longi-tudinal*
Oak heartwood	New	±120	4·8	10·6	0·4
Oak sapwood	New	±145	4·0	9·4	0·4
Beech	New	±115	±6	12	0·3
Oak heartwood	*Bremer Kogge*	120–145	9–10	20	—
Oak heartwood	Utrecht ships	±100	8–9(8)*	17–22(19)	0–2(1)
Oak heartwood	Zwammerdam ships	100–120	12–14(13)	17–23(21)	0–2(0·5)
Oak heartwood	Utrecht ships	375–425	5–13(10)	26–52(33)	9–11(10)
Oak heartwood	Zwammerdam ships	375–400	8–16(11)	38–56(47)	10–17(13)
Oak sapwood	*Bremer Kogge*	600	18–19	54–64	10–13
Oak sapwood	Utrecht ships	670–1000	12–16(14)	41–55(47)	10–16(12)
Silver spruce	Zwammerdam ships	450	11–15(13)	12–49(26)	5–7(6)
Beech	*Bremer Kogge*	590–610	8–9	65–70	15–17

* The figures in parentheses are the mean shrinkage values as calculated from 4–7 experiments.
Source: Partly based on data from Noack [49].

The data in Tables I and II show that even when almost no chemical changes are found, there is still a considerable increase in shrinkage as compared with that of new waterlogged wood. Increased degradation means increased radial and tangential shrinkage. In highly degraded waterlogged wood a longitudinal shrinkage will also occur, possibly of over 10%.

When an outer layer of highly degraded wood is overlying a less degraded inner core, the outer layer shrinks during drying considerably more than the inner core. These differences cause the higher degraded outer layer to chip off from the inner core. Even with only partial chipping-off of this

TABLE II

THE MAIN CHEMICAL CONSTITUENTS IN NEW AND OLD WATERLOGGED WOOD

Wood	*Origin*	*Content as % of oven-dry wood*					
		Water	*Cellu-lose*	*Hemi-cellulose*	*Lignin*	*Ash*	*Fe*
Oak heartwood	New	±120	±65	—	±25	±0·3	—
Oak heartwood	New	—	±50	—	32	0·4	—
Oak sapwood	New	—	±50	—	32	0·6	—
Beech	New	±115	±66	—	±23	0·8	—
Softwood*	New	—	40–50	20–22	28–30	—	—
Heartwood*	New	—	40–50	26–28	20–22	—	—
Oak heartwood	*Wasa*	±150	67	—	29	3·4	—
Oak heartwood	*Wasa*	±800	26	—	62	5·6	—
Oak heartwood	*Bremer Kogge*	—	46	—	36	5·3	—
Oak heartwood	Utrecht ships	131	46·0	8·1	25·1	2·0	0·8
Oak heartwood	Utrecht ships	455	5·5	9·4	50·6	8·9	0·5
Oak heartwood	Zwammerdam ships	98	47·3	11·1	23·1	1·4	0·7
Oak heartwood	Zwammerdam ships	494	5·5	16·9	51·3	7·6	0·5
Oak sapwood	*Bremer Kogge*	600	17	—	62	7·7	—
Oak sapwood	Utrecht ships	760	6·8	3·1	64·1	10·1	0·4
Beech	*Bremer Kogge*	590–610	31	—	64	3·2	—
Silver spruce	Zwammerdam ships	463	11·0	1·6	69·3	2·4	0·1

* After K. Seifert.

Source: Data for the *Wasa* and the *Bremer Kogge* from Noack [6, 49] and Mühlethaler [5]. The data for the Zwammerdam and Utrecht ships were determined by the Forest Products Research Institute, TNO, Delft, The Netherlands.

outer layer, almost all details of the texture of the surface of the wood will be lost (Fig. 3).

The extra-high shrinkage in the outer layer causes crack formation parallel with and perpendicular to the fibre direction, and chipping-off of the outer layer. To prevent these adverse effects, it is necessary to apply a technique of conservation. The aim is to achieve a stabilisation of the dimensions of the find and also an equilibrium between wood and environment, preventing high degrees of shrinkage. In some cases, especially with highly degraded wood, an improvement of the mechanical properties of the material is desirable or necessary. Even new waterlogged wood shrinks

FIG. 3. A dried piece of the bottom planking of Utrecht ship 1, showing the damage in a highly deteriorated surface layer on top of a little degraded inner core of European oak.

when drying, mainly in the radial and tangential directions, and some shrinkage will prove inevitable during almost any conservation process (Table I). The acceptable degree of shrinkage depends largely on the nature of the object and its use after conservation. If the object is fairly small and has a very fine and delicate surface texture (*e.g.* sculpture) and will be used for further research (archaeological or historical), shrinkage should be minimal. If, however, the wood surface is rough and the finds will be used after conservation to give an impression of an entire construction, some shrinkage and surface defects can be accepted.

Deterioration is expected when wood is found in land reclaimed from sea or lakes. During the time these finds are covered by the sea or lake sediments, the wood is attacked anaerobically. After the reclamation of the land, which causes a lowering of the groundwater level, the degradation of the wood is continued by aerobic processes, hence deterioration is speeded up [48]. An indication of the lifetime of wood under such circumstances can be obtained from the durability classes which have been developed for new wood. This division of wood species is based on untreated heartwood during an outdoor exposure and in contact with a humid soil in a temperate climate (Table III). European oak heartwood is, according to this division, placed on the boundary of classes II and III. The expected lifetime will be about 15 years, and for small objects even less.

TABLE III

THE CLASSIFICATION OF WOOD SPECIES ACCORDING TO DURABILITY. THE DIVISION IS BASED ON AN OUTDOOR EXPOSURE OF UNTREATED HEARTWOOD IN CONTACT WITH A HUMID SOIL AND IN A TEMPERATE CLIMATE.

	Class	*Lifetime (years)*	*Wood species (examples)*
I	Very durable	25	Afrormosia, afzelia, azobé, iroko, merbau, teak, wengé
II	Durable	25–15	Californian redwood, red cedar, white cedar, European oak,* American mahogany, sipo mahogany, red meranty (heavy)
III	Moderately durable	15–10	Oregon pine, pitch pine, larch, red meranty (light), sapele mahogany, European walnut
IV	Little durable	10–5	Fir, European pine, Parana pine, American red oak, maple, elm
V	Very little durable	5	Balsa, birch, beech, ash, poplar, alder, lime, willow

* At the boundary of class II and III.

3. METHODS OF CONSERVATION AND RESULTS

3.1. Introduction

Investigations conducted with new wood revealed that stable dimensions could be maintained only under stable environmental conditions, especially of humidity and temperature [3]. So, if the dimensions are to remain constant, temperature and relative humidity values can be allowed to vary only between narrow and predetermined limits. In order to widen this range, techniques have been developed to reduce the hygroscopicity of the wood (*e.g.* acetylation) or to reduce the transport of water vapour and water into and out of the wood (surface coatings).

Since deterioration results in decreased chemical reactivity of the wood (decomposition of celluloses and hemicelluloses), chemical reduction of the hygroscopicity of the material will have a limited effect. Surface coating will merely retard the transport of water and water vapour from waterlogged wood, so that eventually the shrinkage values corresponding to the environmental conditions will be established [3, 10, 11]. If, after conservation, the object proves to be too fragile to carry its own weight, an additional consolidation at and near the surface is necessary to prevent damage when handling the object. In practice, some treatment methods cover both these aspects.

For the treatment of fairly small objects, a number of techniques have been developed giving, in many cases, good results [5]. If, however, a shipwreck or large parts of it have to be treated, even a comparatively small shrinkage may result in the formation of crevices, gaps and tearing apart in an originally tight construction. Brorson Christensen [17], Barkman [1] and Noack [6, 49] have shown that the conservation of very large finds causes many problems. Some shrinkage during and after the conservation process appears to be inevitable. Before starting conservation it is essential to prevent the wood from drying and to control the growth of fungi, bacteria and algae, but the pesticide applied should have no adverse effect on the conservation. If the period of storage is relatively short, storing in a room as cool as possible and periodically changing the water in the storage tanks will be sufficient.

For conservation [8] the first approach has often been to attempt to apply a coating to the surface and the immediate subsurface. The coating must be impermeable to water and water vapour to a very high degree. Unless very thick coatings are applied, this requirement will never be met, drying will occur and cracks readily form. Hence a coating technique will not prevent shrinkage but merely retard the establishment of an equilibrium between the wood and its environment. Only when a very thin deteriorated surface layer overlies a sound and almost undegraded inner core can a surface treatment give satisfactory results. Such was the case for the Kentmere boat where good conservation was achieved [9] by an impregnation and consolidation of the surface of the wood with an epoxide resin. Near Lelystad a nineteenth-century ship has been excavated and left to dry in the open air. The wood was in a very good condition and no damage other than a local slight surface cracking occurred after a few years.

The second approach is to impregnate the wood as thoroughly as possible. Water present in the old waterlogged wood is substituted by non-volatile compounds. If such a compound is strongly absorbed into the cell walls, it can take over the function of a swelling agent from the water which is forced to leave the wood. Hence the wood, though now containing less water, retains its swollen state. Ethyleneglycol and polymers based on this compound are especially suitable for this purpose [35–37, 56].

When the wood is impregnated with compounds with a hygroscopicity higher than the wood itself, such as some salts, sugars and some types of polyethyleneglycol, these compounds may buffer the changes in relative humidity and temperature. When impregnating agents are applied which exert no swelling action on the wood and have a hygroscopicity less than the wood, only their impermeability to water and water vapour, their strength, and their adhesion to the wood will determine the extent to which hygroscopicity might still cause problems. Using a total treatment process, an improvement of the strength of the wood throughout the object can be obtained. The number of techniques available is larger if, after conservation, the object can be stored in a place where temperature and relative humidity are stable (*e.g.* 18–20°C, 65–70% r.h.).

The transport of any agent into the wood capillary system depends largely on the viscosity of the solution, the temperature, the volume and shape of the capillaries and the dimensions of the penetrating molecules [3, 12, 13]. With increasing viscosity and molecular size, penetration speed will decrease. Increasing the temperature and the concentration difference between the outside and the inside of the wood will result in increased penetration speed [3, 12, 13]. In general, compounds with low molecular weights and low viscosity offer the best opportunities for penetration.

3.2. Conservation Methods

Methods for a total conservation of old waterlogged wood can be divided into three groups:

(a) The first group consists of processes based upon dehydration of the wood. The water is removed by freeze-drying or exchanged for solvents like methanol, ethanol, tertiary butylalcohol or acetone. After dehydration, a final surface or total treatment follows. Alternatively, an apolar solvent (*e.g.* diethylether) followed by impregnation with a solution of resins in this ether is used. Freeze-drying techniques where the stabilisation agent is added before freeze-drying give good results.

(b) The second group consists of techniques resulting in an improvement in the strength of the wood to such an extent that the wood can stand a drying-out without suffering considerable damage from shrinkage. Fixation of the water in the wood can also be used, as with the alum method. This type of treatment involves the use of both inorganic compounds (chromates, silicates, arsenates, alum) and organic compounds (urea-formaldehyde and melamine formaldehyde polymers or glycol-methacrylate). In general, no final treatment of the wood is necessary, but aesthetic considerations may demand application of a final surface treatment.

(c) The third group is based on the exchange of water in the wood in a gradually declining water-containing liquid medium for a water-soluble polymer derived from ethyleneglycol. These polyethyleneglycols are capable, when absorbed into the cell walls, of acting as a swelling agent like water. These compounds are non-volatile and, once introduced, remain in the wood under dry circumstances, so giving a good and lasting dimensional stabilisation. Several methods for the impregnation of polyethyleneglycols into old waterlogged wood have been described. If the hard, wax-like polyethyleneglycol (PEG) grades are used and the capillary system is filled with these compounds, there is also an improvement in mechanical strength.

Schematically the methods can be represented as follows:

A. Total treatment methods (results often (very) good):	(a) Dehydration (freeze-drying, liquid exchange). Final treatment, either surface or total, necessary.
	(b) Strength improvement in waterlogged condition; the treatment can be given with both organic and inorganic compounds. Final treatment often based on aesthetic considerations.
	(c) Impregnations with water-soluble polyethyleneglycols. These polymers tend to keep the wood in a swollen condition. Final surface treatment unnecessary but sometimes desirable for aesthetic reasons.
B. Surface treatment methods (results mostly poor):	Surface treatment with compounds, often resins, waxes or polymers, in order to prevent the water from evaporating.

3.2.1. *Dehydration*

(i) *Freeze drying* [7, 14 16]. The waterlogged wood is first frozen and the water then removed by sublimation under vacuum. Ambrose [15, 16] adds PEG 400 to the wood in amounts sufficient to keep it in a swollen condition while avoiding the introduction of any surplus of PEG 400. Depending on its quality, the wood is impregnated with a 5–12% solution of PEG 400 in water. To obtain a good dimensional stabilisation, 20% PEG 400, relative to the dry wood constituents, is regarded as necessary. The more deteriorated the wood, the more diluted the impregnation solution must be. After impregnation, the PEG 400 solution in the wood is frozen and the water removed by sublimation under vacuum.

The results obtained are good if the treated objects are relatively small. After this treatment, Ambrose determined the PEG 400 contents in the wood, and these varied from 15% to 25% of the weight of the dry wood material. One advantage of freeze-drying wood impregnated with these diluted PEG 400 solutions is that the liquids do not expand on freezing and hence cause no damage to the wood while freezing. Brorson Christensen [7] exchanged the water in the wood with tertiary butylalcohol (TBA) and impregnated the wood afterwards with a solution of PEG 4000 in TBA at 50–54°C. Starting with a low PEG 4000 concentration, the PEG in the solution was gradually raised to 50–66%. After this impregnation the wood and the liquid in the wood were frozen and the TBA removed by vacuum sublimation. The shrinkages occurring during this process were small.

(ii) *Liquid exchange.* Water in old waterlogged wood can be exchanged for water-soluble, polar and volatile organic solvents. These solvents can, in turn, be exchanged for apolar, highly volatile solvents. After an impregnation of the wood with a resin solution, the solvent can evaporate off. The alcohol–ether method is the most widely used [19]. Brongers and Wijnman use a vacuum evaporation stage at the end of their alcohol–ether conservation [18]. After cleaning and pretreating, if necessary, such as bleaching in an ammonia–hydrogen peroxide bath (1–2%), the wood is placed in 70–90% ethanol. These baths are renewed until no more water is liberated from the wood. Then, in a series of baths, the alcohol is exchanged for diethylether. The wood is then impregnated with a solution of resins in diethylether and dried [5, 18, 19]. A final surface treatment with microcrystalline waxes is often applied. Mühlethaler [5] emphasises that this method is very suitable for the treatment of small objects of alder, ash, maple, elm, lime, beech and oak sapwood, all porous and easily permeable. Problems with this conservation process arise when treating coniferous wood, especially silver spruce and taxus. Large objects have not yet been conserved by this technique, since the risks in using ethanol and diethylether for large-scale operations are regarded as too high.

McKerrel *et al.* [31] described a method of dehydrating old waterlogged oak. With ancient Roman oak, the appearance of the wood could be greatly improved by pretreating with a 3·5% hydrochloric acid solution, after which the wood was dehydrated with acetone. Both treatments, when combined, improved the permeability of the wood for PEG 4000 considerably. Since the solubility of PEG 4000 in acetone is limited, McKerrel *et al.* looked for a substitute and the natural resin rosin was found to be suitable. The main constituent of rosin is abietic acid.

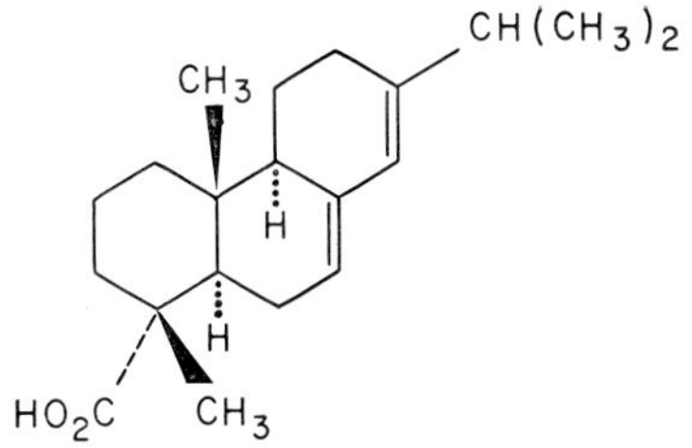

Abietic acid

Rosin is comparatively cheap and is available in large quantities. The disadvantages are that it is often tacky, and it has a low melting point (80°C) and a high acid number (160). Also, it ages fast through air oxidation [61].

Old waterlogged oak treated with diluted hydrochloric acid (3·5%) followed by dehydrating with acetone is easy to impregnate with rosin. In the treated wood, the tangential shrinkages varied between 1·3% and 6·0%, indicating a fairly good dimensional stabilisation.

With the ships found at Zwammerdam and Utrecht, attention was also paid to the dehydration of the wood. When acetone was used both with and without pretreating the wood (European oak heartwood and sapwood) with dilute hydrochloric acid (3·5%), the wood and the acetone solution turned black. The use of methanol and TBA for dehydration of old waterlogged wood during our investigations, however, never caused such a black discoloration. The wood species tested were European oak, beech, silver spruce, elm and ash.

In many cases dehydration can be less than 100%, but needs to be continued only until the azeotropic mixture of the solvent and water has been reached. TBA has distinct practical advantages since the excess of water can be removed by distillation, and wood saturated with the TBA–water azeotrope can be impregnated with different monomers dissolved in this azeotrope, such as methylmethacrylate or butylmethacrylate, or with a solution of PEG 4000 in this azeotrope. Dehydration is now started in an 88% TBA solution in water and the water extracted from the wood by distillation, the distillate being transported back into the dehydration tank. The loss of water and some alcohol vapour is compensated by the addition of TBA (Fig. 4).

The following dehydration experiments have been done with old waterlogged oak samples up to approximately 20 × 20 × 100 cm. During all the experiments the organic solvent was circulated around the samples.

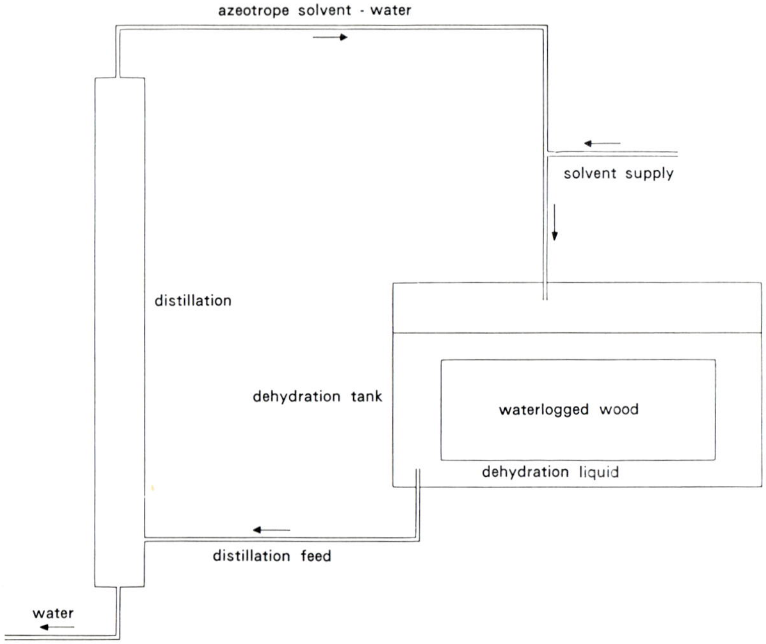

FIG. 4. Scheme for the dehydration of waterlogged wood using azeotropic distillation of the solvent–water mixture.

After dehydration, the dimensions of the samples, which had been measured before the treatment, were measured again and the shrinkages were calculated.

(a) Dehydration with tertiary butylalcohol:

1. Samples were immersed in a 19% TBA solution, after which the concentration of the solution was raised to 100% in 28 days (largest sample 15 × 15 × 20 cm). After 2 days storage in 100% TBA there was no change in water content of the solution and the samples were measured.

2. Samples were immersed in 100% TBA. 10% of that in the bath (10% of the bath volume) was changed daily for new 100% TBA. After 30 days the water content remained almost zero and the samples were measured. The largest samples were 15 × 15 × 20 cm.
3. Samples were immersed in 88% TBA. The water removed from the wood was separated off by azeotropic distillation. The process was continued until no more water was liberated. For the largest samples this took 4 months. Since the available distillation capacity limited the water separation speed, this period may be shorter. The largest samples were 20 × 20 × 95 cm.

(b) Dehydration with methanol. Samples were immersed in 25% methanol. In 26 days the concentration of the methanol was raised to 100%, until no more water was liberated from the wood. After 4 days storage in 100% methanol the samples were measured. The largest samples were 15 × 15 × 20 cm.

(c) Dehydration with acetone. The technique as described for methanol was repeated several times using acetone with and without a pretreatment with 3·5% hydrochloric acid. In all cases the samples and the acetone solution turned black.

The measured shrinkage values (Table IV) indicate that old waterlogged oak wood can be dehydrated with TBA, methanol or acetone without suffering large shrinkages. Since with acetone the wood turns black and the separation of water from methanol is less easy than from TBA, we prefer TBA. The use of TBA results in a slight brightening of the wood's appearance.

(iii) *Final treatment of dehydrated wood.* The wood obtained after freeze-drying or liquid exchange is hygroscopic, hence it will tend to absorb water. When this happens, the familar reactions of the wood with respect to changing environmental conditions will occur. Also, the dehydrated wood can become very brittle and fragile. The problems of hygroscopicity were solved by Ambrose [16] who impregnated the wood with the minimum amount of PEG 400 necessary to obtain dimensional stabilisation followed by freeze-drying.

When wood is dehydrated using the alcohol–ether method [5, 18], or the method described by McKerrel *et al.* [31], or that of Brorson Christensen [7], in all cases a final treatment with natural or synthetic compounds is necessary (dammar resin, rosin, polyvinyl acetate, acrylic resins and polyethyleneglycol). Methylmethacrylate or butylmethacrylate may also be used for the final treatment. After dehydrating wood in

TABLE IV

MEAN SHRINKAGE VALUES FOUND FROM DEHYDRATION EXPERIMENTS WITH OLD WATERLOGGED OAK. TERTIARY BUTYLALCOHOL (TBA), METHANOL AND ACETONE WERE USED TO DEHYDRATE THE WOOD. THE EXPERIMENTS ARE DESCRIBED IN THE TEXT

Dehydration solvent	*No. of experi-ment*	*Wood type*	*Water content*	*Shrinkage after dehydration*		
				Radial	*Tan-gential*	*Longi-tudinal*
TBA	1	Heartwood	100–140%	0·2	0·7	0·1
TBA	1	Heartwood	~700%	0·3	—	—
TBA	2	Heartwood	100–140%	0·4	0·6	0·1
TBA	2	Heartwood	375–425%	—	—	0·5
TBA	2	Heartwood	~700%	—	—	1·2
TBA	3	Heartwood	100–140%	0·5	1·1	0·6
TBA	3	Sapwood	~700%	0·5	1·1	0·6
Methanol	1	Heartwood	100–140%	−0·1	0·1	−0·6
Methanol	1	Sapwood	~700%	—	—	0·2
Acetone	1	Heartwood	100–140%	2·4*	0·6	0·5
Acetone	1	Sapwood	~700%	—	0·3†	1·9

* Strong influence of one value of 12·6% shrinkage out of 7 measurements.
† Result of 2 measurements.

acetone, this solvent can be exchanged for butylmethacrylate. As a polymerisation catalyst, 0·3–1% benzoylperoxide is added. The monomer in the wood is polymerised at 65–95°C for 8–9 hr. The results are said to be good [5].

In laboratory experiments with samples of old waterlogged oak wood in which highly degraded sapwood, deteriorated heartwood and still very sound heartwood were present, good results have also been obtained by means of final treatment with butylmethacrylate. The largest samples treated were 15 × 15 × 15 cm. The wood had been dehydrated until the azeotropic water concentration was reached in TBA, then this solvent was exchanged under vacuum for the butylmethacrylate. As a catalyst, 1–3% benzoylperoxide had been added. When no more TBA and water were liberated from the samples (the solution ceasing to boil), the butylmethacrylate was polymerised for 6 hr at 75°C.

The following shrinkage values have been calculated after a storage period of one month in the laboratory, at 18–20°C, 55–70% r.h.:

Radial shrinkage	1·3%
Tangential shrinkage	4·5%
Longitudinal shrinkage	0·6%

In badly deteriorated heartwood the longitudinal shrinkage was 1·6 %, in highly deteriorated sapwood 2·2 %. These shrinkage values are acceptably low in most cases, though the difference between the longitudinal shrinkage for the three given wood qualities could result in the formation of cracks perpendicular to the fibre direction in large objects. Also, the tangential shrinkage of 4·5 % may result in the formation of some cracks parallel to the fibre direction. Such damage has not been found in the fairly small samples which were used in the laboratory. The treated samples retained the unpleasant smell of butylmethacrylate.

Methylmethacrylate has also been used for final treatment of dehydrated wood. The monomers can be polymerised with the aid of chemical radical generators or with radiation. The technique is comparable to that used for butylmethacrylate, but no vacuum can be applied to raise the exchange speed of the dehydration solvent. Brorson Christensen found a volume shrinkage of approximately 7 % using methylmethacrylate, while Munnikendam [32] also found low shrinkage values at very high impregnation levels. At impregnation levels between 90 % and 100 % the following shrinkage values have been found for old waterlogged oak with an original water content of 163 %:

Radial shrinkage	2–2·3 %
Tangential shrinkage	−0·3–3·5 %
Longitudinal shrinkage	−1·8–0 %

For our experiments, European oak with highly degraded sapwood and deteriorated and still sound heartwood was again used. The samples were dehydrated in TBA until the azeotropic composition of the liquid was achieved. Then this TBA–water mixture was exchanged for methylmethacrylate. No vacuum was applied after the samples were immersed in the monomer to which 1–3 % benzoylperoxide had been added as a catalyst. The exchange time was 7–14 days (18–20°C). The treated samples were polymerised at 65°C for 4–6 hr. The dimensions of the treated samples one month after polymerisation were:

Radial shrinkage	5·6 %
Tangential shrinkage	8·2 %
Longitudinal shrinkage	0·8 %

In the very strongly deteriorated sapwood a longitudinal shrinkage of 5·3 % was found. During the polymerisation, crack formation occurred in the samples both parallel with and perpendicular to the fibre direction. The cracks were formed mainly in the sapwood layer.

3.2.2. *Strength Improvement in Waterlogged Condition*

(i) *The alum method.* The waterlogged wood is immersed in a hot concentrated solution of potassium and aluminium sulphate, following which potassium aluminium sulphate ($K_2SO_4 \cdot Al_2(SO_4)_3 24$ aq.) can be precipitated in the wood on cooling. The crystal mass in the wood prevents shrinkage during evaporation. In Denmark an impregnation of old waterlogged wood with alum and glycerol has been used [7]. After some time, this treatment appeared to result in a disintegration of the objects treated. Alum is very soluble in water and it is necessary to protect the treated objects against high relative humidities. This is even more necessary when the hygroscopic glycerol has also been used.

(ii) *Waterglass.* Munnikendam [20] mentions treatment with waterglass and fluorosilicates. Impregnation with waterglass is used for the fireproofing of new wood. The wood is impregnated with a stabilised solution of sodium waterglass ($Na_2O \cdot nSiO_2$). This sodium waterglass is a viscous liquid which can be diluted with acid-free water. The viscosity of sodium waterglass is at a minimum for $n = 2$.

In air, sodium waterglass solidifies by uptake of carbon dioxide. During this reaction, water-soluble sodium carbonates and insoluble silica oxides are formed. With metal oxides (MgO or ZnO) hard and insoluble silicates are formed during a fast reaction, which can be disadvantageous. Raising the pH in the wood to 10 or 11 can prevent this reaction. Using this technique, a considerable increase in strength and hardness was achieved [58].

(iii) *The electrokinetic method.* The material to be treated is, at one site, saturated with a solution of sodium silicate and, at another site, with a solution of calcium chloride. Passing a direct current in a suitable direction through the material locally then brings about the formation of calcium silicate. In Poland this method has been used for conservations *in situ* [23], but Mühlethaler [5] reports that it has failed to produce satisfactory results.

(iv) *Chromium oxide* [24, 25]. This process is based on the use of a 2–10% solution of chromium oxide, which can be used with or without the addition of 10–20% sodium bichromate. Before starting, it is necessary to extract the wood with a 1% hydrochloric acid solution to remove calcium carbonates. When the wood is found in a freshwater environment, 10% sodium chloride is added to the chromium oxide/bichromate solution. After a complete impregnation, the treated objects can be left to dry in the air. During drying out, only small shrinkages will occur. The treated objects acquire a rust-brown colour and are brittle. They do not (Garrouste [25]) respond with large dimensional changes to variations in the relative humidity of the surrounding air. To obtain the best results [26], final treatment of the surface layers is preferred. Bouis has investigated the use of linseed oil for this purpose [62].

(v) *Water-soluble monomers or prepolymers.* Several methods for the impregnation of waterlogged wood with water-soluble monomers or prepolymers have been studied. The polymerisation of these compounds should result in the desired dimensional stabilisation and strength improvement. The compounds used are glycolmethacrylate and melamine–formaldehyde or urea–formaldehyde prepolymers.

(a) Glycol-(di)-methacrylate. This monomer is a derivative of methacrylic acid, but unlike methyl- or butylmethacrylate it is soluble in water and can be used for direct exchange. Polymerisation can be achieved with the aid of chemical catalysts, such as benzoylperoxide or azobisisobutyronitrile, or radiation. Hydroquinone or hydroquinone methylether can be used for polymerisation inhibition. Thermal polymerisation can be obtained without the addition of a catalyst [26, 30]. The monomer is very easy to polymerise; metal salts can induce the polymerisation, while in solutions containing over 10% of water, a hydrogel can be formed [26]. Conservation experience of old waterlogged wood with glycol-(di)-methacrylate is still limited to small samples.

(b) Melamine–formaldehyde and urea–formaldehyde resins [27, 28]. Arigal-C is a melamine–formaldehyde resin which when in water is present in a resinoid form. The condensation reaction can be started by the addition of a catalyst which lowers the pH, which originally is above 9. Depending on the temperature and the amount of catalyst added, the polymer is formed within 24–48 hr [5, 29].

The disadvantage of the Arigal-C method is the short impregnation and polymerisation time, which prevents a treatment of large objects. In Russia a method has been developed by which this disadvantage is partly eliminated [5]. Waterlogged or dehydrated wood is impregnated with a prepolymer of formaldehyde and urea or melamine. These prepolymers are easily soluble in water or acetone and have a longer polymerisation time than Arigal-C, while their viscosity in the concentration range between 30% and 60% is sufficiently low. As a catalyst, 2% aluminium chloride (based on the amount of polymer) is added.

Lehman in Poland has developed a method resembling the two former methods [5].

(vi) *Impregnation with polyethyleneglycols* [33–38, 50, 51]. In Table V, some properties of polyethyleneglycols of different molecular weights are given. Since polyethyleneglycols with molecular weights less than 1000 are too hygroscopic to be of general practical use, and the diffusion of the high molecular weight types into the wood is too low, in most cases polymers with molecular weights ranging from 1000 to 4000 are used for the dimensional stabilisation of wood. The obtained dimensional stabilisation is based on an exchange of water in the wood for a non-volatile,

TABLE V

SOME PROPERTIES OF POLYETHYLENEGLYCOLS

PEG type	*Molecular weight range**	*Specific weight (20–22°C)*	*Melting point (°C)*	*Solubility in water (parts by weight 20°C)*	*Relative hygroscopicity (glycol = 100)*
200	190–210	1·126	−15	100	80
300	285–315	1·126	−10	100	70
400	380–420	1·126	6	100	60
600	570–630	1·126	20	100	45
1000	950–1050	1·140	38	70	5
4000	3000–3700	1·204	52	60	—
6000	6000–7500	—	58	50	—

* Depending somewhat on the process of manufacture.
Source: Data from Morén and Centerwall [34].

water-soluble compound which also has the property of being a swelling agent for the wood, such as water. The impregnation agent has to keep the wood in a swollen condition while the water is expelled. When treating new wood, excellent results are obtained when, as a general rule—based on the dry wood material—about 35% of polyethyleneglycol is present in the wood [37]. From the investigations of Ambrose [16] it appeared that the dimensions of old waterlogged wood found in swamps in New Guinea were stabilised when the wood contained about 20% PEG with a molecular weight of 400.

Polyethyleneglycols are also soluble in alcohols, acetone, chlorinated aliphatic hydrocarbons and some aliphatic esters, but water is the cheapest solvent. Where it is desirable to reduce the specific gravity, the viscosity or the surface tension of the solution, the addition of alcohol can be considered.

The practical use of polyethyleneglycols is based on the following properties, which they all have in common, namely an excellent solubility in water, a high chemical and thermal stability, and a good resistance against microbiological attack. Polyethyleneglycols are not only used in the timber industry but also in the chemical, pharmaceutical and cosmetic industries.

Tests were made on the use of polyethyleneglycols for the conservation of old waterlogged wood [6, 7, 40–42, 44–46]. Morén and Centerwall [34] describe such a method for the treatment of highly degraded wood with a water content of 400 % or more, but when the water content is below 400 % their method must be suitably modified.

Lefevre [41] found that when painting old Roman waterlogged yew with a 62 % solution of PEG 4000 in water, evaporation was faster than the impregnation of PEG 4000, which resulted in an unacceptably high shrinkage.

Organ [42] simplified the method of Morén and Centerwall. The addition of sodium pentachlorophenolate, the pH control and the addition of sodium carbonate was omitted. PEG 4000 was used again for the impregnation of the wood. When the impregnation liquid turned a dark colour and became resinous, a new bath was substituted. This method has given good results for small pieces of highly deteriorated old waterlogged wood. Coniferous woods require two treatments. Slightly degraded wood cannot be conserved successfully. In such cases the original dimensions which had been changed during the impregnation of the object can be restored by extracting the PEG 4000 from the wood with water. Organ [46] regards PEG 4000 as the most suitable compound for the conservation of old waterlogged wood. The lower molecular weight grades are too hygroscopic, the impregnation of the higher molecular weight grades is too slow.

Albright [43] first dehydrated the wood at 65°C with ethanol, and then immersed it in a 50 % solution of PEG 4000 in ethanol at 65°C. The alcohol slowly evaporates until a bath of molten PEG 4000 is obtained. The impregnation is regarded as complete when the object, after a surface cleaning with water, reaches a weight 25 % greater than after dehydration with ethanol. This method has given good results for small objects.

Brorson Christensen [7] used the following method for the treatment of the Viking ships found at Roskilde in Denmark. After cleaning, the wood was immersed in an aqueous solution of PEG 4000 at 60°C. The bath temperature was sufficiently high to prevent growth of microorganisms. Depending on the size of the objects, the concentration of the PEG 4000 in the bath was raised from 10 % to almost 100 % over a period of 1 month (very small objects) to 24 months (large objects). After cleaning the wood surface, a final surface treatment with an epoxide resin was given.

This method was not successful for conservation of slightly deteriorated old waterlogged European oak wood. A method for a cold impregnation of such wood has been developed by Brorson Christensen [7]. The wood is treated with a 25% solution of PEG 4000. To this solution a 1% borax/boric acid mixture (7 part by weight borax, 4 parts boric acid) is added to prevent the growth of microorganisms. In treating large objects the PEG concentration is raised to 50% for a period of 12 months after which the wood is left to dry slowly. Every day during this drying period the wood is immersed for a short period in the 50% PEG solution in order to obtain a surface saturation as high as possible. The results obtained are acceptable.

Brorson Christensen [7] also developed a method of dehydrating slightly deteriorated oak with tertiary butylacohol (TBA) followed by an impregnation with a solution of PEG 4000 in TBA. After impregnation the TBA was removed from the wood by freeze-drying. This method is now used for the treatment of large objects in Denmark. Results are as yet not available.

Based on this principle, the following experiment was carried out in our laboratory. Wood samples in which highly degraded oak sapwood, deteriorated oak heartwood and sound heartwood were present were dehydrated in 88% TBA. The samples had wet weights from 0·5 kg to 8 kg, with dimensions from about 5 × 10 × 10 cm up to 10 × 15 × 50 cm (radial × tangenital × longitudinal). Dehydration took 2 months.

After dehydration, the wood was impregnated with a solution of PEG 4000 in 88% TBA at 50–55°C. The concentration was raised in 10% steps up to 50% in 4 weeks, after which the concentration was maintained for another 4 weeks at 50%. Occasionally PEG was added to maintain the concentration of the solution at 50%, due to diffusion of PEG into the wood. The final PEG concentration in the bath liquid was 52%. After this impregnation the wood was removed from the bath and cooled. The PEG crystallises from the solution and therefore little or no transport of PEG will occur from the outside of the wood while the solvent evaporates. After 2 weeks the samples were heated for 7 days up to 65°C in order to melt the PEG. After this treatment, which has the advantage that freeze-drying can be omitted and that a second heat treatment is given to ensure the absorption of PEG into the cell walls, it was found that even in samples in which highly degraded sapwood, deteriorated heartwood and sound heartwood were present, no shrinkage damage occurred. Some small fissures along the fibre direction was the only damage observed. The heat treatment caused the wood to lose the familar waxy surface of PEG-treated wood.

The shrinkages have been measured at several stages of the treatment. The final values are determined one month after the heat treatment, with storage in the laboratory at 18–20°C and 55–65% r.h.:

	% Shrinkage		
	Radial	*Tangential*	*Longitudinal*
Little deteriorated oak heartwood	3·0–3·7	2·9–4·1	*
Highly deteriorated oak heartwood	1·8–2·2	2·1–9·6	1·2–1·7

* This value could not be measured with sufficient accuracy but was less than 1%.

Seborg and Inverarity [44] described the treatment of a 200-year-old boat, built from oak and coniferous wood. The thickness of the hull planking varied from 2 to 4 cm. After the boat had been left on the excavation site during the winter under a cover of moist sawdust it appeared that allowing the wood to dry would result in a chipping-off of the highly deteriorated outer layer of the wood. It was decided to impregnate the boat with a 50% solution of PEG 4000 for 3 weeks, after which it was assumed that a sufficient stabilisation of the surface layer had been obtained. No attempt was made to impregnate the sound inner core of the wood. After impregnation and drying, only small cracks in the surface layers developed, and 18 months after this treatment neither shrinkage nor distortions could be observed.

Heslinga [56] observed that ethyleneglycol and its dimer or trimer could be used to swell shrunken and dried old waterlogged wood. The investigations were performed on European oak from a seventeenth-century ship. The wood had a slightly deteriorated hard core (water content 120–185%) and a probably already somewhat dried degraded surface layer (water content 200–260%). Since ethyleneglycol and its dimer or trimer are somewhat volatile, a final treatment of the surface was considered necessary. Since the wood was relatively little deteriorated, water could also have been used as a swelling agent [7, 46]. For the conservation of large objects an impregnation with ethyleneglycol, followed by an impregnation of the surface layers with polyethyleneglycol, is recommended.

From investigations in our laboratory we learned that by impregnating old waterlogged wood with successive baths of PEG 600, PEG 1500 and finally PEG 4000, the initial results look fine and the shrinkages are small. However, after some time, interdiffusion of the different PEG grades occurs and liquefication of the PEG near the surface results. High relative humidities enhance this process.

Barkman [40] describes a method for the conservation of small objects using a PEG bath method and a spraying method for the treatment of large objects, such as the warship *Wasa*. The objects are immersed in a bath at 60°C containing 2% of a borax/boric acid mixture (7 parts borax, 3 parts boric by weight). In a period of 12 months the concentration of PEG is daily raised by $\frac{1}{12}$%, then for 5 months by $\frac{1}{5}$% per day, and finally for 1 month by $\frac{1}{2}$% per day. During the impregnation of the wood, the liquid in the bath is circulated by pumping. After the impregnation period the objects are dried in air at 65–70% r.h. During this drying the objects are surface-treated by spraying with a 35% PEG solution. The surface concentrations of PEG can be much higher than the average value of 40%. Initially PEG 4000 was used, but later PEG 1500 was preferred. However, the high surface concentrations may, if PEG 1500 is used, cause problems in a high relative humidity environment.

For treatment of the hull of the warship *Wasa* a spraying technique was used. The find was too large to be treated by a bath method. Before the actual conservation was started, contaminations such as clay, rust and iron sulphides were removed thoroughly. This cleaning was done while the ship was being sprayed continuously with water. After cleaning, the ship was sprayed with a 15% solution of PEG 1500 containing 4% of the borax/boric acid mixture as a biocide. The spraying of the hull was continued from 1962 to 1972, after which an intermitted spraying was applied until 1975 to saturate the surface of the hull with PEG, and to ensure a uniform drying of the structure. The average PEG concentration in the inner and outer planking is about 28%. The distribution of the PEG, however, was not as uniform as might be desired. During the summer of 1975 the spraying of the hull was continued using PEG 600 [47, 60].

Noack [6, 49] found that the oak of the *Bremer Kogge* could be stabilised with PEG 1000, provided enough of this agent is introduced into the cell walls while applying suitable techniques. To ensure this, the ship will be impregnated in a huge bath (24 × 8 × 8 m). This bath is provided with inspection windows to make it possible to exhibit the ship during conservation. Measures are taken to prevent discoloration of the liquid. During the first years of conservation the concentration of PEG 1000 will be about 5–10%, after which the PEG concentration in the bath liquid will gradually be raised to 60%. The entire conservation will take 20–30 years.

During the final stages of conservation the bath liquid will be somewhat heated (30–35°C). Since the bath liquid must remain clear, it has to be purified continuously with the aid of sand and activated carbon filters. To prevent the growth of microorganisms during the conservation of the ship, boric compounds are used as a biocide.

In 1973, using the Barkman [40] method, the conservation of old waterlogged wood was started in the workshop of the Ship Archaeological Museum at Ketelhaven. PEG 4000 was used and the entire impregnation time was 18 months, after which the objects were stored in the impregnation liquid at 60°C for one month more. After removal they were dried in an environment of initially 70–80% and finally 50–60% r.h. A number of cases of shrinkage, crack formation and warping occurred. The main weight loss of the objects was approximately 30% (Fig. 5(a)). The cracks which developed, mainly parallel to the fibre direction, varied in width from a few millimetres to a few centimetres. These unsatisfactory results led to investigations into the distribution of the PEG 4000 in the treated wood (Fig. 5(b)) [63]. The techniques as used for the determination of PEG 4000 present in wood are described in the Appendix below.

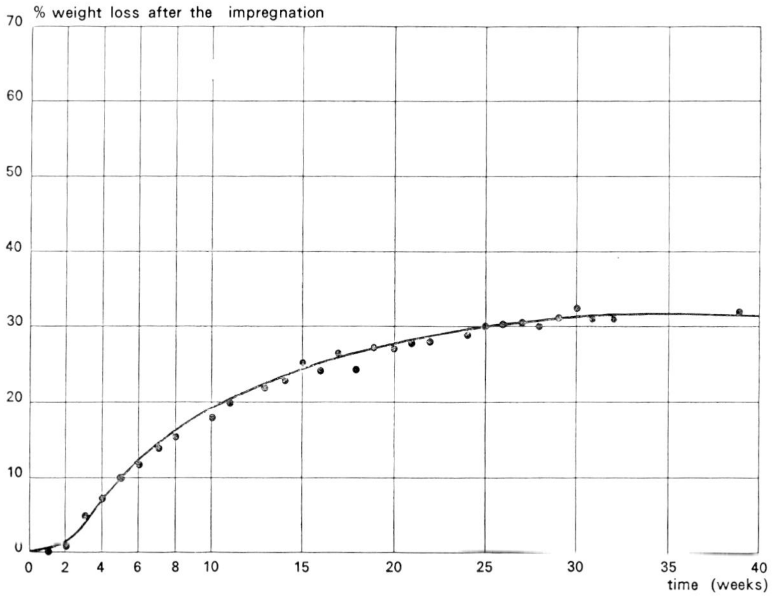

FIG. 5(a). The mean weight loss after conservation with PEG as described in the text. The wood was mainly European oak.

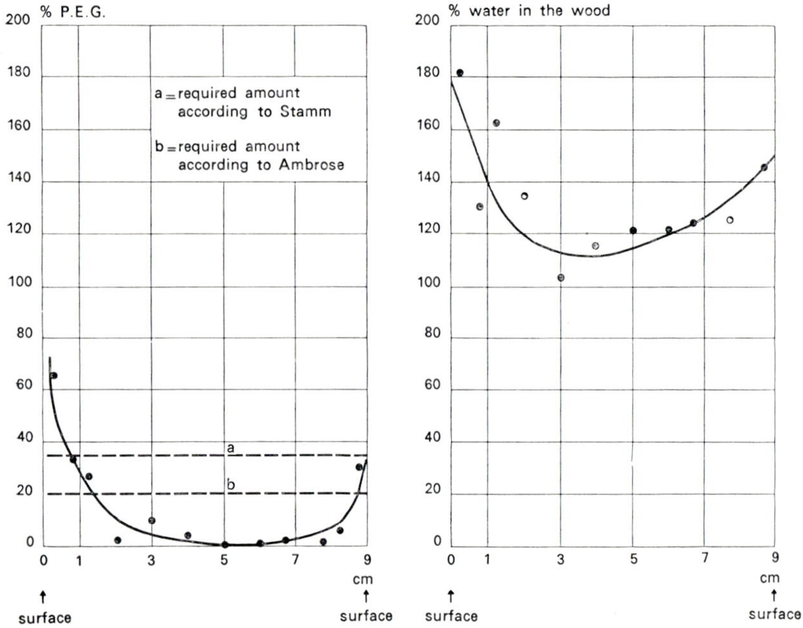

FIG. 5(b). The distribution of PEG 4000 perpendicular to the longitudinal direction in a piece of European oak. Data based on the dry wood material. The conservation process is described in the text.

It was remarkable that the shrinkage problems were caused almost exclusively by oak. The water contents of this wood were found to vary between 100% and 185%. We found that the success of a conservation process also depends on the size of the objects. Results obtained with small test samples can be strongly misleading when conclusions with respect to the applicability of any process for the treatment of large objects are drawn.

The transport of PEG in the longitudinal direction is considerably faster than in the two other directions [52–54]. Hence, in small objects, with surfaces perpendicular to the fibre direction at relatively short distances from each other, the degree of impregnation will largely be determined by the 'fast' transport through those surfaces. For large objects, this 'fast' transport will be too slow to compensate for the much larger distances between the surfaces perpendicular to the fibre direction.

European oak is very difficult to penetrate and to conserve, but other wood species can be impregnated much easier. The penetration of PEG

4000 into European oak test samples with dimensions of 5 × 5 × 10 cm (radial × tangential × longitudinal, water content 120–185%) compared with the data found for shipblocks (up to 30 × 60 × 60 cm) made of elm and ash has been investigated. The impregnation of the even smaller European oak test pieces is much less than that of the shipblocks. The water content of the shipblocks varied from 110% to 160%. The shipblocks contain an amount of PEG 4000 well above the required amount of 35% (or 20% with respect to dry matter) and showed no defect whatsoever.

The observed PEG distribution depends on the following three facts:

1. During impregnation, the concentration of PEG in the bath was raised from zero to about 75% in a period of 18 months, after which the impregnation was stopped.
2. When the wood is dried, evaporation of the water starts at the surface; this evaporation causes a transport of water with dissolved PEG from the interior of the wood towards the surface where the PEG will remain after evaporation of the water.
3. The amounts of PEG are expressed as amounts relative to the dry matter. In general, the wood near the surfaces is more deteriorated than that in the interior per unit of volume, and the amount of dry matter at the surface is less than in the inside. Even when the same amount of PEG is present per unit of volume, this method of calculating the PEG concentration as related to the dry wood will give higher amounts for the surface regions.

In practice, a surplus of PEG is found near the outside of the wood, while, in the inside, often insufficient PEG to ensure a good dimensional stabilisation is found (Fig. 6).

In the surface areas of the wood a partly unabsorbed surplus of PEG is present which is certain to absorb water from the surrounding air whenever possible. Hence problems may arise when using the low molecular weight grades of PEG, if the treated objects are not kept under conditions of low relative humidity.

The PEG distribution along the longitudinal direction has been determined in a piece of the hull planking of a seventeenth-century merchant ship found in the Lake IJssel polders. The thickness of the hull planking was approximately 5 cm. Parallel longitudinal direction layers have been sampled. In these layers the amount of PEG was determined at distances of 1, 3, 5, 9 and 13 cm perpendicular to the fibre direction from a freshly cut surface. In all cases the water content of the wood was approximately

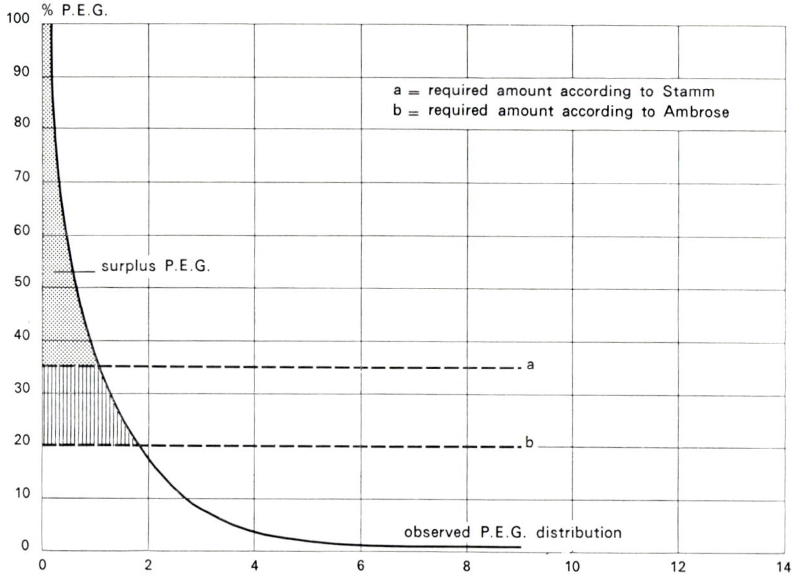

FIG. 6. The practical distribution of PEG 4000 in wood and the theoretically preferable distribution, both expressed as a percentage of the dry weight of the wood.

100–120%. In the surface layers (0–0·5 and 4·5–5·0 cm) it was found that the amount of PEG is sufficiently high. In the underlying layers the amount of PEG decreases with the distance from the outer plane. The amount of PEG decreases rapidly with increasing distance from the outer surface and goes well below the level required for a good dimensional stabilisation. The transport in a longitudinal direction is considerably better than in a radial direction. In sound European oak the transport of PEG 4000 is insufficient to ensure sufficient penetration in reasonable time; hence dimensional stabilisation with the process used is insufficient and other techniques should be investigated.

Results similar to those found for the hull planking have been observed for a large prefabricated 'knee' (weight about 500 kg). This knee of European oak was also still relatively slightly deteriorated, having a water content of 120–140%. After impregnation with PEG 4000 it was found that the impregnation was limited to the surface layers of the knee and the areas around the holes caused by teredos.

It can be concluded that old waterlogged wood can be successfully conserved with polyethyleneglycols provided there is a thorough knowledge of the wood to be treated. Knowledge about the different stages of deterioration present in the find and possible conservation techniques is essential. More fundamental research into methods for improving the speed of penetration of PEG into some wood species, *e.g.* European oak, is needed.

4. CONSERVATION OF THE ZWAMMERDAM AND UTRECHT SHIPS

4.1. Introduction

In February 1973 the IJsselmeerpolders Development Authority established a board of Dutch experts in the field of wood conservation. The task of this board was to advise on the conservation of the large finds of waterlogged wood found in the Lake IJssel polders. Early in 1973 the Zwammerdam ships, dating from Roman times, were found and it was decided that the IJsselmeerpolders Development Authority should excavate, salvage and conserve these ships. In Utrecht in 1974 two medieval ships were also found and it was again decided that the IJsselmeerpolders Development Authority should salvage and conserve these finds. From these developments it will be clear that the attention of the advisory board has been focused almost entirely on the possibilities of conserving these finds.

In both cases the condition of the wood was too bad and the damage sustained by the wood in the soil too heavy to enable salvaging and restoration of the ships as a whole. It was therefore decided to take the timber of the ships apart and to put the separated parts on boards of new wood upon which the old waterlogged wood could be transported, cleaned and conserved. During the storage in water an attempt was made to prevent the growth of microorganisms by the addition of thymol to the water [5]. However, this chemical proved to be insufficient during the summer and it was decided that the water should be renewed regularly.

Although, as described in Section 3, the conservation of old waterlogged wood with polyethyleneglycols has not always been successful, and a number of drawbacks can be listed, this method was selected as the best for the treatment of the ships found at Zwammerdam and Utrecht. The following considerations led to this conclusion:

(a) The experience obtained in conserving large quantities of old waterlogged wood is limited to conservation with PEG (Roskilde ships and *Wasa*). Since both the Zwammerdam and the Utrecht ships have been taken apart during the excavation, the bath method seems to be the most effective technique. The ships can then be reconstructed with the conserved wood. The conservation of a previously reconstructed ship has the disadvantage that, especially with the highly deteriorated timber of the Zwammerdam ships, damage will increase and a very large bath will be needed for the conservation. Another solution is the spraying method, which will result in longer impregnation times than with the bath method.

(b) The PEG bath method is a technically simple process and has the lowest estimated process costs.
(c) The process is safe with respect to the wood to be conserved.
(d) The process can be automated and requires very little labour.
(e) The process is non-hazardous from an environmental point of view.

In Krefeld, West Germany, a boat 13·5 m long and 2·6 m wide was found and the possibilities for excavation and conservation studied. Individual pieces of timber will be conserved with polyethyleneglycol for reconstruction of the boat later [57].

Using polyethyleneglycols for conservation of waterlogged wood, it is essential to know the wood species and the stages of degradation to avoid failures. Based on classifications given in the literature [7, 15, 16, 34], the following classification for the treatment of European oak with PEG 4000 has been made and is used now:

I Wood containing over 400% water. This wood has lost almost all of its hard core.
II Wood containing 185–400% water. A hard core is present but is comparatively small.
III Wood containing less than 185% water. A large hard, little deteriorated core is present below a thin deteriorated surface layer (Fig. 7).

Wood belonging to classes I and II can be conserved with PEG 4000; large objects consisting of class III wood cannot successfully be conserved with solutions of PEG 4000 in water. Class I wood can be conserved by means of the warm bath method with PEG 4000, while for class II wood the cold bath method with PEG 4000 can be used.

Treatment of class III wood with lower molecular weight types of PEG is possible, but requires many years (*Wasa* spraying with PEG 1500 for 13 years, *Bremer Kogge* treatment with PEG 1000 for an expected period of 20–30 years). Conservation methods which are less time-consuming still have to be developed and research in this field needs to be stimulated.

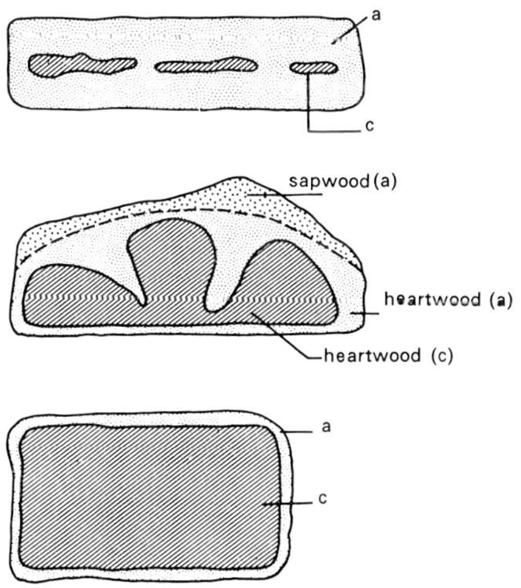

FIG. 7. Three typical wood quality types. In the first, found at Zwammerdam, highly degraded wood is predominant. In the second type, found in the Utrecht ships, there is a rather large sound inner core present under a layer of highly deteriorated wood. In the third type, found in a seventeenth-century ship, a thin layer of highly deteriorated wood is present on top of a large sound inner core.

4.2. Conservation Technique

For the conservation of the wood of the Zwammerdam and Utrecht ships, four concrete conservation tanks have been built. The inner dimensions of the tanks are 9·1 × 2·4 × 1·6 m. All tanks are coated on the inside with an epoxide resin to prevent chemical attack. All parts of the tanks are made of wood, stainless steel or brass. Two tanks will be used for a warm process, and the other two for the cold process. To reduce the energy consumption during the warm process, much attention has been paid to the thermal insulation, with the result that 10 kW of electrical power is more than sufficient to maintain a bath temperature of 80°C when the room temperature is 5°C. The time needed for heating the bath is rather long, about two weeks, but this is no serious drawback since the expected process time is about two years. Two concrete storage tanks measuring 5·2 × 9·1 × 1·1 m and 9·1 × 2·1 × 2·2 m have also been made.

For very large objects a shower installation is available for the prevention of drying-out. Outside the workshop a resin-coated concrete storage tank for approximately 50 m^3 of PEG solution has been constructed. Figure 8 shows the layout of the workshop of the Museum for Ship Archaeology at Ketelhaven.

For the warm process the treatment of the wood is as follows. After a thorough cleaning, the wood is immersed in a 10% solution of PEG 4000.

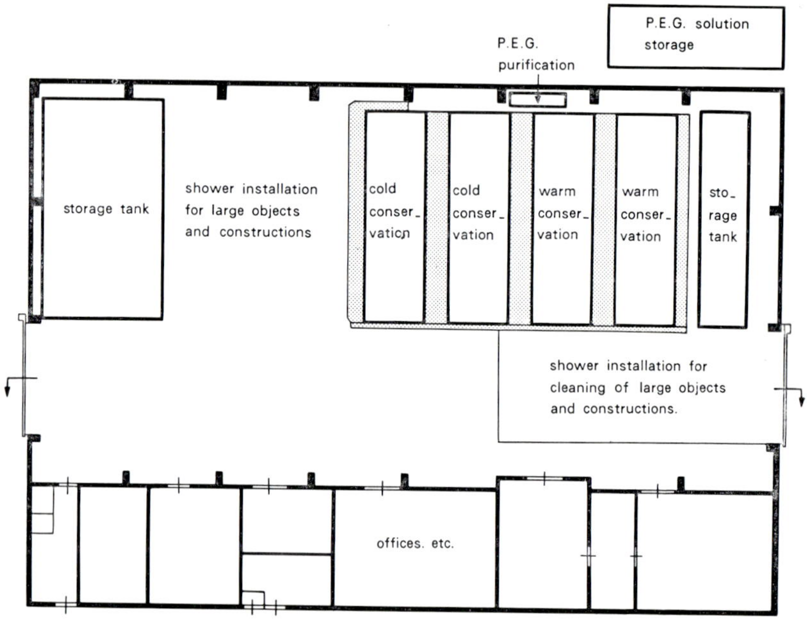

FIG. 8. The layout of the workshop of the Museum for Ship Archaeology at Ketelhaven.

Approximately 30% of the tank volume can be filled with waterlogged wood; ±10% of the volume is consumed by the devices needed for the support of the waterlogged wood, the rest of the tank volume can be filled with the solution of PEG 4000. After filling the tank with the 10% solution of PEG 4000, the temperature is raised to 60°C and, over a period of two years, the concentration of PEG 4000 in the bath will gradually be raised to 90% or higher. The solution in the tank will be circulated by pumping. The end of the process will be determined by sampling the wood and analysing it for the amount of PEG present and its distribution in the wood.

To increase the PEG concentration, water has to be removed from the conservation tank and PEG added. To avoid sudden changes in concentration and to have an easy PEG supply system, this chemical will be added dissolved in water. By using a 50% PEG 4000 solution in water, the chances of PEG crystallising from the solution during storage outside the workshop will be small. The removal of water from the conservation tank will be achieved with the aid of a forced air current flowing over the surface of the liquid, and water vapour will be transported through a vent to the outside of the building. This forced air current will be operated by means of a level control in the conservation tank. When evaporation is too fast and the level of the PEG 4000 solution in the tank becomes too low, the air current will stop and tap water can be added. This also will be operated through the level control. Using this system it is possible to maintain the level of the impregnation liquid in the conservation tank within predetermined limits.

The PEG 4000 concentration in the tank will be determined weekly by analysing samples from the tank, and the required amount of PEG 4000 to be added can be calculated. With the aid of an adjustable pump, this amount can be added in a period of 6 days.

During the conservation of the wood, some impurities and dispersed solid materials will always contaminate the impregnation liquid. The possibility of the formation of thermal degradation products of PEG has also been mentioned [5], although we have found that PEG 4000 is thermally very stable. We heated this chemical for 3 months and kept the temperature between 140° and 160°C. After this we compared the melting point with that of an unheated sample; both melted at from 56·0–59·5°C. A mixed sample showed no melting point depression. The impurities appearing in an impregnation liquid will be mainly extractives from the wood and from coatings applied to it. To avoid a dark surface colour of the wood after conservation, these impurities have to be removed. Noack [49] has shown that a good purification of PEG can be obtained by filtration of the impregnation liquid through sand and activated carbon. In the laboratory we used this method for cleaning PEG 4000 used for 18 months at 60°C and that had become dark brown. The liquid was filtered and treated with activated carbon twice. After the second treatment a pale-yellow PEG 4000 was obtained from the solution which had a melting point just below that of a fresh, unused sample (Table VI).

Since impurities in PEG 4000 can be removed with the aid of activated carbon, it was decided to build PEG purification units into the new designed PEG impregnation plant (Fig. 9).

TABLE VI

THE MELTING POINTS OF FRESH PEG 4000 AND USED PEG 4000 AFTER SUCCESSIVE TREATMENTS WITH ACTIVATED CARBON

Sample	*Melting point (°C)*
Fresh PEG 4000	56·0–59·5
Used PEG 4000	35·0–42·0
Used PEG 4000, after 1 treatment with activated carbon	53·3–56·0
Used PEG 4000, after 2 treatments with activated carbon	55·0–59·0

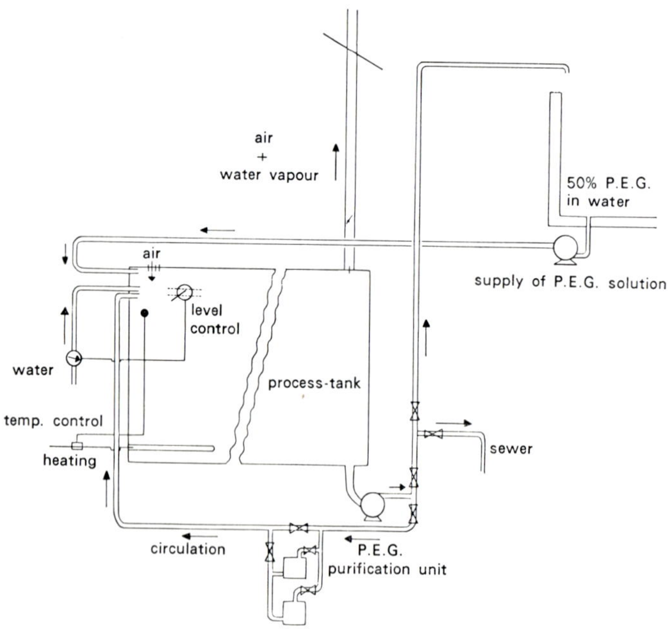

FIG. 9. The process schema for the conservation of old waterlogged wood using warm impregnation with PEG 4000.

Detailed knowledge of the quality and state of degradation of the individual pieces of timber of a boat is required in order to select the correct conservation method. Each individual part of the ships was therefore sampled once or twice; the water content in the sample's surface layers and core was determined. Next, the presence or absence of large hard cores of little degraded wood was determined with the aid of a thin, strong needle. Based on the results of the water content and the hardness assessment, the type of treatment was decided upon. If the wood belonged to class III it had to be stored and left untreated until a safe conservation process will have been developed. All ships found at Zwammerdam and Utrecht have been investigated in this way. In Table VII the number of

TABLE VII

THE NUMBERS OF SAMPLES TAKEN FOR WATER CONTENT DETERMINATION AND THE DISTRIBUTION OF THE RESULTS OVER THE THREE WOOD QUALITY TYPES

Ships	*No. of samples*	*Distribution of wood quality types, in percentages of the number of samples*		
		Class I	*Class II*	*Class III*
Zwammerdam 2*	64	31	38	31
Zwammerdam 4*	131	21	51	28
Zwammerdam 6*	77	35	52	13
Utrecht 1	20	15	25	60
Utrecht 2	12	25	25	50

* Zammerdam ships 1, 3 and 5 are comparatively small and highly degraded. The wood is of class I quality.

samples which have been taken and the division (expressed as a percentage of the number of samples) over the three different wood quality types is given. Based on these results, drawings of the ships were made, showing the distribution of the different wood quality types in the ships (Figs. 10 and 11).

From the results summarised in Table VII and the two examples shown in Figs. 10 and 11, it can be seen that every wood quality type is present in all the large ships and that only their relative quantities differ. Even between ships of equal age these differences can exist. In general, the Utrecht ships contain less degraded wood than the Zwammerdam ships. It will also be clear that as long as no general and safe conservation process for the treatment of large finds of old waterlogged wood has been proved successful,

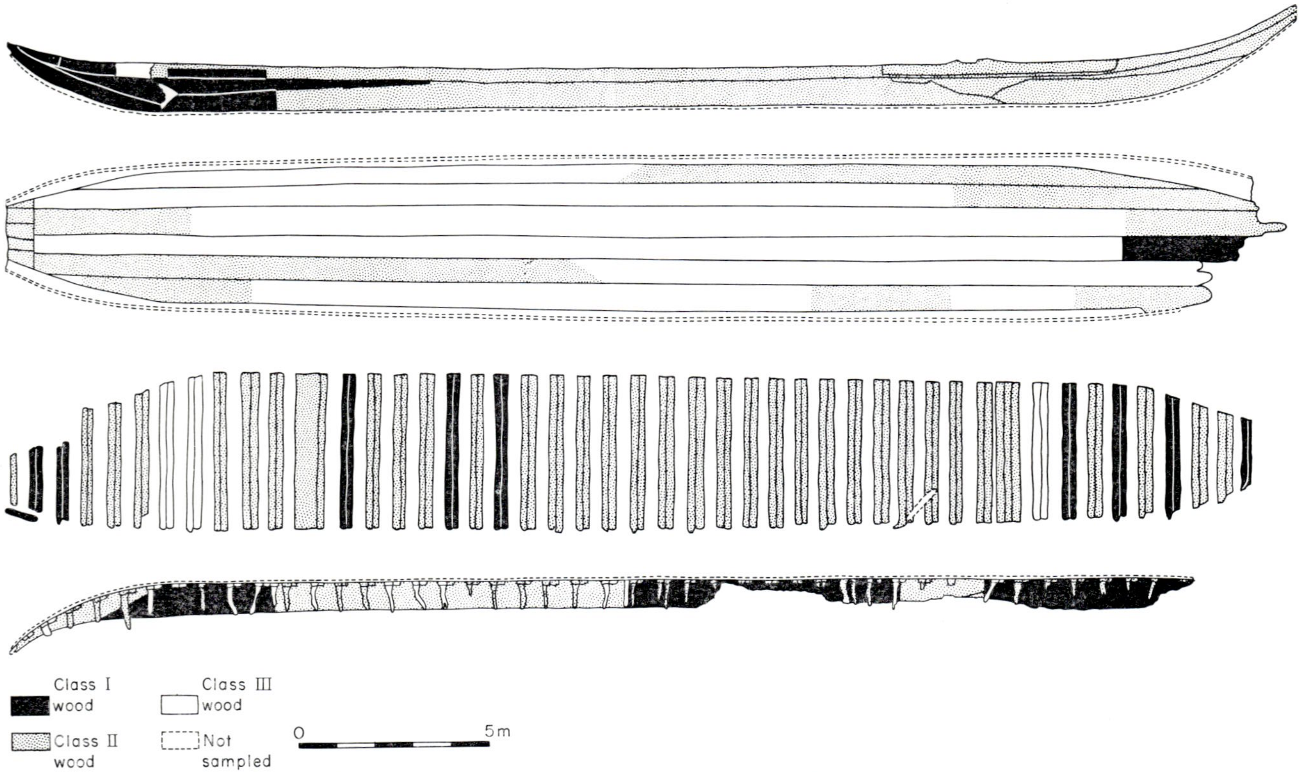

FIG. 10. The distribution of wood of class I, II and III in Zwammerdam ship 4.

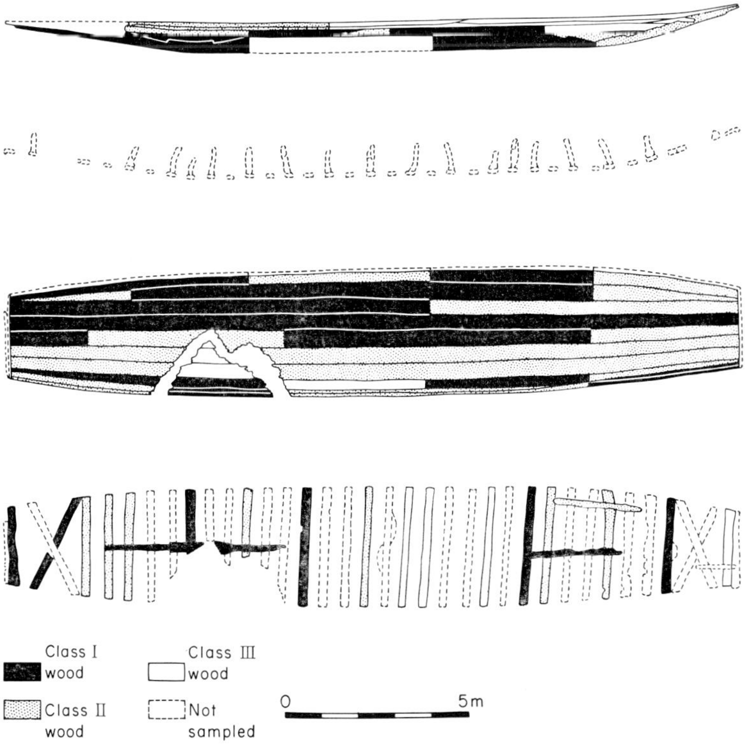

FIG. 11. The distribution of wood of class I, II and III in Zwammerdam ship 6.

it will be necessary to know these quality distributions within one find and to select the most appropriate treatment method for a given piece of timber. A prerequisite for this approach is that the ship is taken apart and the individual pieces of timber reconstructed after conservation.

ACKNOWLEDGEMENTS

Thanks are due to Mr W. Eenkhoorn, Mr A. Wevers, Mr L. J. Sollie and Mr G. van Panhuis for their contribution to the laboratory research, and to Mr W. Eenkhoorn, Mr A. Wevers, Drs H. R. Reinders and Mrs J. van der Elburg for carefully reading the manuscript and suggesting improvements.

REFERENCES AND BIBLIOGRAPHY

1. Barkman, L. (1965). The preservation of the *Wasa. Wasastudier*, No. 5. Statens Sjöhistoriska Museum, Stockholm.
2. Boutelje, L. B. and Kiessling, H. (1964). On water-stored oak timber and its decay by fungi and bacteria. *Arch. Mikrobiol.*, **49** (3), 305–14.
3. Stamm, A. J. (1964). *Wood and Cellulose Science*. Ronald Press, New York.
4. Hashimoto, Y. and Nisizawa, K. (1963). In: *Advances in Enzymic Hydrolysis of Cellulose and Related Materials* (ed. E. T. Reese). Pergamon Press, Oxford.
5. Mühlethaler, B. (1973). *Conservation of Waterlogged Wood and Leather*. ICOM Travaux et Publications, No. XI. Ed. Eyrolles, Paris.
6. Noack, D. (1965). Der gegenwärtige Stand der Dimensionsstabilisierung von Holz und Schlussfolgerungen für die Konservierung der *Bremer Kogge. Bremisches Jahrb.*, **50,** 43–72.
7. Brorson Christensen, B. (1970). *The Conservation of Waterlogged Wood in the National Museum of Denmark*. Copenhagen.
8. Risi, J. and Arseneau, D. F. (1957). Dimensional stabilization of wood. *Forest Prod. J.*, **7,** 210–13.
9. Werner, A. E. A. Technical Notes 218. CIBA Ltd, Duxford, Cambridge, England.
10. Seborg, R. M., Tarkow, H. and Stamm. H. (1953). Effect of heat on the dimensional stabilization of wood. *J. Forest Prod. Res. Soc.*, **3** (3), 59–67.
11. Stamm, A. J. (1959). Dimensional stabilization of wood by thermal reaction and formaldehyde-crosslinking. *Tappi*, **42** (1), 39–44.
12. Burr, H. K. and Stamm, A. J. (1947). Diffusion in wood. *J. Phys. Chem.*, **51,** 240–57.
13. Behr, E. A., Briggs, D. R. and Kaufert, F. H. (1953). Diffusion of dissolved material through wood. *J. Phys. Chem.*, **57,** 476–80.
14. Rosenquist, A. M. (1959). *Studies Conserv.*, **4,** 62–72.
15. Ambrose, W. (1970). Freeze-drying of swamp degraded wood. IIC Conference, New York.
16. Ambrose, W. (1972). The treatment of swamp degraded wood by freeze-drying. Report to the ICOM, Madrid.
17. Brorson Christensen, B. Personal communication.
18. Brongers, J. A. and Wijnman, H. F. (1971). The conservation of waterlogged wood in the Netherlands by means of the alcohol-ether method. *Berichten van de Rijksdienst voor het Oudheidkundig Bodemonderzoek*, **20–21,** 321–2.
19. Kramer, W. and Mühlethaler, B. (1967–68). Über die Erfahrungen mit der Alkoholäther die Konservierung von Nassholz am Schweizerischen Landesmuseum. *Zschr. für Schweizerische Archäologie und Kunstgeschichte*, **25,** 78–88.
20. Munnikendam, R. A. (1971). Conservering van stenen en houten voorwerpen. *Chemisch Weekblad*, **67** (20), 23–6.
21. Malz, A. D.P. 704,836, 6 Mar. 1941.
22. Willstatter, R., Kraut, H. and Lobingen, K. D.P. 520,970, 12 Dec. 1928.

23. Wielicka, H. (1959–60). Research on the preservation of timber from Biskupin and Gdansk excavations. *Wiadowmosci Archeologiczne*, **26,** 288–96.
24. Bouis, J. (1967). Principes de conservation et de restauration des bois antiques immergés. Report to the ICOM, Brussels.
25. Garrouste, L. Procédé de conservation des bois antiques décomposés dans l'eau douce ou l'eau de mer. Fr. Pat. 1,392,835, 1965.
26. Munnikendam, R. A. Personal communication.
27. CIBA. Angaben betr. Ausübung der Arigal-Konservierung unter Verwendung von Arigal C und Arigal Katalysator C. CIBA, Basel.
28. Mueller-Beck, H. J. and Haas, A. (1960). A method for wood preservation using Arigal C. *Studies Conserv.*, **5,** 150–7.
29. Bikales, N. M. (1972). *Encyclopedia of Polymer Science and Technology.* Interscience, New York.
30. Munnikendam, R. A. (1972). The conservation of waterlogged wood with glycolmethacrylate. Report to the ICOM, Madrid.
31. McKerrel, H., Roger, E. and Versanyi, A. (1972). The acetone/rosin method for conservation of waterlogged wood. *Studies Conserv.*, **17,** 111–25.
32. Munnikendam, R. A. (1967). Conservation of waterlogged wood using radiation polymerisation. *Studies Conserv.*, **12,** 70–5.
33. Morén, R. E. (1964). Some practical applications of polyethyleneglycol for the stabilization and presentation of wood. Modo Technical Information, No. 82.
34. Morén, R. E. and Centerwall, B. (1960). The use of polyglycols in the stabilizing and preservation of wood.Modo Technical Information No. 64.
35. Stamm, A. J. (1956). Dimensional stabilization of wood with carbowaxes. *Forest Prod. J.*, **6,** 201–4.
36. Stamm, A. J. (1959). Effect of polyethyleneglycol on the dimensional stability of wood. *Forest Prod. J.*, **9,** 375–81.
37. Stamm, A. J. (1964). Factors affecting the bulking and dimensional stabilization of wood with polyethyleneglycols. *Forest Prod. J.*, **14,** 403–8.
38. Tarkow, H. (1971). In: *Cellulose and Cellulose Derivatives*, vol. V (ed. N. M. Bikales and L. Segal). Interscience, New York.
39. Tarkow, H., Feist, W. C. and Sutherland, C. F. (1966). Interaction of wood with polymeric materials: penetration versus molecular size. *Forest Prod. J.*, **16,** 61–5.
40. Barkman, L. (1973). The preservation of the warship *Wasa*. Paper presented to the Symposium on the Problems of the Conservation of Waterlogged Wood, Greenwich, England.
41. Lefevre, R. (1961). Conservation treatment with polyethyleneglycol of a wooden Roman bucket excavated at Wemmel. *Bull. Inst. Royal du Patrimoine Artistique*, **4,** 97–108.
42. Organ, R. M. (1959). Carbowax and other materials in the treatment of waterlogged paleolithic wood. *Studies Conserv.*, **4,** 96–105.
43. Albright, A. B. (1966). The preservation of small waterlogged wood specimens with polyethyleneglycol. *Curator*, (9) 3. American Museum of Natural History, New York.

44. Seborg, R. M. and Inverarity, R. B. (1962). The preservation of old water-logged wood by treatment with polyethyleneglycol. *Science*, **136**, 649–50.
45. Seborg, R. M. and Inverarity, R. B. (1962). Conservation of 200 years old waterlogged boats with polyethyleneglycol. *Studies Conserv.*, **7**, 111–19.
46. Organ, R. M. (1960). Polyethyleneglycols in the treatment of waterlogged wood: some precautions. *Studies Conserv.*, **5**, 161–2.
47. Barkman, L. Personal communication.
48. van der Elburg, J. Personal communication.
49. Noack, D. (1969). *Die Bremer Hanse Kogge. Fund, Konservierung, Forschung*, pp. 127–56. Friedrich Röver, Bremen.
50. Morén, R. (1965). Die polyäthylenglycol Imprögnierung von Holz und ihre Auswirkungen bei Holztrocknung und Holzbearbeitung. *Holz als Roh- und Werkstoff*, **23**, 142–52.
51. Morén, R. (1964). Some practical applications of polyethyleneglycol for the stabilization and preservation of wood. Modo Technical Information No. 82.
52. Behr, E. A., Briggs, D. R. and Kaufert, F. H. (1953). Diffusion of dissolved materials through wood. *J. Phys. Chem.*, **57**, 476–80.
53. Kumar, S. and Jain, V. K. (1973). Diffusion through wood: diffusion of boric acid. *Holzforschung und Holzverwertung*, **25**, 21–4.
54. Stamm, A. J. (1964). *Wood and Cellulose Science*. Ronald Press, New York.
55. Rosenquist, A. M. Personal communication.
56. Heslinga, F. J. M. (1973). The impregnation and conservation of deteriorated oak using ethyleneglycol and polyethyleneglycol. *Museuologia*, (1), 24–39.
57. Pirling, R. and Buchwald, G. (1974). Ein Schiff aus karolingischer Zeit und seine Konservierung. *Naturwissenschaften*, **61**, 396–8.
58. Heertjes, P. M., Loxham, M. and Snijder, P. J. Personal communication.
59. Roberts, J. D. and Caserio, M. (1965). *Basic Principles of Organic Chemistry*. W. A. Benjamin Inc., New York.
60. Crèvecoeur, R. Personal communication.
61. Fieser, L. F. and Fieser, M. (1956). *Organic Chemistry*. Reinhold. New York.
62. Bouis, M. J. (1975). Report to the ICOM, Venice.
63. de Jong, J. (1975). Report to the ICOM, Venice.

APPENDIX: THE DETERMINATION OF PEG 4000 IN WOOD

Pretreatment

The wood is sampled with the aid of a hollow drill producing cylindrical samples of 2·5 cm diameter. The samples are divided into layers, for example from 0–0·5, 0·5–1, 1–2, 2–3, 3–5 cm, starting from both surfaces. The different pieces of wood are dried and weighed and then extracted with water in a Soxhlet apparatus in order to remove the PEG from the wood. The extraction takes approximately 40 hr. Large samples are divided into smaller pieces before extraction. The extract is dried, weighed and dissolved in 50 ml 0·1 N NaOH and filtered (MN 619¼) until a clear filtrate is obtained. The filtrate is neutralised with concentrated HCl, then transferred in a volumetric flask of 250 ml and diluted up to the mark with water. In this solution the PEG is precipitated with an excess of potassium hexacyanoferrate(II). The excess of potassium hexacyanoferrate(II) is determined either titrimetrically, using cerium ammonium sulphate as reagent and ferroin as indicator (A), or spectrophotometrically (B).

The extracted wood is stored in water to prevent drying, until the water content is measured by weighing and drying.

A. Titrimetric Determination

1. Approximately 0·02 mol cerium ammonium sulphate solution. Approximately 0·1 mol cerium ammonium sulphate solution is prepared by dissolving 63·0 g cerium ammonium sulphate in a mixture of 30 ml concentrated sulphuric acid and 500 ml distilled or deionised water.
   The solution is filtered and diluted up to 1000 ml. It is then stored for 14 days and filtered again (MN 619¼). 200 ml of the filtrate and 25 ml conc. sulphuric acid are diluted to obtain the 0·02 molar solution.
2. Ferroin (1,10-phenantroline-ferrosulphate complex). Dilute 50 ml trade solution (0·025 M) to 2000 ml with distilled or deionised water.
3. Potassium hexacyanoferrate(II) solution. Dissolve 10 g $K_4Fe(CN)_6 \cdot 3H_2O$ (p.a.) and 5 g water-free sodium carbonate in distilled or deionised water and dilute to 1000 ml.
4. Sodium chloride solution. Dissolve 300 g NaCl p.a. in water and dilute to 1000 ml.
5. Hydrochloric acid, iron-free. Dilute concentrated HCl in water (1:1).
6. NaOH solution about 0·1 N.
7. Washing liquid. Dissolve 240 g NaCl and 80 ml hydrochloric acid (conc.) in water and dilute to 1000 ml.
8. Stock solution of PEG 4000. Dissolve 2500 g PEG 4000 to 1000 ml.

PEG Determination

Transfer into conical flasks of 300 ml with a pipette:

(a) four times (two as samples and two as blanks) 25 ml of extract solution, or less, if 25 ml exceeds 50 mg of solids in the extract;
(b) two times 20·0 ml of PEG stock solution;
(c) two times 25 ml of water (for the titration of 10·0 ml potassium hexacyanoferrate(II) solution).

Add to all the flasks 50 ml of the sodium chloride solution, 30 ml of HCl (1 + 1) solution and dilute with water to 100–105 ml. Then add to all the flasks 10·0 ml of potassium hexacyanoferrate(II) solution except to the blanks. Add to all the flasks 10 ml of water. Shake and allow to stand for 5 min.

Filter off the precipitate (MN 619¼) and collect the filtrate in a conical flask of about 700 ml. Wash the precipitate thoroughly with 3 × 30 ml of washing liquid.

Add 20 ml of ferroin indicator solution to the filtrate and titrate with 0·02 M cerium(IV) ammonium sulphate solution until the indicator changes from orange to clear green.

Calculation

The amount of PEG present in the sample is:

$$500 \times \frac{(E1 + E4 - E3)}{(E1 - E2) \times D \times M} = \% \text{ PEG}$$

where E1 = mean titration of 10·0 ml potassium hexacyanoferrate(II) treated like the samples;

E2 = mean titration of 60 mg PEG from the stock solution and 10·0 ml of the hexacyanoferrate solution treated like the samples;

E3 = mean titration of the samples treated with 10·0 ml potassium hexacyanoferrate(II);

E4 = mean titration of the samples with the addition of 10·0 ml potassium hexacyanoferrate(II);

D = dry weight of the wood;

M = number of ml of the sample (25 or less).

Remarks

1. The solution of potassium hexacyanoferrate is relatively stable in water, but is unstable in acidic solution, in which a dark-blue colour quickly develops. The titration should therefore be done within 1 hr of the addition of the potassium hexacyanoferrate.
2. When reducing agents are extracted from the wood, these are also oxidised during the titration. A correction can be made by determining the reagent consumption without the addition of potassium hexacyanoferrate. Since the colour change of ferroin may be difficult to observe, a potentiometric end-point indication may be preferred.

Iron(III), if present, will cause the formation of Berlin blue with the hexacyanoferrate ion and give a dark-blue precipitate which is difficult to remove by filtration. Therefore the dried extract is redissolved in NaOH and filtered.

3. Because of the low concentration of the reagent, an accurate amount of indicator must be used. Therefore we dilute the ferroin and add 20 ml.

B. Spectrophotometric Determination of PEG

Reagents and solution (in all cases distilled or deionised water should be used):

1. Dissolve 300 g NaCl p.a. in water and dilute to 1000 ml.
2. Dissolve 10 g potassium hexacyanoferrate ($K_4Fe(CN)_6 \cdot 3H_2O$) and 5 g water-free sodium carbonate in water and dilute to 1000 ml.
3. Dissolve 4 g NaOH (chemical pure grade) in water and dilute to 1000 ml (0·1 N solution).
4. Dissolve 2 parts concentrated HCl (chemical pure) into 1 part (by volume) water.
5. Dissolve 50 g hydroxyl ammonium chloride in 1000 ml of the hydrochloric acid (2 + 1) solution.
6. For the preparation of a stock solution of PEG, dissolve 2500 g PEG 4000 in water and dilute to 1000 ml.

Determination of PEG 4000 *Content in Impregnated Old Waterlogged Wood*

Transfer with pipette:

(a) four times (two as samples, two as blanks) 25·0 ml of the extract solution into a calibrated 100 ml volumetric flask (if 25 ml of the extract solution contains more than 50 mg PEG, less should be taken);
(b) two times 20·0 ml of the PEG 4000 stock solution into calibrated 100 ml flasks;
(c) two times 25 ml water into calibrated 100 ml flasks.

To all flasks 40 ml of the sodium chloride solution is then added and 20 ml of a freshly prepared hydroxyl ammonium chloride/hydrochloric acid solution is added to each flask except for the two blanks. All flasks are then filled up to the 100 ml mark, shaken and left to stand for 5 min. The precipitate is filtered through a S & S 588 filter. The first 30 ml of filtrate is rejected, the rest is used for the spectrophotometric determination. The extinction of the filtrate is measured with the aid of a spectrophotometer at 314 nm. In automatic sampling, all the samples, the PEG standard solution and 10 ml potassium hexacyanoferrate(II) should be taken first, and then the blanks.

The amount of PEG present in the sample is calculated from the extinctions of the samples, the potassium hexacyanoferrate solutions, the PEG solutions and the blank, using the following formula:

$$\text{PEG as \% of dry wood} = \frac{500 \times (E1 + E4 - E3)}{(E1 - E2) \times D \times M}$$

where E1 = mean extinction of 10·0 ml potassium hexacyanoferrate solution treated like the samples;
E2 = mean extinction of 60 mg PEG from the stock solution and 10·0 ml of the potassium hexacyanoferrate solution treated like the samples;
E3 = mean extinction of the samples treated with 10·0 ml potassium hexacyanoferrate;
E4 = mean extinction of the samples without the addition of 10·0 ml potassium hexacyanoferrate;
D = dry weight of the wood;
M = number of ml of the sample (25 or less).

Remarks

1. Potassium hexacyanoferrate(II) is unstable in an acid solution, so hydroxyl ammonium chloride is added as a reducing agent to prevent the oxidation of Fe(II) into Fe(III) and the measurement should be completed within 1 hr of filtration.
2. When sufficient potassium hexacyanoferrate has been added, a cloudy precipitate is formed, a clear filtrate is obtained and the filter becomes light blue.

C. Determination of the Water Content in the Samples

The samples which had been stored in water are blotted dry to remove any water attached to the surface and weighed. They are then dried for 8–12 hr at 105°C and weighed again. The water content is calculated from the weight loss.

Index